THE 2ND EDITION MATHEMATICS COMPANION

ANTHONY C. FISCHER-CRIPPS
Fischer-Cripps Laboratories Pty Ltd
Sydney, Australia

CRC Press is an imprint of the
Taylor & Francis Group, an **Informa** business

Cover design by Ray Cripps.

CRC Press
Taylor & Francis Group
6000 Broken Sound Parkway NW, Suite 300
Boca Raton, FL 33487-2742

Printed on acid-free paper
Version Date: 20140709

International Standard Book Number-13: 978-1-4665-1587-1 (Paperback)

Visit the Taylor & Francis Web site at
http://www.taylorandfrancis.com

and the CRC Press Web site at
http://www.crcpress.com

This book is dedicated to
Mr McGuire of the former
Seaforth Technical College in
Sydney, who, when presented
with a class of unsuccessful
high school students, rekindled
our interest in mathematics by
telling us which procedures
were a "WOT" (waste of time)
and which were worth
knowing.

Contents

Preface xv
Part 1. Mathematics Essentials
1.1 Numbers, Trigonometry and Analytical Geometry 2
1.1.1 Real Numbers 3
1.1.2 Complex Numbers 4
1.1.3 Coordinate Systems 6
1.1.4 Vectors 7
1.1.5 The Unit Vectors 8
1.1.6 Trigonometry 9
1.1.7 Straight Line 10
1.1.8 Circle and Ellipse 11
1.1.9 Parabola 12
1.1.10 Hyperbola 13
1.2 Limits and Functions 14
1.2.1 Functions 15
1.2.2 Quadratic Function 16
1.2.3 Limits 17
1.2.4 Theorems on Limits 18
1.3 Differentiation 19
1.3.1 Derivative 20
1.3.2 Rules for Calculating the Derivative 21
1.3.3 Higher Order Derivatives 23
1.3.4 Maxima and Minima 24
1.3.5 The Second Derivative 25
1.3.6 Curve Sketching 26
1.3.7 Time Rate of Change 27
1.3.8 Anti-Derivatives 28
1.4 Integration 29
1.4.1 Definite Integral 30
1.4.2 Fundamental Theorem of Calculus 31
1.4.3 Properties of the Definite Integral 32
1.4.4 Indefinite Integral 33
1.4.5 Numerical Integration 34
1.5 Exponential and Logarithmic Functions 35
1.5.1 Logarithms 36
1.5.2 The Natural Logarithm 37

1.5.3 The Natural Exponential 38
1.5.4 Differentiation and Integration of e^x 39
1.5.5 Exponential Law of Growth and Decay...................40
1.6 Trigonometric and Hyperbolic Functions...........................41
1.6.1 Circular Measure 42
1.6.2 Derivatives and Integrals of Trigonometric Functions 43
1.6.3 Inverse Trigonometric Functions............................44
1.6.4 Derivatives of Trigonometric Functions 45
1.6.5 Hyperbolic Functions...46
1.6.6 Properties of Hyperbolic Functions........................47
1.6.7 Derivative of Hyperbolic Functions 48
1.6.8 Inverse Hyperbolic Functions 49
1.7 Methods of Integration 50
1.7.1 Integration by Substitution....................................51
1.7.2 Integration by Parts...52
1.7.3 Trigonometric Substitutions..................................53
1.7.4 Integration by Partial Fractions 54
1.7.5 Quadratic Expressions 55
1.7.6 Indeterminate Forms...56
1.7.7 Improper Integrals...57
1.8 Waves...58
1.8.1 Simple Harmonic Motion.......................................59
1.8.2 Waves 60
1.8.3 Wave Equation 62
1.8.4 Sign Conventions 63
1.8.5 Complex Functions 64
1.8.6 Complex Representation of a Wave.......................65
1.8.7 Superposition in Complex Form 66
1.8.8 Energy in a Wave 67
1.9 Infinite Series...68
1.9.1 Sequences...69
1.9.2 Series 70
1.9.3 d'Alembert's Ratio Test 71
1.9.4 Power Series...72
1.9.5 Binomial Series 73

1.10 Probability 74
1.10.1 Mean, Median and Mode 75
1.10.2 Permutations and Combinations 76
1.10.3 Probabilities, Odds and Expectation 77
1.10.4 Probability Distribution 78
1.10.5 Expected Value 79
1.10.6 Binomial Distribution 80
1.10.7 Normal Distribution 81
1.10.8 Sampling 83
1.10.9 *t* Distribution 84
1.10.10 Chi-Squared Distribution 85
1.11 Matrices 86
1.11.1 Matrices 87
1.11.2 Determinants 88
1.11.3 Systems of Equations 89
1.11.4 Eigenvalues and Eigenvectors 90
1.11.5 Cayley–Hamilton Theorem 91

Part 2 Advanced Mathematics
2.1 Ordinary Differential Equations 94
2.1.1 Ordinary Differential Equations 95
2.1.2 Separation of Variables 96
2.1.3 Homogeneous Equations 97
2.1.4 Exact Equations 99
2.1.5 Linear Equations 100
2.1.6 Linear Equations with Constant Coefficients 102
2.1.7 Method of Undetermined Coefficients 104
2.1.8 Systems of Equations 106
2.1.9 Complex Eigenvalues 109
2.1.10 Power Series 110
2.2 Laplace Transforms 111
2.2.1 Laplace Transform 113
2.2.2 Laplace Transform of Derivatives 114
2.2.3 Step Functions 115
2.2.4 Laplace Transforms and Differential Equations 116
2.2.5 Laplace Transforms and Partial Fractions 117

2.3 Vector Analysis 120
2.3.1 Vectors 121
2.3.2 Direction Cosines 122
2.3.4 Vector Dot Product 123
2.3.5 Equation of a Line in Space 124
2.3.6 Equation of a Plane 125
2.3.7 Distance from a Point to a Plane 126
2.3.8 Vector Cross Product 127
2.3.9 Distance from a Point to a Line 128
2.3.10 Distance between Two Skew Lines 129
2.3.11 Vector Differentiation 130
2.3.12 Motion of a Body 131
2.4 Partial Derivatives 132
2.4.1 Partial Differentiation 133
2.4.2 Chain Rule for Partial Derivatives 134
2.4.3 Increments and Differentials 135
2.4.4 Directional Derivatives 136
2.4.5 Tangent Planes and Normal Vector 137
2.4.6 Gradient, Divergence and Curl 138
2.4.7 Maxima and Minima 140
2.4.8 Lagrange Multipliers 141
2.4.9 Multiple Least Squares Analysis 142
2.4.10 Constraints 144
2.5 Multiple Integrals 146
2.5.1 Line Integrals 147
2.5.2 Electrical Potential 150
2.5.3 Work Done by a Force 151
2.5.4 Double Integral 152
2.5.5 Triple Integral 153
2.5.6 Surface Integrals 154
2.5.7 Gauss' Law 157
2.5.8 Divergence Theorem 158
2.5.9 Stokes' Theorem 159
2.5.10 Green's Theorem 160
2.5.11 Vector Representations of Green's Theorem 161
2.5.12 Application of Green's Theorem 162

2.5.13 Maxwell's Equations (integral form) 163
2.5.14 Maxwell's Equations (differential form) 164
2.6 Fourier Series 165
2.6.1 Fourier Series 166
2.6.2 Fourier Transform 167
2.6.3 Sampling 168
2.6.4 Discrete Fourier Transform 169
2.6.5 Odd and Even Functions 170
2.6.6 Convolution 171
2.7 Partial Differential Equations 173
2.7.1 Partial Differential Equations 174
2.7.2 General Wave Equation 175
2.7.3 Solution to the General Wave Equation 176
2.7.4 d'Alembert's Solution to the Wave Equation 179
2.7.5 Heat Conduction Equation 180
2.7.6 Solution to the Heat Conduction Equation 181
2.7.7 Heat Equation for a Thin Rod of Infinite Length ...182
2.8 Numerical Methods 183
2.8.1 Newton's Method 184
2.8.2 Interpolating Polynomial 185
2.8.3 Linear Least Squares 186
2.8.4 Non-Linear Least Squares 187
2.8.5 Error Propagation through Equations 188
2.8.6 Cubic Spline 189
2.8.7 Differentiation 190
2.8.8 Integration 191
2.8.9 First Order Ordinary Differential Equations 192
2.8.10 Runge–Kutta Method 193
2.8.11 Finite Element Method 194

Part 3. Applications

3.1 Capacitance 196
3.1.1 Permittivity 197
3.1.2 Complex Permittivity 198
3.1.3 Series Impedance 199
3.1.4 Parallel Impedance 200

3.1.5 AC Bridge 201
3.1.6 Capacitor Equivalent Circuit 202
3.1.7 Schering Bridge..203
3.1.8 Measurement of Relative Permittivity 204
3.2 Solid Mechanics 205
3.2.1 Hooke's Law...206
3.2.2 Stress 207
3.2.3 Strain 208
3.2.4 Poisson's Ratio...209
3.2.5 Stress Tensor 210
3.2.6 Triaxial Stresses and Strains 211
3.2.7 Principal Stresses and Strains 212
3.2.8 Equilibrium 213
3.2.9 Calculation of Stresses and Displacements 214
3.2.10 Moment of Inertia 215
3.2.11 Stresses and Displacements in a Beam 216
3.3 Signal Processing 218
3.3.1 Transfer Function 219
3.3.2 Transforms and Operators 220
3.3.3 Low Pass Filter – Integrator.................................221
3.3.4 Integrator 222
3.3.5 Differentiator...223
3.3.6 Mechanical Property Measurements 224
3.4 Fourier Optics 225
3.4.1 Diffraction...226
3.4.2 Fourier Transform – Graphical Approach 227
3.4.3 Fourier Transform – Single Frequency 228
3.4.4 Diffraction Grating...229
3.4.5 Single Slit 230
3.4.6 Double Slit 231
3.4.7 Fourier Transform – Mathematical Approach......232
3.4.8 Fourier Transform – Continuous Function 233
3.4.9 Lenses 234
3.4.10 Spatial Filtering...235
3.5 Quantum Mechanics 236
3.5.1 Quantum Mechanics 237

3.5.2 Solution to the Schrödinger Equation 238
3.5.3 Interpretation of the Wave Function 239
3.5.4 The Time-Dependent Equation 240
3.5.5 Normalisation and Expectation 241
3.5.6 Zero Potential 242
3.5.7 Particle in a Box 245
3.5.8 Infinite Square Well 246
3.5.9 Step Potential 250
3.5.10 Finite Square Well 251
3.5.11 Potential Barrier 252
3.5.12 Harmonic Oscillator 253
3.5.13 Coulomb Potential – Bohr Atom 254
3.5.14 Superposition 256
3.5.15 Orthogonal Eigenfunctions 257
3.5.16 Operator Notation 258
3.5.17 Commutators 259
3.5.18 Perturbation Theory 260
3.5.19 Perturbation Theory – Infinite Square Well 261
3.5.20 Harmonic Oscillator – First Order Perturbation 262
3.5.21 Harmonic Oscillator – Second Order Perturbation 263
3.5.22 Harmonic Oscillator – Eigenfunctions 264
3.5.23 Harmonic Oscillator – Exact Solution 265
3.5.24 Variational Method 266
3.5.25 Variational Method – Example 267
3.5.26 Helium Atom 269
3.5.27 Transitions 270
3.5.28 Transition Rate 271

Appendix 273
A.1 Useful Information 274
A.2 Some Standard Integrals 275
A.3 Special Functions 276

Index 277

Preface

Scientists use the language of mathematics to both describe and communicate physical principles. A significant portion of the content of university science degrees, particularly in physics, is therefore devoted to the study of mathematics. In most cases, mathematics in such courses is taught by mathematicians with the result that science students are taught as if they were mathematics majors. It is not surprising perhaps, therefore, to find that the pass rate for students in science courses in mathematics subjects is usually much lower than that for their other subjects.

This book has been written by a physicist. The book is intended to serve two purposes: (i) to provide a useful learning aid for science students by providing an alternative point of view to the recommended text for any particular course or subject in mathematics, and (ii) to serve as a handy reference for the professional physicist who just wants to refresh his knowledge, or find an equation or method of solution without having to wade through a formal mathematics text.

My own experience as a student learning mathematics was rather poor, mainly due to the enormous number of side issues, caveats, phrases such as "clearly" and "it can be shown," etc., that seemed to get in the way of any enjoyment of the subject. In this book, I have written what I believe is the essential core of mathematical principles needed for a useful knowledge of the subject. I hope that you will find the book useful and that it imparts a sense of beauty and wonder at this extraordinary field of human endeavour.

Finally, I thank John Navas for his sponsorship of the first edition and Francesca McGowan at Taylor & Francis for her continued support for this second edition.

Tony Fischer-Cripps,
Killarney Heights, Australia

Part 1

Mathematics Essentials

1.1 Numbers, trigonometry and analytical geometry
1.2 Limits and functions
1.3 Differentiation
1.4 Integration
1.5 Exponential and logarithmic functions
1.6 Trigonometric and hyperbolic functions
1.7 Methods of integration
1.8 Waves
1.9 Infinite series
1.10 Probability
1.11 Matrices

1.1 Numbers, Trigonometry and Analytical Geometry

Summary

$$z = a + bi$$ Complex number

$$r = |z| = \sqrt{a^2 + b^2}$$

$$e^{i\theta} = \cos\theta + i\sin\theta$$

$$(\cos\theta + i\sin\theta)^n = \cos n\theta + i\sin n\theta$$ de Moivre's theorem

$$e^{(a+bi)x} = e^{ax}(\cos bx + i\sin bx)$$ Euler's formula

$$d = \sqrt{(x_2 - x_1)^2 + (y_2 - y_1)^2 + (z_2 - z_1)^2}$$ Distance formula

$$\left(\frac{x_1 + x_2}{2}, \frac{y_1 + y_2}{2}, \frac{z_1 + z_2}{2}\right)$$ Mid-point formula

$$(x - h)^2 + (y - k)^2 = r^2$$ Equation of a circle

$$\frac{(x - h)^2}{a^2} + \frac{(y - k)^2}{b^2} = 1$$ Equation of an ellipse

$$x^2 = 4ay$$ Equation of a parabola

$$\frac{(x - h)^2}{a^2} - \frac{(y - k)^2}{b^2} = 1$$ Equation of a hyperbola

1.1.1 Real Numbers

Real numbers are all the numbers, positive and negative, integers and fractions from negative infinity to positive infinity, including zero. Real numbers are called real to distinguish them from **imaginary numbers**. Imaginary numbers also cover the range from minus infinity to plus infinity but have a different dimension from real numbers.

A **rational number** can be expressed as a fraction

$$\frac{a}{b}$$

where a and b are integers and b does not equal zero.

Numbers that are not rational numbers are **irrational numbers**. Examples are:

$$\sqrt{2};\pi;e$$

Rational numbers can be expressed as decimals which are either terminating or repeating. Irrational numbers can only be represented by non-terminating, non-repeating decimals. **Transcendental numbers** are irrational numbers that cannot be used to solve an algebraic equation.

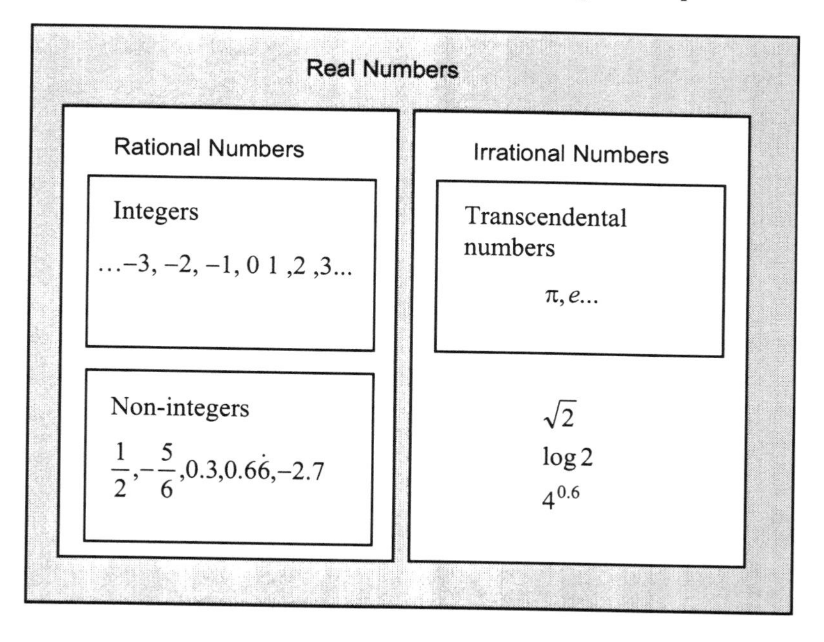

1.1.2 Complex Numbers

Imaginary numbers consist of rational and irrational numbers but are kept separated from the set of **real numbers**. There is nothing "imaginary" about imaginary numbers, this is just a label that distinguishes them from real numbers. Real and imaginary numbers are related by axes on the coordinate plane.

We customarily indicate imaginary numbers on the vertical axis, and real numbers on the horizontal axis.

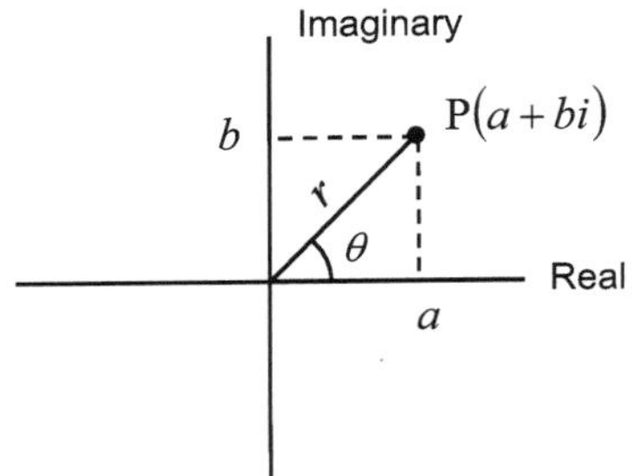

A combination of a real and an imaginary number is called a **complex number**.

A complex number has the form $z = a + bi$ where a is the real part and b is the imaginary part of the number. The symbol $\boldsymbol{i}$ is used to identify the imaginary part of a complex number.

A complex number is represented by a single point on the complex plane. The **magnitude of a complex number** is given by:

$$\boxed{r = |z| = \sqrt{a^2 + b^2}}$$

and represents the distance from the origin to the point. The magnitude of a complex number is sometimes called the **modulus** of the number. For mathematical reasons, the symbol i can be seen to be equal to the square root of −1. Most often, it is $i^2 = -1$ that is used in calculations.

It is readily seen that a complex number can be expressed in trigonometric terms:

$$\begin{aligned} a &= r\cos\theta \\ b &= r\sin\theta \\ z &= (r\cos\theta) + i(r\sin\theta) \\ &= r(\cos\theta + i\sin\theta) \end{aligned}$$

If $a + bi$ represents a complex number, then $a - bi$ is called the **complex conjugate** of the number. It can be seen that:

$$\begin{aligned} (a+bi)(a-bi) &= a^2 - abi + abi + b^2 \\ &= a^2 + b^2 \quad \text{a real number} \\ &= |z|^2 \end{aligned}$$

Complex numbers may be added and subtracted by adding and subtracting the real and imaginary parts separately.

$$(a+bi)+(c+di)=(a+c)+(b+d)i$$
$$(a+bi)-(c+di)=(a-c)+(b-d)i$$

Complex numbers may be multiplied by multiplying the terms as in a normal multiplication of factors as shown (where $i^2 = -1$):

$$\begin{aligned}(a+bi)(c+di) &= ac+adi+cbi+bdi^2 \\ &= (ac-bd)+(ad+bc)i\end{aligned}$$

or, in trigonometric terms:

$$\begin{aligned}z_1 &= r_1(\cos\theta_1 + i\sin\theta_1) \\ z_2 &= r_2(\cos\theta_2 + i\sin\theta_2) \\ z_1 z_2 &= r_1 r_2(\cos(\theta_1+\theta_2) + i\sin(\theta_1+\theta_2))\end{aligned}$$

Complex numbers may be divided by making use of the **complex conjugate** to transfer the imaginary part from the denominator to the numerator:

$$\begin{aligned}\frac{(a+bi)}{(c+di)} &= \frac{(a+bi)}{(c+di)}\frac{(c-di)}{(c-di)} \\ &= \frac{(a+bi)(c-di)}{(c^2+d^2)}\end{aligned}$$

or, in trigonometric terms:

$$\frac{z_1}{z_2} = \frac{r_1}{r_2}(\cos(\theta_1-\theta_2) + i\sin(\theta_1-\theta_2))$$

Complex numbers may be raised to a power. In trigonometric terms, this is known as **de Moivre's** theorem:

$$\boxed{(\cos\theta + i\sin\theta)^n = \cos n\theta + i\sin n\theta}$$

Euler's formula expresses a complex number in terms of an exponential function:

$$\boxed{e^{(a+bi)x} = e^{ax}(\cos bx + i\sin bx)}$$

$$C_1 e^{(a+bi)x} + C_2 e^{(a-bi)x} = e^{ax}(A\cos bx + iB\sin bx) \qquad \begin{aligned}A &= C_1 + C_2 \\ B &= C_1 - C_2\end{aligned}$$

1.1.3 Coordinate Systems

Linear types of coordinate systems are referred to as **Cartesian coordinate systems** and points in space are identified by the coordinates (x,y,z).

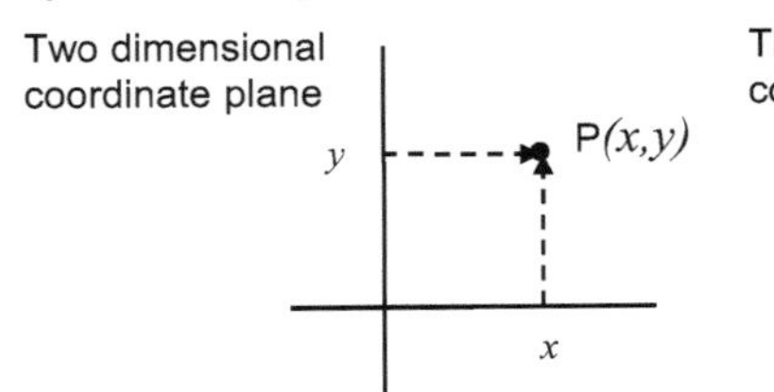

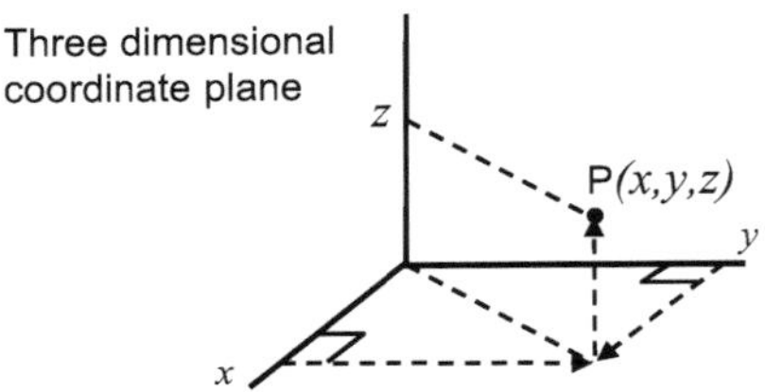

Distance formula:

$$d = \sqrt{(x_2 - x_1)^2 + (y_2 - y_1)^2 + +(z_2 - z_1)^2}$$

Midpoint formula:

$$\left(\frac{x_1 + x_2}{2}, \frac{y_1 + y_2}{2}, \frac{z_1 + z_2}{2}\right)$$

Where there is axial symmetry in a particular problem, a **cylindrical coordinate system** can be used. A point in space is given by the coordinate (r, θ, z).

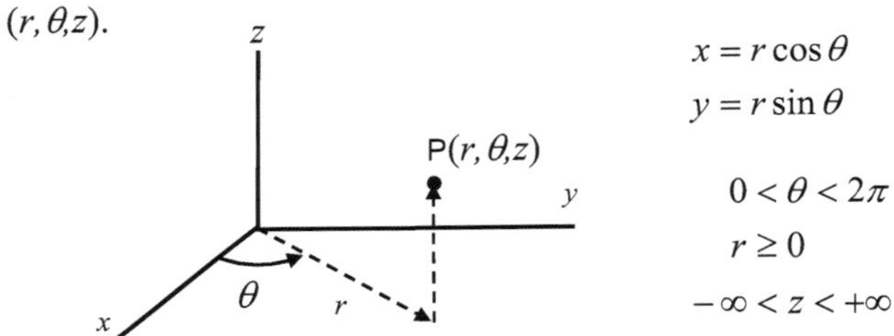

$$x = r\cos\theta$$
$$y = r\sin\theta$$

$$0 < \theta < 2\pi$$
$$r \geq 0$$
$$-\infty < z < +\infty$$

Where there is spherical symmetry in a particular problem, a **spherical coordinate system** can be used. A point in space is given by the coordinate (ρ, θ, ϕ).

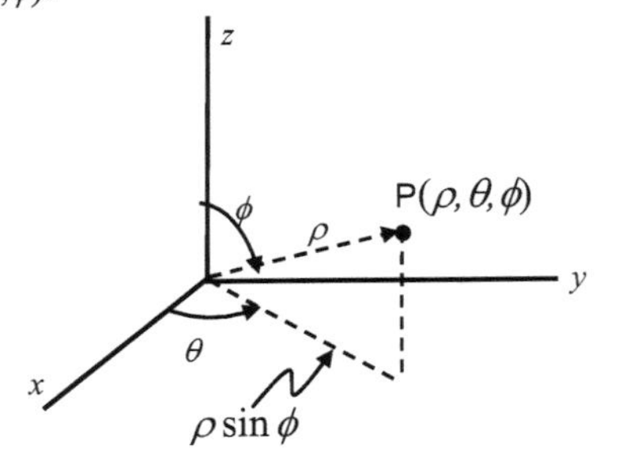

$$x = \rho\sin\phi\cos\theta$$
$$y = \rho\sin\phi\sin\theta$$
$$z = \rho\cos\phi$$

$$0 \leq \theta < \pi$$
$$\rho \geq 0$$
$$0 \leq \phi \leq 2\pi$$

$$\rho = \sqrt{x^2 + y^2 + z^2}$$

1.1.4 Vectors

Physical quantities that have a direction associated with them are called **vectors**.

Examples are displacement, velocity, acceleration, angular acceleration, torque, momentum, force, weight.

Physical quantities that do not have a direction associated with them are called **scalars**.

Examples are temperature, energy, mass, electric charge, distance.

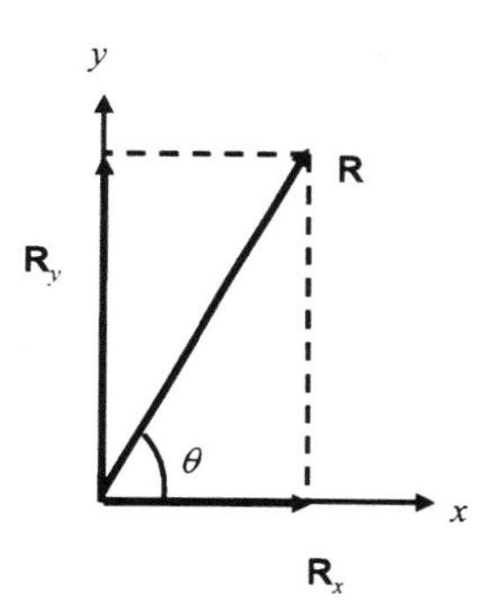

A vector acting at some angle to the coordinate axes can be represented by component vectors that lie along these axes.

$$|\mathbf{R}_x| = |\mathbf{R}|\cos\theta$$

$$|\mathbf{R}_y| = |\mathbf{R}|\sin\theta$$

The magnitude of resultant **R** of these two components is given by **Pythagoras' theorem**:

$$|\mathbf{R}| = \sqrt{|\mathbf{R}_x|^2 + |\mathbf{R}_y|^2}$$

The direction, or angle, of the resultant is found from:

$$\tan\theta = \frac{|\mathbf{R}_y|}{|\mathbf{R}_x|}$$

The division of a vector into horizontal and vertical components is a very useful concept. The idea also applies in three dimensions, where a vector has components along the x, y and z axes.

That is, the direction of the components of the vector is that of the corresponding coordinate axis. Let the magnitude of these components be A, B and C.

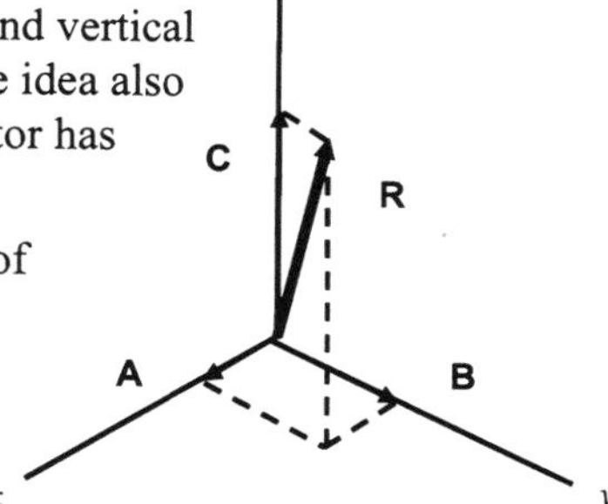

$$|\mathbf{R}| = \sqrt{A^2 + B^2 + C^2}$$

1.1.5 The Unit Vectors

It is often convenient to separate the magnitude and direction characteristics of a vector. This is done by using unit vectors, that is, vectors which lie along the coordinate axes and have a magnitude of one unit. Thus, let the vectors **i**, **j** and **k** have a magnitude of 1 unit, and have directions along the x, y and z axes, respectively. The vector **R** consists of the components:

$$\mathbf{R} = A\mathbf{i}, B\mathbf{j}, C\mathbf{k}$$

Why do this? It is a good way to keep the components of a vector organised. Indeed a great way of writing a vector in terms of its components is:

$$\mathbf{R} = A\mathbf{i} + B\mathbf{j} + C\mathbf{k}$$

magnitude ← (points to A)
direction → (points to $\mathbf{i}$)

This is a **vector equation**. We express a vector as the resultant of its component vectors. The magnitudes of the component vectors are the scalars A, B and C. When the unit vector **i** (which points in the x direction) is multiplied by the scalar magnitude A, we obtain $A\mathbf{i}$, which is the vector component in the x direction and so on for the product $B\mathbf{j}$ and $C\mathbf{k}$. The magnitude of **R** is given by:

$$|\mathbf{R}| = \sqrt{A^2 + B^2 + C^2}$$

When two vectors are to be added, we simply add together the corresponding magnitudes of the **i unit vector**, the **j** and the **k** unit vectors.

$$\mathbf{A} = A_1\mathbf{i} + B_1\mathbf{j} + C_1\mathbf{k}$$

$$\mathbf{B} = A_2\mathbf{i} + B_2\mathbf{j} + C_2\mathbf{k}$$

$$\begin{aligned}\mathbf{A} + \mathbf{B} &= A_1\mathbf{i} + B_1\mathbf{j} + C_1\mathbf{k} + A_2\mathbf{i} + B_2\mathbf{j} + C_2\mathbf{k} \\ &= (A_1 + A_2)\mathbf{i} + (B_1 + B_2)\mathbf{j} + (C_1 + C_2)\mathbf{k}\end{aligned}$$

> Different ways of writing vectors: $\mathbf{R}, \langle R \rangle, \overline{R}, \vec{R}, \hat{R}$

1.1.6 Trigonometry

The **trigonometric ratios sine, cosine,** and **tangent** are defined as:

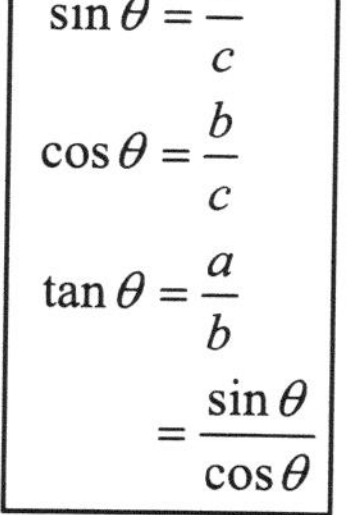

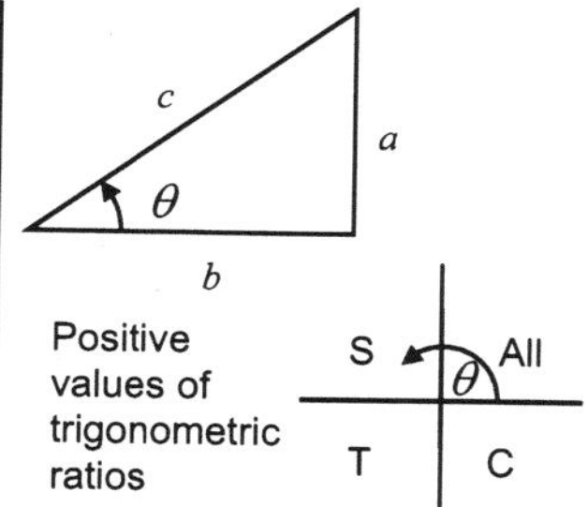

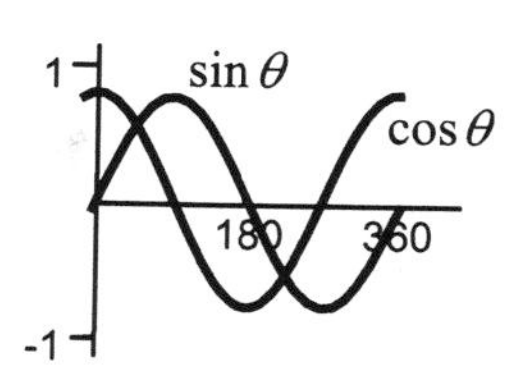

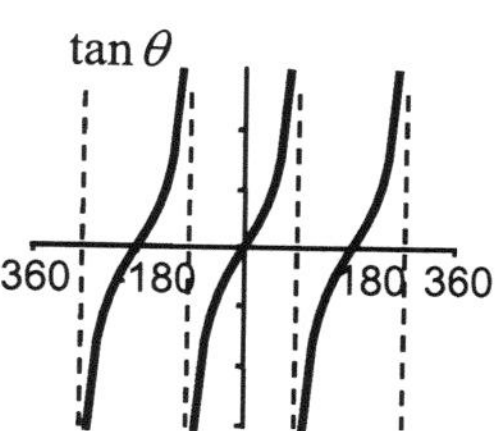

The **reciprocal trigonometric ratios cosecant, secant** and **cotangent** are defined as:

$$\csc\theta = \frac{1}{\sin\theta}; \sec\theta = \frac{1}{\cos\theta}; \cot\theta = \frac{1}{\tan\theta}$$

Trigonometric relationships:

$$\sec\theta = \operatorname{cosec}(90-\theta)$$
$$\tan\theta = \cot(90-\theta)$$
$$\sin\theta = \cos(90-\theta)$$

$$\sin(a \pm b) = \sin a\cos b \pm \cos a\sin b$$
$$\cos(a \pm b) = \cos a\cos b \mp \sin a\sin b$$
$$\tan(a \pm b) = \frac{\tan a \pm \tan b}{1 \mp \tan a\tan b}$$

$$\sin^2\theta + \cos^2\theta = 1$$
$$1 + \tan^2\theta = \sec^2\theta$$
$$1 + \cot^2\theta = \operatorname{cosec}^2\theta$$

$$\sin 2\theta = 2\sin\theta\cos\theta$$
$$\cos 2\theta = \cos^2\theta - \sin^2\theta$$
$$\tan 2\theta = \frac{2\tan\theta}{1-\tan^2\theta}$$

$$\sin^2\frac{\theta}{2} = \frac{1-\cos\theta}{2}$$
$$\cos^2\frac{\theta}{2} = \frac{1+\cos\theta}{2}$$
$$\tan\frac{\theta}{2} = \frac{1-\cos\theta}{\sin\theta}$$

$$\sin a \pm \sin b = 2\cos\frac{a \mp b}{2}\sin\frac{a \pm b}{2}$$
$$\cos a + \cos b = 2\cos\frac{a+b}{2}\cos\frac{a-b}{2}$$
$$\cos a - \cos b = 2\sin\frac{a+b}{2}\sin\frac{b-a}{2}$$

Sine rule and **cosine rule**:

$$a^2 = b^2 + c^2 - 2bc\cos A$$
$$b^2 = c^2 + a^2 - 2ca\cos B$$
$$c^2 = a^2 + b^2 - 2ab\cos C$$
$$\frac{a}{\sin A} = \frac{b}{\sin B} = \frac{c}{\sin C}$$

1.1.7 Straight Line

The equation of a straight line is called a **linear equation** and has the form:

$$y = mx + b$$

where m is the **gradient**, or **slope**, and b is the y axis **intercept**.

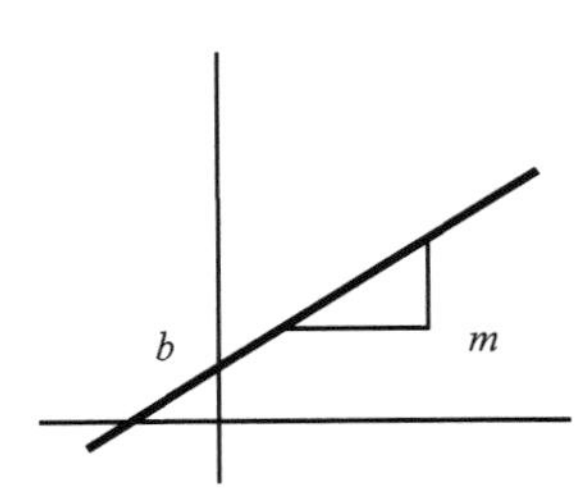

Given the coordinates of any two points on the line, the slope can be calculated from:

$$m = \frac{y_2 - y_1}{x_2 - x_1}$$

Two lines are parallel if $m_1 = m_2$. Two lines are perpendicular if $m_1 m_2 = -1$.

The angle between two lines of slopes m_1 and m_2 is given by:

$$\tan\theta = \frac{m_2 - m_1}{1 + m_2 m_1}$$

Different forms of equation of a straight line:

$$y - y_1 = \frac{y_2 - y_1}{x_2 - x_1} x_2 - x_1 \qquad \text{Two-point form}$$

$$y - y_1 = m(x - x_1) \qquad \text{Point-slope form}$$

The **general form of equation for a straight line** has the form:

$$Ax + By + C = 0$$

where the gradient is $m = -\frac{A}{B}$ and the intercept is $b = -\frac{C}{B}$

The perpendicular distance from a point $P(x, y)$ to a straight line is found from:

$$d = \frac{Ax + By + C}{\pm\sqrt{A^2 + B^2}}$$

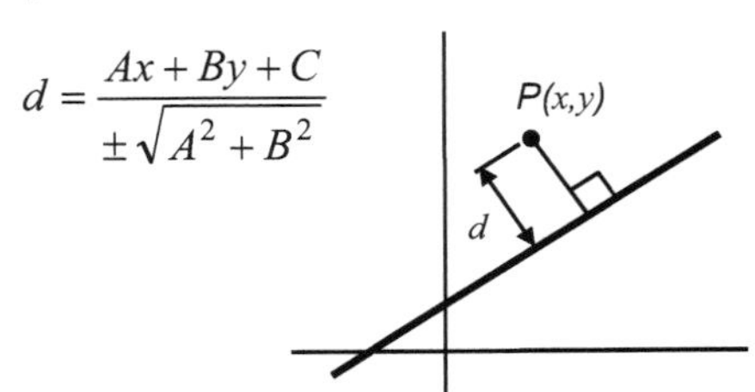

1.1.8 Circle and Ellipse

A **circle** is formally defined as the set of points on a plane such that the distance of each point to a given fixed point is a constant.

The **standard equation of a circle** with centre at C(h,k) and radius r is given by:

$$(x-h)^2+(y-k)^2=r^2$$

Alternately, the **general equation of a circle** can be found by expanding the terms to obtain:

$$x^2+y^2+ax+by+c=0$$

An **ellipse** is formally defined as the set of points on a plane such that the sum of the distance of each point to two fixed points (foci) is a constant.

The **standard equation of an ellipse** whose centre is at (h,k) is given by:

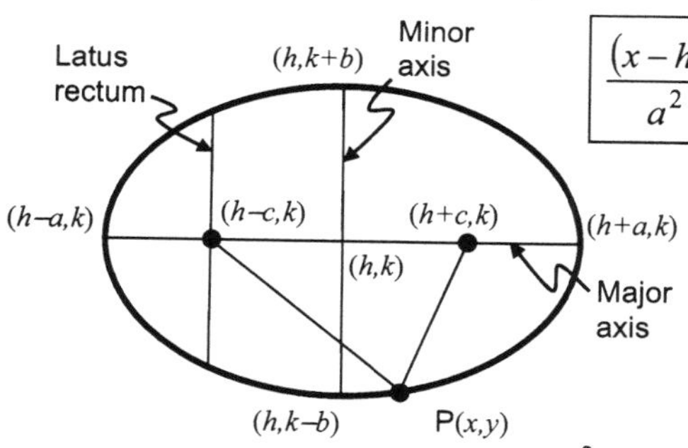

$$\frac{(x-h)^2}{a^2}+\frac{(y-k)^2}{b^2}=1$$ where $a>b$

The distance from $(-c,0)$ to P added to the distance from $(c,0)$ to P is a constant. The distances a, b and c are related by:

$$c^2=a^2-b^2$$

The length of the **latus rectum** is $\dfrac{2b^2}{a}$.

The ratio c/a is called the **eccentricity** of the ellipse.

When the **major axis** of the ellipse is in the direction of the y axis, we keep $a>b$ and write:

$$\frac{(y-k)^2}{a^2}+\frac{(x-h)^2}{b^2}=1$$

1.1.9 Parabola

A **parabola** is formally defined as the set of points on a plane such that the distance of each point to a given fixed point (the **focus**) and a fixed line (the **directrix**) is a constant.

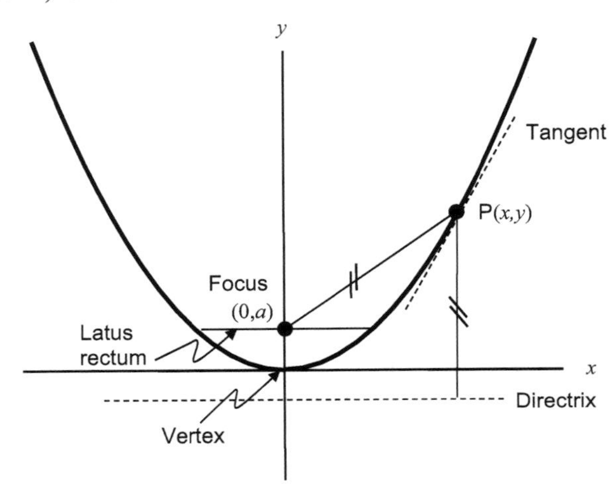

When the vertex of the parabola is at (0,0), then the coordinates of all points on the parabola satisfy the **standard equation of a parabola**:

$$\boxed{x^2 = 4ay}$$

or

$$y = \frac{x^2}{4a}$$

When a is > 0, the parabola opens upwards. When $a < 0$, the parabola opens downwards.

When the **vertex** of the parabola is at coordinates (h,k) with the axis parallel to the y axis, then the standard equation of a parabola becomes:

$$(x-h)^2 = 4a(y-k)$$

$$x^2 - 2xh + h^2 = 4ay - 4ak$$

$$y = \frac{x^2 - 2xh + h^2 + 4ak}{4a}$$

$$\boxed{y = Ax^2 + Bx + C}$$

1.1.10 Hyperbola

An **hyperbola** is formally defined as the set of points on a plane such that the difference of the distance of each point to two fixed points (foci) is a constant.

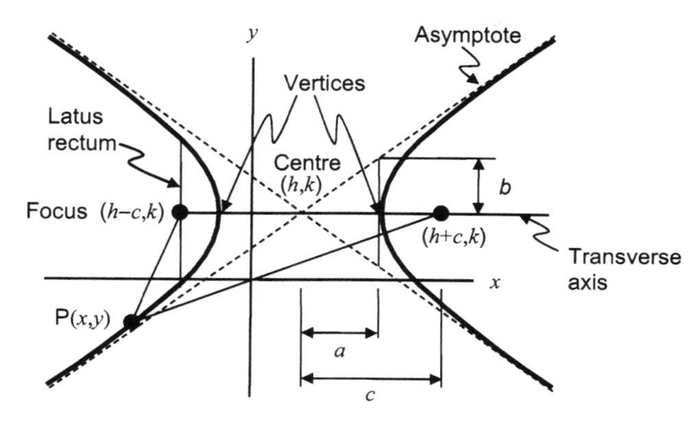

The difference in distance between the point P and one focus and the other focus is $2a$.

When the **centre** of the parabola is at coordinates (h,k) with the axis parallel to the x axis, then the **standard equation of a hyperbola** is:

$$\boxed{\frac{(x-h)^2}{a^2}-\frac{(y-k)^2}{b^2}=1} \qquad \text{where } c^2=a^2+b^2$$

The length of the **latus rectum** is $\dfrac{2b^2}{a}$.

The ratio c/a is called the **eccentricity** of the hyperbola.

The asymptotes are given by:

$$y-k=\pm\frac{b}{a}(x-h)$$

When the transverse axis is parallel to the y axis, we have:

$$\frac{(y-k)^2}{a^2}-\frac{(x-h)^2}{b^2}=1$$

1.2 Limits and Functions

Summary

$y = ax^2 + bx + c$	Quadratic polynomial
$\dfrac{-b \pm \sqrt{b^2 - 4ac}}{2a}$	Quadratic formula
$\lim\limits_{x \to a} f(x) = L$	Limit
$\lim\limits_{x \to a^+} f(x) = L$ $\lim\limits_{x \to a^-} f(x) = L$	One-sided limits

1.2.1 Functions

A **function** is an ordered set of pairs of numbers in which one number, y, depends in a particular way upon the value of the other number, x. When y is a function of x, we say that y is the **dependent variable** and x is the **independent variable**. The value of y at a particular value of x is written:

$$y = f(x)$$

where f can be thought as an **operator** which acts upon x to give y.

The **domain** of a function is the set of all values of x. For example, consider the function $y = x^2$. The domain of this function is the set of all real numbers.

The **range** of a function is the set of values for y. In the above example, the range of the function is zero and all positive numbers.

Functions can be described by equations and graphs. Graphs are useful for visualising how y changes as x varies.

A function is said to be **odd** if: $f(-x) = -f(x)$

A function is said to be **even** if: $f(-x) = f(x)$

A function which has the form:

$$y = a_n x^n + a_{n-1} x^{n-1} + \dots a_1 x + a_o$$

where n is a positive integer (or zero) and a_n are constants is called a **polynomial function**. The highest value of n gives the degree of the polynomial. For the case of $n = 1$, a graph of $y = ax + b$ is a straight line and is thus called a **linear function**.

It is important to note that the word function implies a one to one correspondence between the dependent and independent variables.

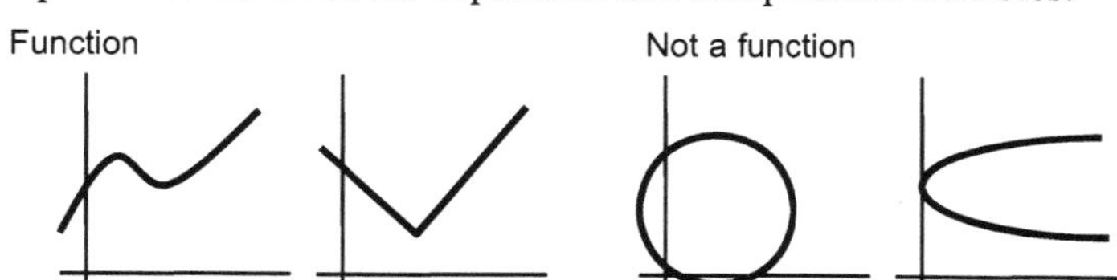

A function is an ordered set of pairs of numbers in which no two pairs of numbers have the same value of x. If the function is such that no two pairs have the same value of y, then there exists an **inverse function** $g(y)$.

Since $g(y)$ is also a function, then we say that $g(y)$ is the inverse function of $f(x)$. A function can have at most only one inverse function. Not every function has an inverse function.

1.2.2 Quadratic Function

A quadratic function is a polynomial function of degree 2 and has the form:

$$y = ax^2 + bx + c$$

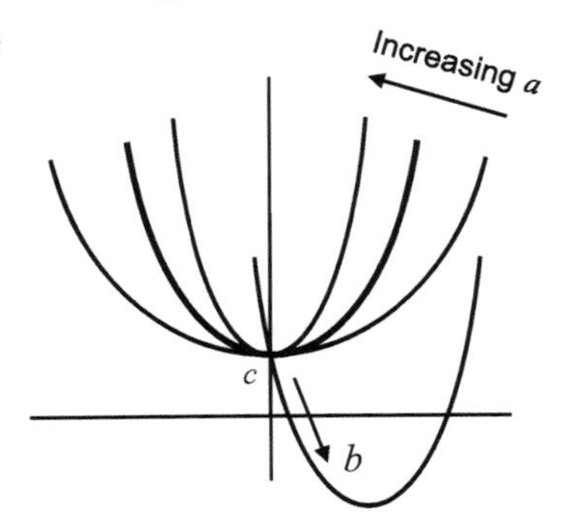

A graph of $y = f(x)$ in this case is a **parabola**. The value of a determines the rate of change of the parabola. A positive value for a represents a curve which is concave upwards, a negative value concave downwards. For large values of x, the term ax^2 dominates the value of the function. The value of c determines the vertical position of the parabola. The value of b tends to linearise the function and shifts the minimum value off-centre.

A **quadratic equation** arises when we wish to find where the function crosses the x axis.

$$ax^2 + bx + c = 0$$

The **roots** of the quadratic equation are the values of x that satisfy the quadratic equation. The roots are found by factorisation, completing the square, or by the **quadratic formula**:

$$\frac{-b \pm \sqrt{b^2 - 4ac}}{2a}$$

The quantity $b^2 - 4ac$ has special significance and is called the **discriminant**:

$b^2 - 4ac > 0$ crosses x axis at two points, 2 roots.

$b^2 - 4ac = 0$ touches the x axis at a single point.

$b^2 - 4ac < 0$ lies completely above or below the x axis.

Later we shall see that complex roots can be defined for the case of $b^2-4ac < 0$.

A quadratic equation can have no roots (lies completely above or below the x axis), one (touches the x axis) or two roots but no more than two since it can only cross the x axis ($y = 0$) no more than twice.

Given the roots α, β of a quadratic equation, we can construct the equation from:

$$\alpha + \beta = -\frac{b}{a} \quad \text{and} \quad \alpha\beta = \frac{c}{a}$$

1.2.3 Limits

The **limit** L of a function is that number that the function approaches as the independent variable approaches (but does not necessarily become equal to) a certain value a.

$$\lim_{x \to a} f(x) = L$$

Often, the function $f(x)$ is undefined at $x = a$. In general, there may or may not be a limit as x approaches a.

For example, the function:

$$f(x) = \frac{1}{x}$$

has no limit as x goes to 0.

When the function $y = f(x)$ increases or decreases without limit as x approaches some value a, then the line $x = a$ is a vertical **asymptote** of the function. Horizontal and oblique asymptotes are also possible.

A function may have a **left limit** or a **right limit.** For example, a particular function may approach a limit when x approaches a but x is always greater than a (i.e., from the right). Or, when x approaches a but x is always less than a (i.e., from the left). In such cases, we write the following notation for **one-sided limits**: $\lim_{x \to a^+} f(x) = L$ and $\lim_{x \to a^-} f(x) = L$

Determining the existence of a limit or the value of a limit can require significant ingenuity and algebraic manipulation of the form of the function. Consider the following examples:

$$\lim_{x \to 2} \frac{x+3}{x+2} = \frac{2+3}{2+2} = \frac{5}{4}$$

$$\lim_{x \to 1} \frac{x^2 - 1}{x - 1} = \lim_{x \to 1} \frac{(x-1)(x+1)}{(x-1)} = 2$$

$$\lim_{x \to -3} \frac{x^2 + 4x + 3}{x + 3} = \lim_{x \to -3} \frac{(x+3)(x+1)}{(x+3)} = -2$$

$$\lim_{x \to \infty} \frac{x^2 - 2x + 3}{2x^2 + 5x - 3} = \lim_{x \to \infty} \frac{1 - \frac{2}{x} + \frac{3}{x^2}}{2 + \frac{5}{x} - \frac{3}{x^2}} = \frac{1}{2}$$

It is important to note that this result (and similar treatments) say that the function as it is originally written approaches a value of 2 as x approaches 1 but $x = 1$ is excluded since, from the point of view of the original function, the function is undefined at $x = 1$. The discontinuity at $x = 1$ is said to be removable since the function can be rearranged and expressed in an alternate form where the value of $f(x)$ is defined. However, the limit need not be equal to $f(a)$. If this is the case, then the function is said to be continuous at $x = a$.

Oftentimes limits exist for $x \to a$ even if the function is undefined at $x = a$. The concept of a limit is partly a philosophical one and an understanding of the nature of limits is often something that takes time and experience to achieve. Much like irrational numbers, limits sometimes appear physically intuitive even though they are difficult to express numerically.

1.2.4 Theorems on Limits

Determination of a limit of a function can be facilitated by making use of standard **theorems on limits** which have been shown to be true:

$$\lim_{x\to a} c = c$$ the limit of a constant.

$$\lim_{x\to a}(mx+b) = ma+b$$ the limit of a linear equation is found by direct substitution.

If $f(x)$ is a polynomial function, then

$$\lim_{x\to a} f(x) = f(a)$$

$$\lim_{x\to a}[f(x)+g(x)] = \lim_{x\to a} f(x) + \lim_{x\to a} g(x)$$ Limit of a sum

$$\lim_{x\to a}[f(x)g(x)] = \lim_{x\to a} f(x) \lim_{x\to a} g(x)$$ Limit of a product

$$\lim_{x\to a}\left[\frac{f(x)}{g(x)}\right] = \frac{\lim_{x\to a} f(x)}{\lim_{x\to a} g(x)}$$

Limit of a quotient

To evaluate the limit of a quotient which has an indeterminate form 0/0 or ∞/∞ at x equal to some number a, **L'Hôpital's rule** may be used:

$$\lim_{x\to a}\frac{f(x)}{g(x)} = \lim_{x\to a}\frac{f'(x)}{g'(x)}$$

$$\lim_{x\to a}[f(x)]^n = \left[\lim_{x\to a} f(x)\right]^n$$ Limit of a function raised to power

Sandwich theorem

$$\text{If } f(x) \le h(x) \le g(x)$$
$$\text{and } \lim_{x\to a} f(x) = L = \lim_{x\to a} g(x)$$
$$\text{then } \lim_{x\to a} h(x) = L$$

The **formal definition of a limit** is $\lim_{x\to a} f(x) = L$

For every $\varepsilon > 0$, there exists $\delta > 0$ such that: if $0 < |x-a| < \delta$

then $|f(x)-L| < \varepsilon$

A function $f(x)$ approaches the limit L as x approaches some value a if, when ε is a given positive number (however small), a positive number δ can be found such that $|f(x) - L| < \varepsilon$ whenever $|x-c| < \delta$. A function $f(x)$ is **continuous** at $x = a$ if $f(a)$ is defined, the limit of the function as x approaches a exists, and the limit as x approaches a is given by $f(a)$.

1.3 Differentiation

Summary

$$\frac{dy}{dx} = \lim_{\Delta x \to 0} \frac{f(x+\Delta x) - f(x)}{\Delta x}$$ Derivative

$$y = ax^n; \frac{dy}{dx} = nx^{n-1}$$ Power rule

$$\frac{dy}{dx} = f(x)\frac{dg}{dx} + g(x)\frac{df}{dx}$$ Product rule

$$\frac{dy}{dx} = \frac{g(x)\frac{df}{dx} - f(x)\frac{dg}{dx}}{(g(x))^2}$$ Quotient rule

$$\frac{dy}{dx} = \frac{dy}{du}\frac{du}{dx}$$ Chain rule

$$\frac{d}{dx}f(g(x)) = \frac{df}{dg}\frac{dg}{dx}$$ Function of a function

$$\frac{dy}{dx} = n(f(x))^{n-1}\frac{d}{dx}f(x)$$ Power rule for functions

Local max	Local min	Points of inflection	
$f'(a-\varepsilon) > 0$	$f'(a-\varepsilon) < 0$	$f'(a-\varepsilon) < 0$	$f'(a-\varepsilon) > 0$
$f'(a) = 0$	$f'(a) = 0$	$f'(a) = 0$	$f'(a) = 0$
$f'(a+\varepsilon) < 0$	$f'(a+\varepsilon) > 0$	$f'(a+\varepsilon) < 0$	$f'(a+\varepsilon) > 0$

$$f(x) = ax^n; F(x) = \frac{a}{n+1}x^{n+1} + C$$

Power rule for anti-derivatives

1.3.1 Derivative

Differential calculus is concerned with the calculation of the rate of change of a function with respect to the independent variable.

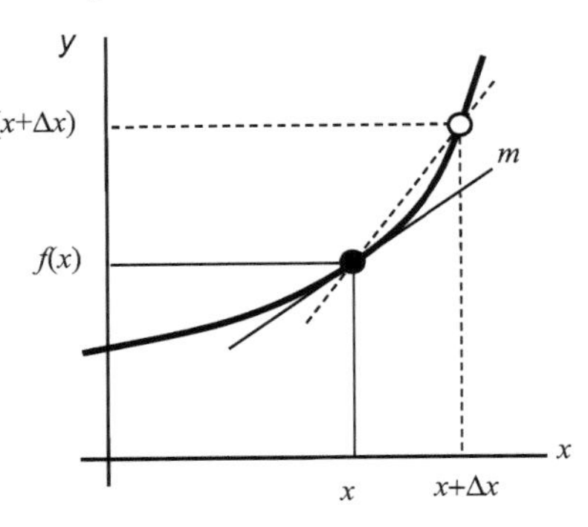

Consider a function $y = f(x)$. The **instantaneous rate of change** of $f(x)$ with respect to x is the slope of the **tangent** to the graph of the function at x. If we select two points on the curve, and draw a straight line between them (a **secant** line), then the slope of that line is given by:

$$m = \frac{y_2 - y_1}{x_2 - x_1}$$

$$= \frac{f(x+\Delta x) - f(x)}{\Delta x}$$

The smaller we make Δx, the closer the calculation of slope is to the slope of the tangent at x. That is, expressed as a limit:

$$m = \lim_{\Delta x \to 0} \frac{f(x+\Delta x) - f(x)}{\Delta x}$$

The slope, or the rate of change of y with respect to x, is called the **derivative** of y with respect to x. There are various forms of notation for expressing the derivative:

$$\frac{dy}{dx};\, f'(x);\, \dot{y};\, D_x(x);\, \frac{d}{dx}f(x)$$

It is evident from the above notation that $\dfrac{dy}{dx} = f'(x)$ or $dy = f'(x)dx$

Expressed in this way, Δy and Δx are **differentials** of the variables y and x.

Not all functions are differentiable. For the derivative of a function to exist at x, the function must be **continuous** at x. However, not all functions which are continuous can be differentiated.

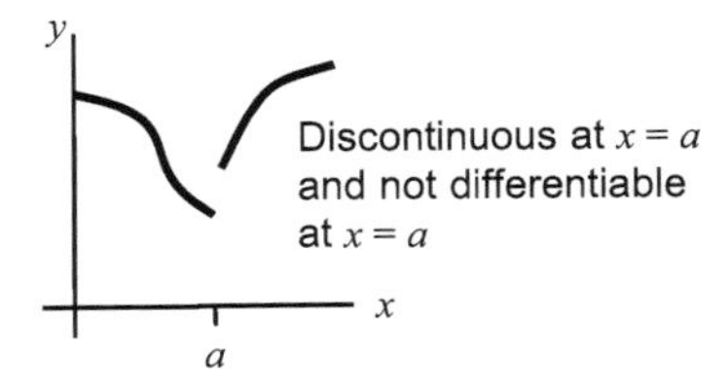

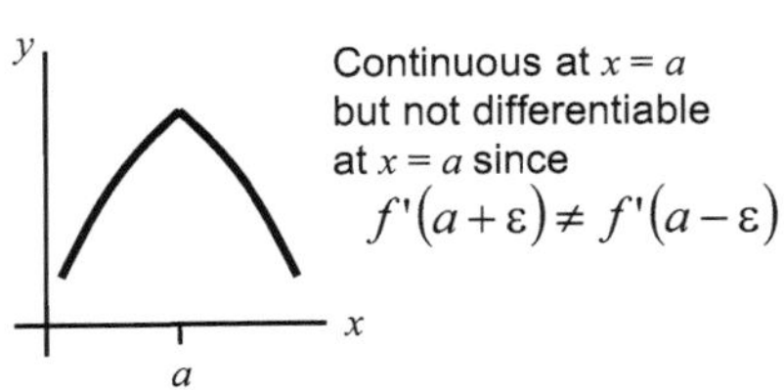

1.3.2 Rules for Calculating the Derivative

It is evident that calculation of the **derivative** of a function involves finding the limit of a function, which is oftentimes inconvenient. Standard results for a range of functions simplify the procedure. More complicated functions can be treated by expressing them as combinations of standard forms.

If $y = c$, then the rate of change of the function with respect to x is zero. The **derivative of a constant** is zero.

$$y = c$$
$$\frac{dy}{dx} = 0$$

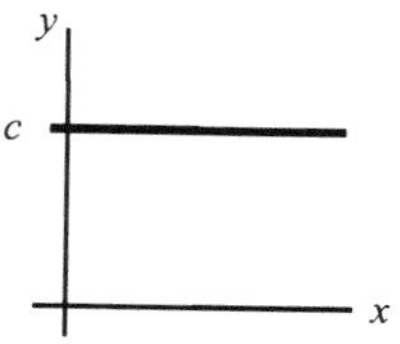

If $y = x$, it can be seen that the slope of the function $y = x$ is 1. More formally, we can write:

$$y = x$$
$$\frac{dy}{dx} = 1$$

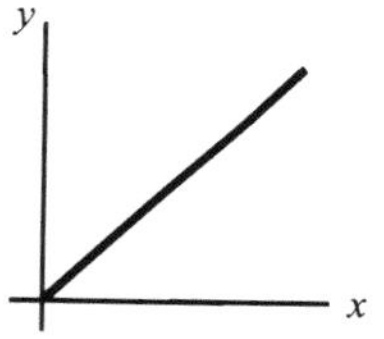

$$\begin{aligned}\frac{dy}{dx} &= \lim_{\Delta x \to 0} \frac{f(x+\Delta x) - f(x)}{\Delta x} \\ &= \lim_{\Delta x \to 0} \frac{x + \Delta x - x}{\Delta x} \\ &= \lim_{\Delta x \to 0} 1 \\ &= 1\end{aligned}$$

One of the most widely used standard results for finding derivatives is that for powers of x:

$$y = x^n$$
$$\frac{dy}{dx} = \lim_{\Delta x \to 0} \frac{(x+\Delta x)^n - x^n}{\Delta x}$$

By the **binomial theorem**:

$$(x+\Delta x)^n = x^n + nx^{n-1}\Delta x + \frac{n(n-1)}{2!}x^{n-2}\Delta x^2 + \ldots + nx\Delta x^{n-1} + \Delta x^n$$

Thus, the limit becomes:

$$\frac{dy}{dx} = \lim_{\Delta x \to 0}\left[nx^{n-1} + \frac{n(n-1)}{2!}x^{n-2}\Delta x^1 + \ldots + nx\Delta x^{n-2} + \Delta x^{n-1}\right]$$

As $\Delta x \to 0$, all the terms but the first remain and so:

$$\boxed{\frac{dy}{dx} = nx^{n-1}}$$ **Power rule**

Sums and differences:

$$y = (f(x) + g(x))$$
$$\frac{dy}{dx} = \frac{df}{dx} + \frac{dg}{dx}$$

Product rule:

$$y = (f(x)g(x))$$
$$\boxed{\frac{dy}{dx} = f(x)\frac{dg}{dx} + g(x)\frac{df}{dx}}$$

The derivative is given by the first times the derivative of the second plus the second times the derivative of the first.

Quotient rule:

$$y = \frac{f(x)}{g(x)}$$
$$\boxed{\frac{dy}{dx} = \frac{g(x)\frac{df}{dx} - f(x)\frac{dg}{dx}}{(g(x))^2}}$$

The derivative is given by the bottom times the derivative of the top minus the top times the derivative of the bottom all over the bottom squared.

Chain rule:

$$y = f(u)$$
$$u = g(x)$$
$$\boxed{\frac{dy}{dx} = \frac{dy}{du}\frac{du}{dx}}$$

Function of a function:

$$\boxed{\frac{d}{dx}f(g(x)) = \frac{df}{dg}\frac{dg}{dx}}$$

Power rule for functions:

$$y = (f(x))^n$$
$$\boxed{\frac{dy}{dx} = n(f(x))^{n-1}\frac{d}{dx}f(x)}$$

The derivative is given by the derivative of the brackets times the derivative of what's inside.

Example:
In some cases, y might be a function of x but cannot be explicitly written in this way. In such cases, y is called an **implicit function** of x. The derivative dy/dx can still be found. Find dy/dx of the following function:

$$2x^3 + x^2y + y^3 = 1$$

Solution:

$$\frac{d}{dx}2x^3 + \frac{d}{dx}x^2y + \frac{d}{dx}y^3 = \frac{d}{dx}1$$
$$6x^2 + (x^2)\frac{dy}{dx} + y(2x) + 3y^2\left(\frac{dy}{dx}\right) = 0$$
$$\frac{dy}{dx} = \frac{-6x^2 - 2xy}{x^2 + 3y^2}$$

1.3.3 Higher Order Derivatives

In general, the derivative of a function produces another function which itself may be differentiable. This second function is called the **second derivative** of the original function. There are several ways of writing the second derivative:

$$\frac{d^2y}{dx^2}; f''(x); \ddot{y}; D^2{}_x(x); \frac{d^2}{dx^2}f(x); \frac{d}{dx}\left(\frac{dy}{dx}\right)$$

Physically, the second derivative gives a measure of the rate of change or the rate of change of the original function. For example, for a linear function, we have:

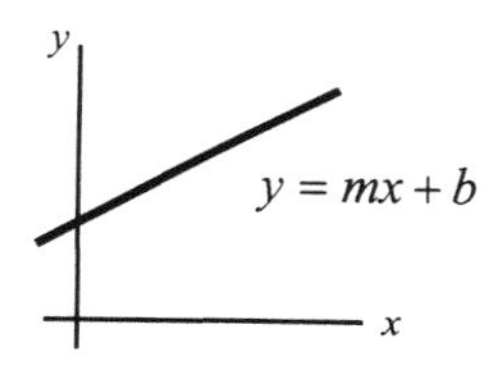

The slope or first derivative at any point on this function is a constant (since it is a linear function) so: $m = \frac{dy}{dx}$

In this case, since the rate of change of the function is a constant, the second derivative is zero: $\frac{d^2y}{dx^2} = \frac{d}{dx}m = 0$

For a second degree **polynomial function**, we have:

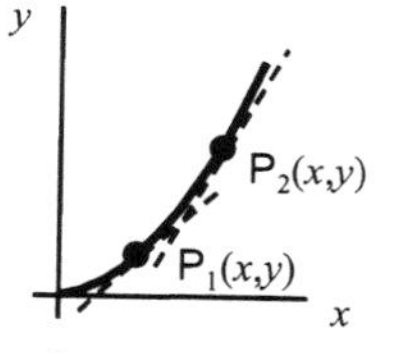

$$y = ax^2 + bx + c$$

In this case, the slope of the tangent is a function of x. That is, the slope of the tangent at any point x depends on where x is. As x increases, the slope of the tangent increases. The value of the slope of the tangent is given by dy/dx, which is a linear function in x. The rate of change of the slope of the tangent is given by the second derivative, which, in this case, is a constant. That is, the slope of the tangent increases at a uniform rate.

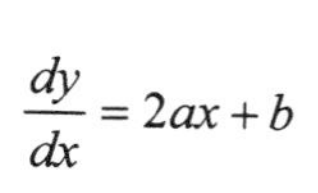

$$\frac{dy}{dx} = 2ax + b$$

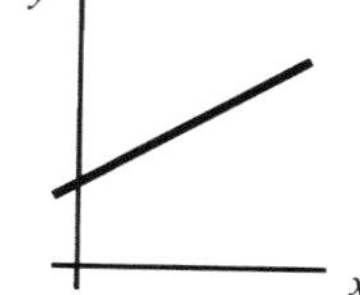

$$\frac{d^2y}{dx^2} = 2a$$

Higher order derivatives are found by taking the derivative of the preceding order. The notation is:

$$\frac{dy}{dx}; \frac{d^2y}{dx^2}; \frac{d^3y}{dx^3}; \frac{d^4y}{dx^4} \cdots \frac{d^ny}{dx^n}$$

1.3.4 Maxima and Minima

The first derivative of a function evaluated at a point provides a measure of the **instantaneous rate of change** of the function at that point. When the instantaneous rate of change of a function is equal to zero, we have what is called a **stationary point**. Some stationary points are local maxima or minima, while others are horizontal **points of inflection**.

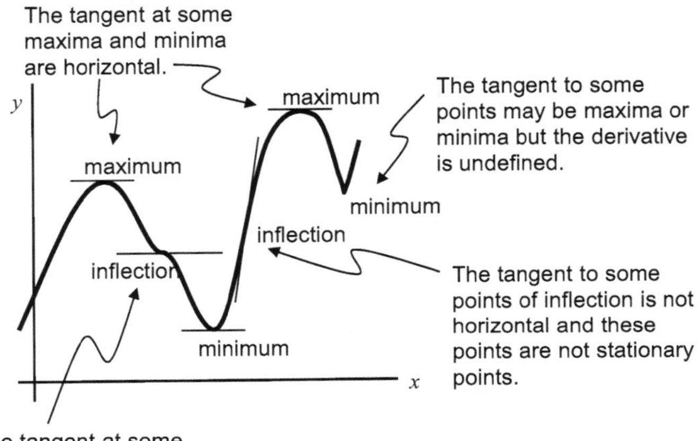

Maxima and minima in a function are called **turning points** or **extrema**. The points at which the first derivative is zero or is undefined are called **critical points**. Extrema of the function may occur at the end points of the function, in which case they are referred to as **end point extrema**. In many cases, stationary points of a function can be determined when $dy/dx = 0$. This is called the **first derivative test** – which is reliable. Further tests are required to determine the nature of the stationary point. If the first derivative is evaluated at $x = a$, for $x = a + \varepsilon$ and $x = a - \varepsilon$, then:

Local maximum	Local minimum	Horizontal points of inflection	
$f'(a-\varepsilon)>0$	$f'(a-\varepsilon)<0$	$f'(a-\varepsilon)<0$	$f'(a-\varepsilon)>0$
$f'(a)=0$	$f'(a)=0$	$f'(a)=0$	$f'(a)=0$
$f'(a+\varepsilon)<0$	$f'(a+\varepsilon)>0$	$f'(a+\varepsilon)<0$	$f'(a+\varepsilon)>0$

More information about the nature of stationary points is given by the second derivative of the function.

Caution: There are some instances where the tests above do not work.

1.3.5 The Second Derivative

Some information about the nature of stationary points in a function can be obtained by examining the sign of the **second derivative**.

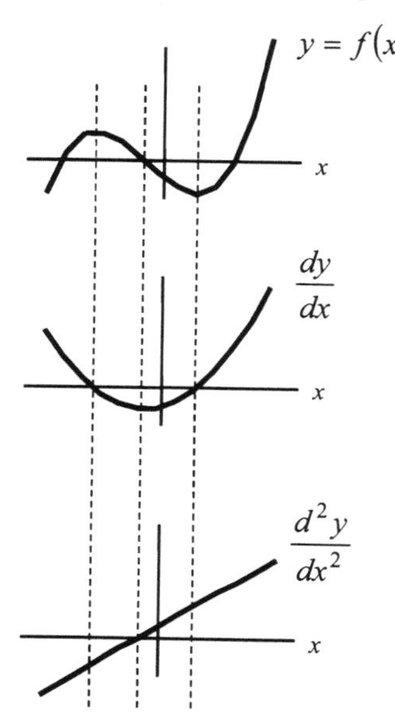

The second derivative gives the rate of change of the first derivative. The physical significance of this is that if the first derivative is zero at any particular point $x = a$, then the sign of the second derivative indicates whether the stationary point is a **maximum, minimum** or **point of inflection**.

The **second derivative test** also gives information about the **concavity** of the original function. When the second derivative is greater than zero, the function is concave upwards. When the second derivative is less than zero, the function is concave downwards.

Local maximum

$f'(a) = 0$
$f''(a) < 0$

Function is concave downwards.

Local minimum

$f'(a) = 0$
$f''(a) > 0$

Function is concave upwards.

Any point of inflection

$f''(a-\varepsilon) < 0 \quad f''(a-\varepsilon) > 0$
$f''(a) = 0 \quad f''(a) = 0$
$f''(a+\varepsilon) < 0 \quad f''(a+\varepsilon) > 0$

Change in concavity of the function. At points of inflection, the tangent crosses the curve. Points of inflection can be of any slope, including horizontal or vertical. For vertical points of inflection, $f''(a)$ may not exist.

Some caution is needed for the second derivative test. The rules given here apply for many cases but in some cases, they do not work. For example, one can have a local minimum when $f'(a) = f''(a) = 0$ e.g., $y = -x^4$ at $x = 0$

$$f'(a-\varepsilon) > 0; f'(a+\varepsilon) < 0$$

1.3.6 Curve Sketching

Curve sketching leads to physical intuition, which can help solve the problem at hand.

Properties and features of a curve:

Extent: Describes the domain (in the x direction) and the range (in the y direction). Usually, only real values of x and y are used in curve sketching.

Symmetry: $f(x,y) \equiv f(x,-y)$ Symmetrical about the x axis

$f(x,y) \equiv f(-x,y)$ Symmetrical about the y axis

$f(x,y) \equiv f(-x,-y)$ Symmetrical about the origin

$f(x,y) \equiv f(y,x)$ Symmetrical about the line $y = x$

Intercepts: Find where the function $f(x,y)$ touches or crosses the coordinate axes. x intercept when $y = 0$, y intercept when $x = 0$.

Vertical and horizontal **asymptotes**: When the function $y = f(x)$ increases or decreases without limit as x approaches some value a, then the line $x = a$ is a **vertical asymptote** of the function. If $y = f(x)$ approaches a value a as x increases or decreases without limit, then the line $y = a$ is a **horizontal asymptote**.

Oblique asymptotes:

If $y = f(x)$ can be written in the form $y = ax + b + h(x)$ and $\lim_{x\to\infty} h(x) = 0$ or $\lim_{x\to-\infty} h(x) = 0$ then the function is asymptotic to $y = ax + b$

Critical points: When: $f'(x) = 0$

$f'(x) = \text{undefined}$

Extrema at maxima, minima or end points.

Concave up: $f''(x)$ is +

Concave down: $f''(x)$ is –

Points of inflection: $f''(0) = 0$ and $f''(x)$ changes sign through x

1st derivative test when $f'(x) = 0$ and observe change of sign through x

2nd derivative test when $f''(x) = $ –ve, +ve or 0

If $f'(x) = 0$ and $f''(x) = 0$, then x may be a horizontal point of inflection (subject to 2nd derivative test).

1.3.7 Time Rate of Change

The derivative of a function plays a very important role in determining the rate of change of physical quantities with respect to time. For example, the **average velocity** for a body in motion can be determined from the total distance travelled divided by the time taken:

$$v_{av} = \frac{\Delta s}{\Delta t}$$

The smaller the time interval selected, the more representative is the calculation of the **instantaneous velocity** at a particular time t.

$$v = \lim_{\Delta t \to 0} \frac{\Delta s}{\Delta t} = \frac{ds}{dt}$$

If the velocity is not a constant but itself varies with time, then the second derivative gives the **instantaneous acceleration** of the body at a particular time t.

$$a = \lim_{\Delta t \to 0} \frac{\Delta v}{\Delta t} = \frac{dv}{dt}$$

Expressed in a slightly different way, we have:

$$v = \frac{ds}{dt} \qquad \text{and} \qquad a = \frac{dv}{dt}$$

$$dt = \frac{ds}{v} \qquad\qquad = \frac{dv}{ds}v$$

Therefore: $\boxed{a\,ds = v\,dv}$ This is called a **differential equation** because it contains **differentials** (ds and dv).

Example:

The volume of liquid in a spherical tank which has a depth h from the bottom of the tank is given by: $V = \frac{1}{3}\pi h^2(3a - h)$

If the tank is being filled with water at a rate of Q litres per minute, determine an expression for the rate of rise of the level of liquid in the tank for a given value of h.

Solution:

Let $u = h^2(3a - h)$

$$V = \frac{1}{3}\pi u$$

$$\frac{dV}{dt} = \frac{1}{3}\pi\frac{du}{dt}$$

$$\frac{du}{dt} = \frac{du}{dh}\frac{dh}{dt}$$

$$= \left(6ah - 3h^2\right)\frac{dh}{dt}$$

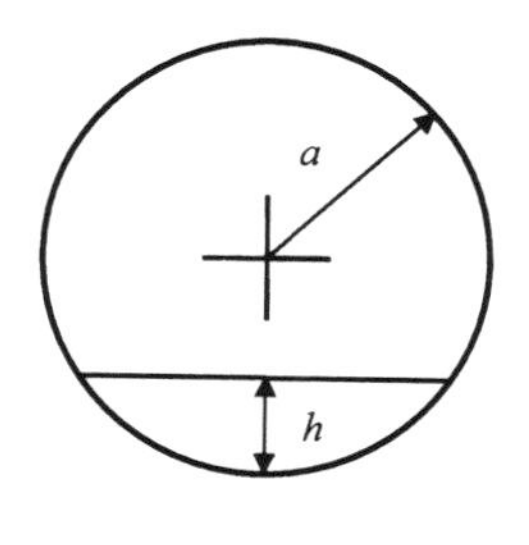

Thus: $Q = \frac{1}{3}\pi\left(6ah - 3h^2\right)\frac{dh}{dt}$ from which dh/dt can be found.

1.3.8 Anti-Derivatives

The derivative of a function represents the rate of change of that function with respect to the independent variable. The derivative usually is a function of that variable on its own. In many situations, we are given the derivative but wish to find the original function. A function whose derivative is given is called the **anti-derivative**.

Formally stated, $F(x)$ is the anti-derivative of $f(x)$ if:

$$\frac{dF}{dx} = f(x)$$

Because the derivative of a constant is zero, it is possible that there are many anti-derivatives of a function all being different by a constant term. For example, consider the function $f(x) = 2x^3$

Possible anti-derivatives of $f(x)$ are: $F(x) = 0.5x^4 + 2$

$$F(x) = 0.5x^4 + 6\times10^4$$

$$F(x) = 0.5x^4 + 0.001$$

$$F(x) = 0.5x^4$$

In general, we write: $F(x) + C$ as being the anti-derivative of $f(x)$ where C is an **arbitrary constant**.

The procedure for finding the anti-derivative of a function is often a reversal of the rules for finding the derivatives of functions (but being careful to add the arbitrary constant). The most commonly used rule is the **power rule for anti-derivatives**:

$$f(x) = ax^n$$

$$\boxed{F(x) = \frac{a}{n+1}x^{n+1} + C} \qquad n \neq -1$$

In many cases, finding the anti-derivative of a function involves a manipulation of the function into a form in which the power rule can be employed.

Anti-derivatives are closely connected with the concept of **integration** of the function $f(x)$.

1.4 Integration

Summary

$A=\int_a^b f(x)dx=\lim_{\Delta x\to 0}\sum_{i=1}^{\infty} f(x_i)\Delta x$	Definite integral
$A=\int_a^b f(x)dx=f(z)(b-a)$	Mean value theorem
$A=\int_a^b f(x)dx=[F(x)]_a^b=F(b)-F(a)$	Fundamental theorem of calculus
$\int f(x)dx=F(x)+C$	Indefinite integral
$\int_a^b cf(x)dx=c\int_a^b f(x)dx$	Multiplication by constant
$\int_a^b (f(x)+g(x))dx=\int_a^b f(x)dx+\int_a^b g(x)dx$	Addition and subtraction
$\int_a^c f(x)dx=\int_a^b f(x)dx+\int_b^c f(x)dx$	Partition into two integrals
$\int_a^b f(x)dx=-\int_b^a f(x)dx$	Reversal of sign
$\int_a^b f(x)dx=\int_a^b f(t)dt$	Change of variable

1.4.1 Definite Integral

Integral calculus is concerned with finding the limit of a sum. The most common application of the integral is to calculate the area underneath the curve of a function. Consider the following graph of a function:

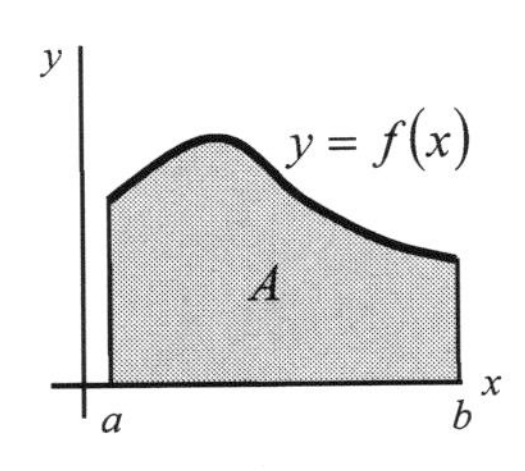

We desire to calculate the **area under the curve**. This area might represent, for example, the work done as a variable force (y) is plotted against the distance (x).

The procedure we might take is to divide the area into a series of rectangles and calculate the sum of the areas of each.

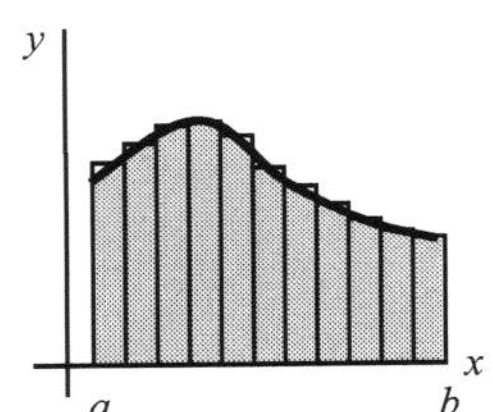

The area of each rectangle is given by:

$$A_i = f(x_i)\Delta x$$

The total area is thus the summation of all the individual areas:

$$A \approx \sum_{i=1}^{N} f(x_i)\Delta x$$

Of course if we pick very large rectangles, features of the function may not be captured very well. So for best results, we make Δx as small as possible. The exact value of the area under the curve occurs when Δx approaches zero and hence N approaches infinity. The **definite integral** is thus defined as the limit of the sum:

$$A = \int_a^b f(x)dx = \lim_{\Delta x \to 0} \sum_{i=1}^{N} f(x_i)\Delta x$$

where a and b are called the limits (or extent) of the integration (not to be confused with the limit $\Delta x \to 0$). The expression $f(x)$ is called the **integrand**.

Since A is the area under the curve, it must be possible to draw a rectangle between the limits of integration such that its height gives the same value of A.

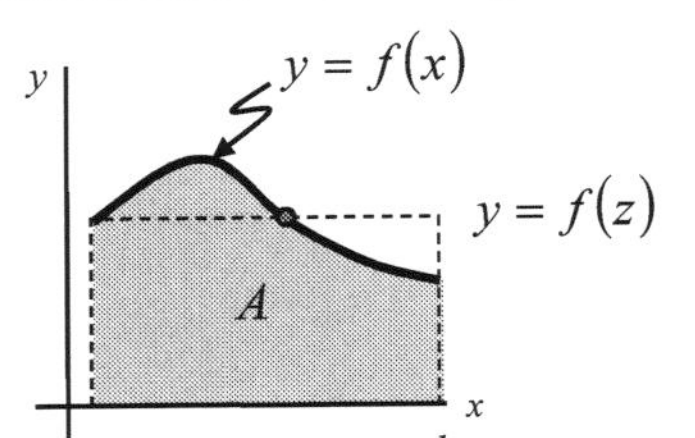

$$\boxed{A = \int_a^b f(x)dx = f(z)(b-a)}$$

This is the **mean value theorem**. The number z may not be unique it may occur for various values of x.

1.4.2 Fundamental Theorem of Calculus

Integration of a function $f(x)$ with respect to x where $a \le x \le b$ gives the area under the curve from a to b. If we hold a constant, then the area A swept out depends on how far x is along the path from a to b.

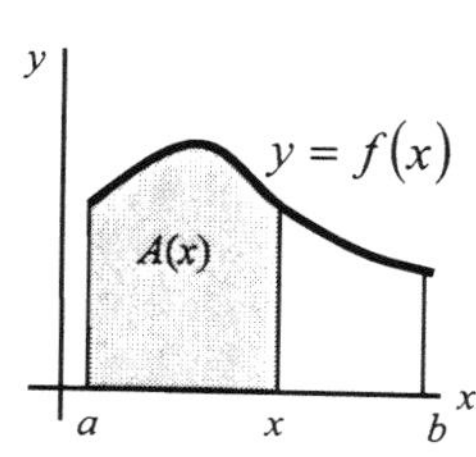

$$A(x) = \int_a^x f(x)dx$$

Consider the movement of x going from a to b. $A(x)$ gives the area under the curve as x increases. Now consider an incremental increase in x, that is, $x + \Delta x$.

The incremental area ΔA is found from:

$$\Delta A = A(x + \Delta x) - A(x)$$

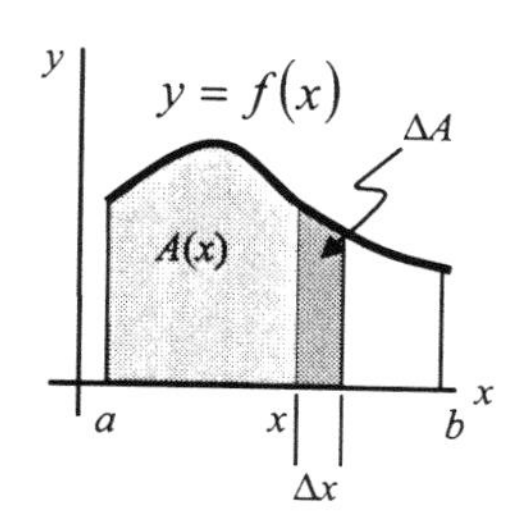

Now, by the **mean value theorem**, there exists a number z between x and $x + \Delta x$ such that:

$$\Delta A = f(z)\Delta x$$

Thus, we can say:

$$f(z)\Delta x = A(x + \Delta x) - A(x)$$

$$f(z) = \frac{A(x + \Delta x) - A(x)}{\Delta x}$$

As Δx approaches zero, $f(z)$ approaches $f(x)$, and so we can write:

$$f(x) = \lim_{\Delta x \to 0} \frac{A(x + \Delta x) - A(x)}{\Delta x}$$

That is, $f(x)$ must be the derivative of $A(x)$ with respect to x: $f(x) = A'(x)$

The function $A(x)$ must therefore be one **anti-derivative** of $f(x)$. Let $F(x)$ be any anti-derivative of $f(x)$. Thus $A(x) = F(x) + C$

Since $A(a) = 0$, then $C = -F(a)$ and so $A(x) = F(x) - F(a)$

The total area A is $A(b)$ and so $A(b) = F(b) - F(a)$

Thus, in summary, to find the area under the curve of $f(x)$ between a and b, we find the difference between the anti-derivatives of $f(x)$ evaluated at a and b:

$$A = \int_a^b f(x)dx = [F(x)]_a^b = F(b) - F(a)$$

This is a remarkable result and is called the **fundamental theorem of calculus**.

1.4.3 Properties of the Definite Integral

The **definite integral** of $f(x)$ with respect to x between $x = a$ and $x = b$ is written:

$$\int_a^b f(x)dx = [F(x)]_a^b$$

where $F(x) + C$ is the anti-derivative of $f(x)$. The relationship is called definite integral because the sum is taken over a definite interval: a to b.

To calculate the integral of a function, it is often necessary to express it in a form in which the anti-derivative can be found using the power rule. **Properties of the definite integral** are useful for these manipulations.

Multiplication by a constant $$\int_a^b cf(x)dx = c\int_a^b f(x)dx$$

Addition and subtraction $$\int_a^b (f(x) + g(x))dx = \int_a^b f(x)dx + \int_a^b g(x)dx$$

Partition into two integrals $$\int_a^c f(x)dx = \int_a^b f(x)dx + \int_b^c f(x)dx$$

Reversal of sign $$\int_a^b f(x)dx = -\int_b^a f(x)dx$$

Change of variable $$\int_a^b f(x)dx = \int_a^b f(t)dt$$

If $f(x) > 0$ then the integral is positive and is the area under the curve from a to b.

$$A = \int_a^b f(x)dx$$

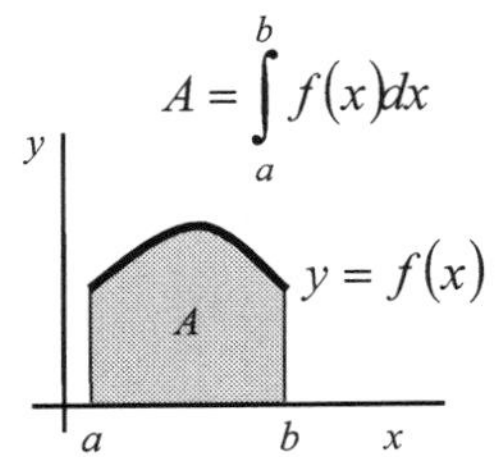

If $f(x) < 0$ then the integral is negative and is minus the area under the curve from a to b (where $b > a$).

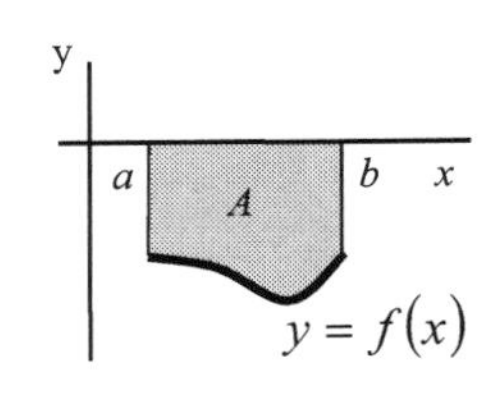

If $f(x) < 0$ and $f(x) > 0$ at different places within the interval a,b then the integral may be positive or negative and is the difference in areas $A_1 - A_2$ under the curve from a to b (where $b > a$).

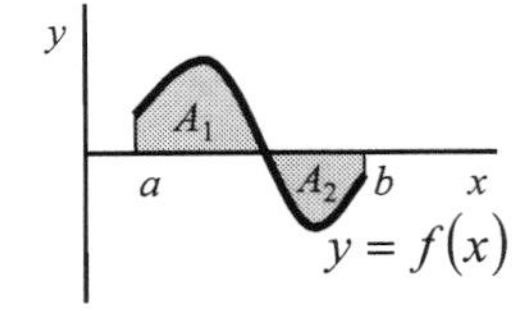

1.4.4 Indefinite Integral

The definite integral is determined from the limits of the integration applied to the anti-derivative of the function being integrated. Often, we wish to be more general about the process of integration and not specify the limits explicitly. We thus do not write the limits of integration, but write the **indefinite integral** as:

$$\boxed{\int f(x)dx = F(x) + C}$$

The result of the process of integration is another function $F(x)$ rather than a number as in the case of the definite integral.

Two of the most important rules of finding the integral of a function are the **power rule**

$$\int x^n dx = \frac{1}{n+1}x^{n+1} + C \qquad \text{where n} \neq -1$$

and the **change of variable** rule:

$$\begin{aligned}\int f(g(x))g'(x)dx &= \int f(u)du \quad \text{where } u = g(x)\\ &= F(u) + C\\ &= F(g(x)) + C\end{aligned}$$

The change of variable rule is particularly important and most frequently used in the evaluation of integrals.

It is often convenient to manipulate expressions in terms of differentials. For example, consider the expression of **velocity** as a function of distance s and time t:

$$\begin{aligned}v &= \frac{ds}{dt}\\ ds &= v\,dt\\ s &= \int v\,dt \qquad \text{Note: } \int ds = s\end{aligned}$$

Similarly for a constant **acceleration**, we have:

$$\begin{aligned}a &= \frac{dv}{dt}\\ dv &= a\,dt\\ v &= \int a\,dt\\ &= at + C\end{aligned}$$

If we now apply some boundary conditions, we obtain:

$$\begin{aligned}&\text{at } t = 0, v = u\\ &\therefore C = u\\ &v = u + at \quad \text{a familiar formula in kinematics}\end{aligned}$$

1.4.5 Numerical Integration

In some cases, evaluation of a definite integral cannot be accomplished very easily using analytical methods. In these cases, a numerical approach is needed. Consider the division of area under a curve into a series of N subintervals of width Δx:

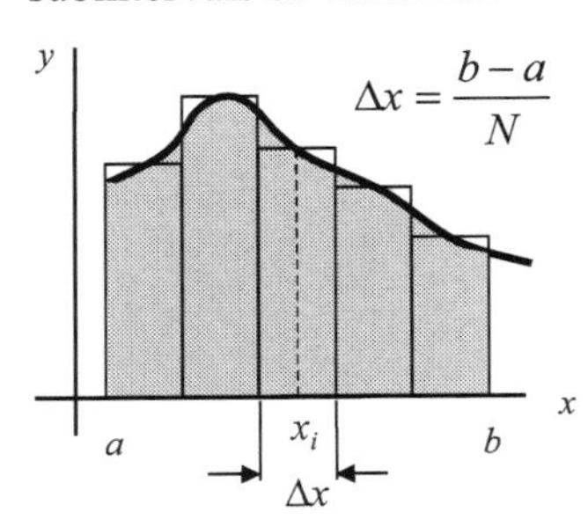

The area under the curve from $x = a$ to b is given by the sum of the rectangles of width Δx and height equal to $f(x_i)$ where $f(x_i)$ is taken at the midpoint of the rectangles.

$$\int_a^b f(x) \approx \sum_{i=1}^{N} f(x_i)\Delta x$$

The error in counting the area of the rectangle above the curve is negated by the area not swept by the rectangle.

This procedure is known as the **midpoint rule** or **mid-ordinate rule**. The percentage error is reduced by the square of the interval Δx.

A better estimate of the area can be found by the **trapezoidal rule** in which the height of the rectangle is given by the average if $f(x)$ evaluated at x and Δx.

$$\int_a^b f(x)dx \approx \frac{b-a}{2N}\left(f(x_o) + 2f(x_1) + 2f(x_2)\ldots + 2f(x_{N-1}) + f(x_N)\right)$$

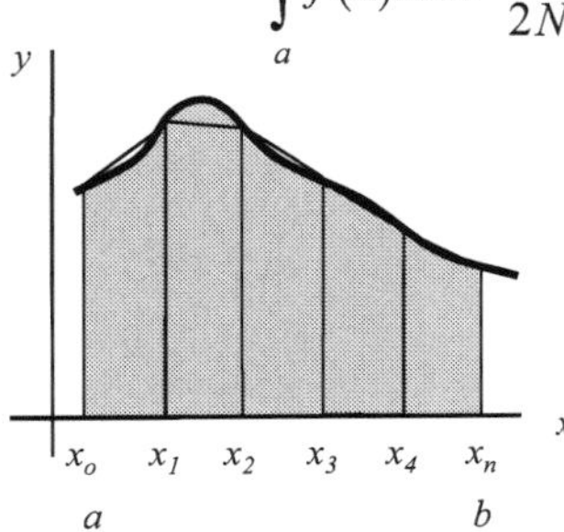

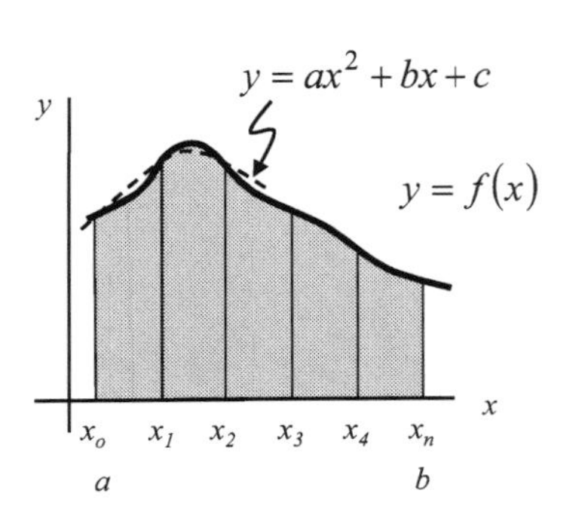

An even better estimate is found by polynomial fit to the regions of the function in each interval. This is **Simpson's rule**. Dividing the region of integration into N sub-intervals, we obtain:

$$\int_a^b f(x)dx \approx \frac{b-a}{3N}\begin{bmatrix}(f(x_o) + f(x_N)) + 4(f(x_1) + f(x_3) + f(x_5) + \ldots + f(x_{N-1})) + \\ 2(f(x_2) + f(x_4) + f(x_6) + \ldots + f(x_{N-2}))\end{bmatrix}$$

1.5 Exponential and Logarithmic Functions

Summary

$y = e^x$ $x = \ln y$	Natural exponential
$\ln x = \int_1^x \frac{1}{x} dx$	Natural logarithm
$\frac{d}{dx} e^{f(x)} = e^{f(x)} f'(x)$	Derivative of e^x
$\frac{d}{dx} a^{f(x)} = a^{f(x)} (\ln a) f'(x)$	Derivative of a^x
$\int e^x dx = e^x + C$	Integral of e^x
$\int e^{f(x)} f'(x) dx = e^{f(x)} + C$	Integral of $e^{f(x)}$
$\int a^x dx = \frac{1}{\ln a} a^x + C$	Integral of a^x

1.5.1 Logarithms

The **logarithm** is the index to which the base must be raised to equal the number.

Thus, if: $y = \log_a x$

then: $x = a^y$

Laws of logarithms:

$$\log_a pq = \log_a p + \log_a q$$

$$\log_a \frac{p}{q} = \log_a p - \log_a q$$

$$\log_a p^n = n \log_a p$$

$$\log_a N = \frac{\log_b N}{\log_b a}$$

$$\log_b a = \frac{1}{\log_a b}$$

Laws of exponents:

$$a^m a^n = a^{m+n}$$

$$(a^m)^n = a^{mn}$$

$$(ab)^n = a^n b^n$$

$$\frac{a^m}{a^n} = a^{m-n}$$

$$\left(\frac{a}{b}\right)^n = \frac{a^n}{b^n}$$

$$a^0 = 1$$

$$a^{-n} = \frac{1}{a^n}$$

$$a^{\frac{p}{q}} = \sqrt[q]{a^p}$$

1.5.2 The Natural Logarithm

The indefinite integral for a power law function can be written:

$$\int x^n dx = \frac{1}{n+1}x^{n+1} + C$$

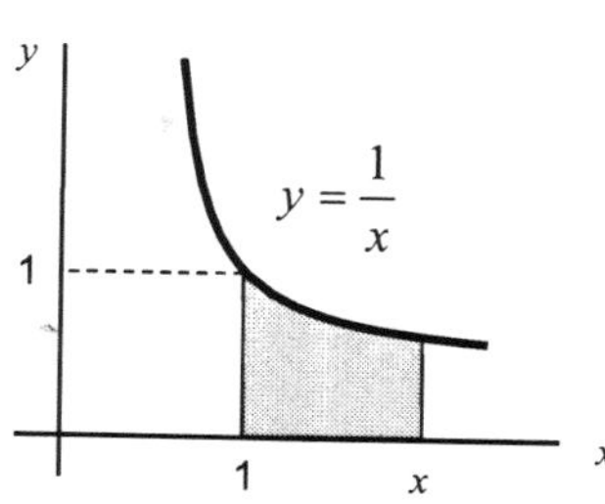

with the condition that $n \neq -1$. However, it is evident from the graph of $y = 1/x$ that it should be possible to integrate this function. Let us select one limit of integration to be $x = 1$. Thus, the integral between $x = 1$ and some other point x is:

$$[F(x)]_1^x = \int_1^x f(x)dx \text{ and thus also } F'(x) = \frac{d}{dx}F(x) = f(x) = \frac{1}{x}$$

Let us examine some interesting properties of this special function $F(x)$:

(i) Let $g(x) = kx$

$$\frac{d}{dx}F(g(x)) = \frac{1}{kx}k = \frac{1}{x}$$

and $\frac{d}{dx}F(x) = \frac{1}{x}$

$$\frac{d}{dx}F(kx) - \frac{d}{dx}F(x) = 0$$

$$\frac{d}{dx}F(kx - x) = 0$$

$$F(kx) - F(x) = C$$

let $x = 1$

$$F(k) - F(1) = C$$

$$F(1) = 0$$

$$C = F(k)$$

$$F(kx) = F(k) + F(x)$$

(ii) Let $g(x) = x^n$

$$\frac{d}{dx}F(g(x)) = \frac{1}{x^n}nx^{n-1} = \frac{n}{x}$$

$$\frac{d}{dx}F(x) = \frac{1}{x}$$

$$\frac{d}{dx}F(x^n) - n\frac{d}{dx}F(x) = 0$$

$$\frac{d}{dx}F(x^n - nx) = 0$$

$$F(x^n) - nF(x) = C$$

let $x = 1$

$$F(1) = 0$$

$$C = 0$$

$$F(x^n) = nF(x)$$

These two tests, plus other similar ones, give the impression that $F(x)$ is a logarithm since the function obeys the **laws of logarithms**. We do not know what base of logarithm yet, but we can define a logarithmic function ln such that $F(x) = \ln x$ and so:

$$\ln x = \int_1^x \frac{1}{x}dx$$

Natural logarithm

$x > 0$

1.5.3 The Natural Exponential

The **natural logarithm** is defined such that:

$$F(x) = \ln x = \int_1^x \frac{1}{x}dx$$

where $\ln 1 = 0$ and $x > 0$.

It is of interest to determine the value of the base of this logarithm. Thus, let us give the base the symbol ***e*** and attempt to evaluate it. From the point of view of a limit, the function $1/x$ can be expressed as:

$$\begin{aligned}
f(x) &= \lim_{\Delta x \to 0} \frac{F(x+\Delta x) - F(x)}{\Delta x} \\
\frac{1}{x} &= \lim_{\Delta x \to 0} \frac{\ln(x+\Delta x) - \ln x}{\Delta x} \\
&= \lim_{\Delta x \to 0} \frac{1}{\Delta x} \ln \frac{x+\Delta x}{x} \\
&= \lim_{\Delta x \to 0} \ln\left(\frac{x+\Delta x}{x}\right)^{\frac{1}{\Delta x}}
\end{aligned}$$

Let $x = 1$

$$\begin{aligned}
1 &= \lim_{\Delta x \to 0} \ln(1+\Delta x)^{1/\Delta x} \\
e^1 &= \lim_{\Delta x \to 0} (1+\Delta x)^{1/\Delta x} \\
&= e
\end{aligned}$$

The logarithm is the index to which the base must be raised to equal the number.

Taking some trial values of Δx closer and closer to zero we find that e takes on a value between 2.7 and 2.8. Indeed, it is not possible to obtain a precise value for e; it is an irrational number.

Using the **binomial theorem**, we can write:

$$\begin{aligned}
e &= 1 + \frac{1}{1} + \frac{1}{2(1)} + \frac{1}{3(2)(1)} + \frac{1}{4(3)(2)(1)} + \ldots \\
&= 1 + \frac{1}{1!} + \frac{1}{2!} + \frac{1}{3!} + \frac{1}{4!} \ldots \\
&= 2.71828\ldots
\end{aligned}$$

Note: The product: $(n)(n-1)(n-2)\ldots(1)$ is more compactly written $n!$ and is called **factorial** n.

1.5.4 Differentiation and Integration of e^x

The **natural logarithm**:

$$\ln x = \int_1^x \frac{1}{x} dx \qquad x > 0$$

$$\ln f(x) = \int_1^x \frac{1}{f(x)} f'(x) dx$$

$$\frac{d}{dx} \ln x = \frac{1}{x}$$

$$\frac{d}{dx} \ln f(x) = \frac{1}{f(x)} f'(x)$$

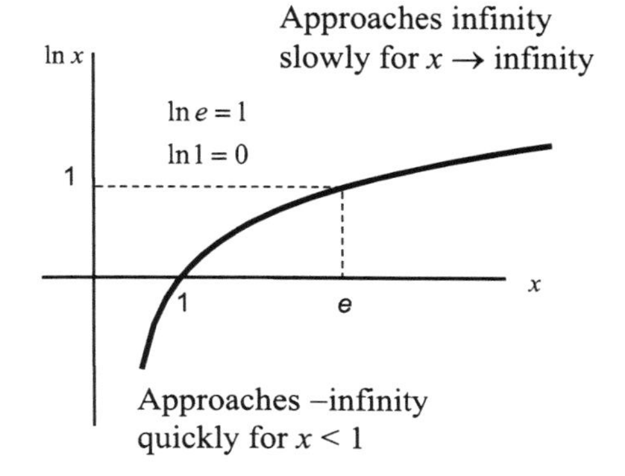

Derivative of e^x

Let $y = e^x$

Thus $x = \ln y$

$$\frac{dx}{dy} = \frac{d}{dy} \ln y = \frac{1}{y}$$

$$= \frac{1}{e^x}$$

$$\boxed{\frac{d}{dx} e^x = e^x}$$

$$\frac{d}{dx} e^{f(x)} = e^{f(x)} f'(x)$$

Let $y = a^x$

Thus $\ln y = x \ln a$

$$x = \frac{\ln y}{\ln a}$$

$$\frac{dx}{dy} = \frac{1}{\ln a} \frac{1}{y}$$

$$\boxed{\frac{d}{dx} a^x = a^x \ln a}$$

$$\frac{d}{dx} a^{f(x)} = a^{f(x)} (\ln a) f'(x)$$

Integral of e^x

$$\int e^x dx = e^x + C$$

$$\int e^{f(x)} f'(x) dx = e^{f(x)} + C$$

$$\int a^x dx = \frac{1}{\ln a} a^x + C$$

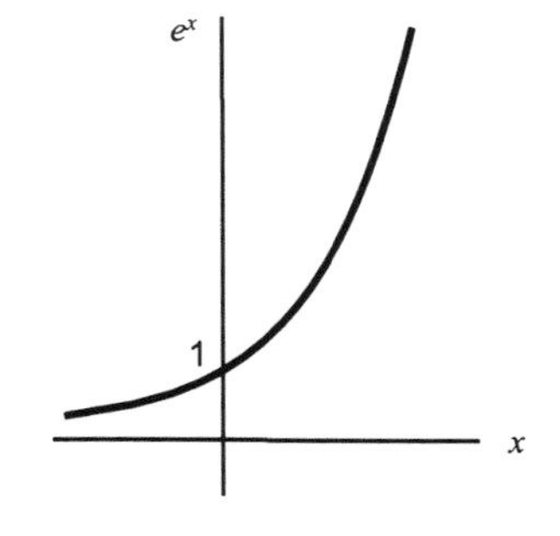

1.5.5 Exponential Law of Growth and Decay

In many physical phenomena, the **rate of change** of a variable is proportional to the value of the variable at any given time t. Such processes are the rate of cooling of a body from a high temperature, the number of atoms that disintegrate per second in a radioactive material, and the growth of bacteria in a food culture. Mathematically, we express this as:

$$\frac{dN}{dt} = \lambda N$$

where λ is a constant which is positive for **growth** and negative for **decay** processes. We can separate the variables in this type of equation to express the relationship in terms of differentials:

$$\frac{1}{N} dN = \lambda dt$$

Both sides of the equation are now integrated:

$$\int_{N_o}^{N_t} \frac{1}{N} dN = \lambda \int_0^t dt$$

$$\ln N_t = \lambda t + \ln N_o$$

$$\boxed{N_t = N_o e^{\lambda t}}$$

In this equation, N_o is the value of the variable at $t = 0$, and N_t is the value of the variable at time t. Thus, since dN/dt is proportional to N_t, a plot of $\ln(dN/dt)$ has a slope λ.

A convenient measure of the rate of growth or decay is the time taken for the variable to increase or decrease by a factor of 2. This is called the **half life** $t_{1/2}$.

$$\frac{N_t}{N_o} = e^{\lambda t_{1/2}} = \frac{1}{2}$$

$$\ln \frac{1}{2} = \lambda t_{1/2}$$

$$\boxed{t_{1/2} = -\frac{0.69}{\lambda}}$$

1.6 Trigonometric and Hyperbolic Functions

Summary

$$\frac{d}{dx}\sin x = \cos x$$

Derivative of trigonometric functions

$$\frac{d}{dx}\cos x = -\sin x$$

$$\frac{d}{dx}\tan x = \sec^2 x$$

$$\sinh x = \frac{e^x - e^{-x}}{2}$$

Hyperbolic functions

$$\cosh x = \frac{e^x + e^{-x}}{2}$$

$$\tanh x = \frac{e^x - e^{-x}}{e^x + e^{-x}}$$

$$\frac{d}{dx}\sinh x = \cosh x$$

Derivative of hyperbolic functions

$$\frac{d}{dx}\cosh x = \sinh x$$

$$\frac{d}{dx}\tanh x = \frac{1}{\cosh^2 x} = \text{sech}^2 x$$

1.6.1 Circular Measure

The **radian** is the angle swept out by an arc of length equal to the radius of the circle.

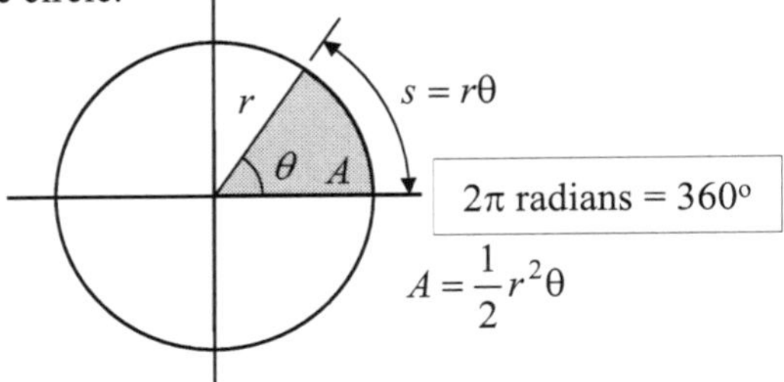

Consider a circle of radius = 1 centred on the xy coordinate axes.

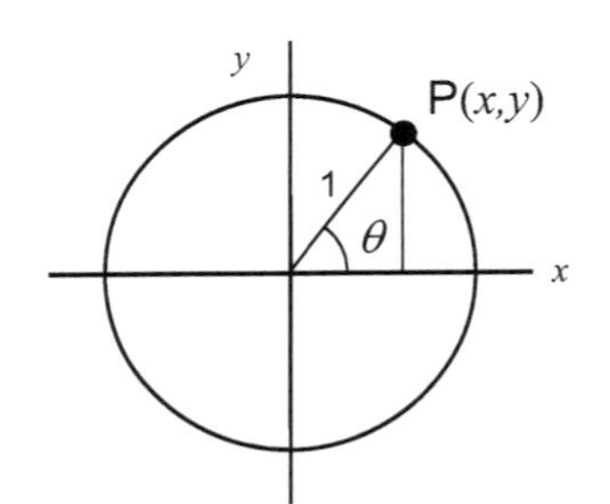

Now, $x = \cos\theta$

and $y = \sin\theta$

A plot of sin θ vs θ shows the variation of the y coordinate of P as a function of θ as the point P travels around the circle.

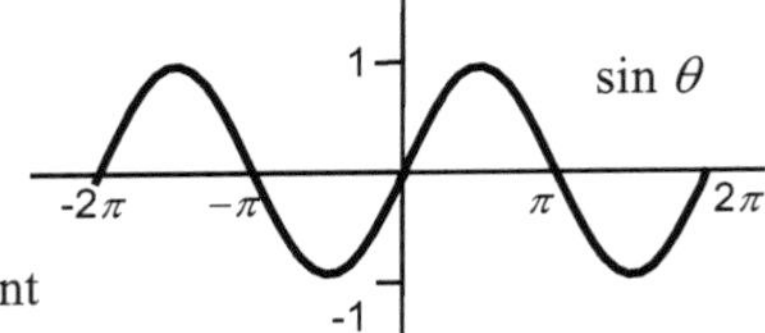

Because the sine, cosine and tangent of an angle depend upon the motion of the point P as it travels around a circle, these quantities are often termed circular functions. Indeed, the coordinates of P are (cos θ, sin θ).

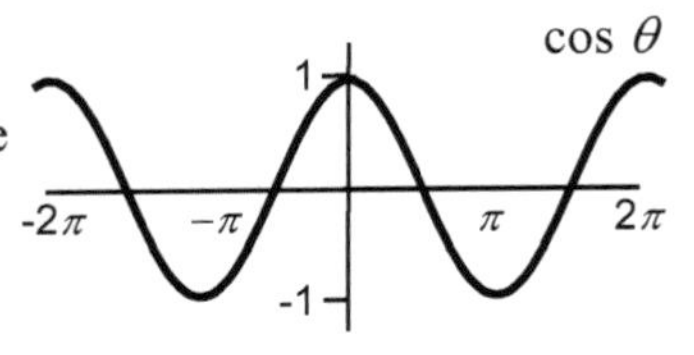

For small values of θ (in radians), we have:

$$\sin\theta \approx \theta$$
$$\cos\theta \approx 1$$
$$\tan\theta \approx \theta$$

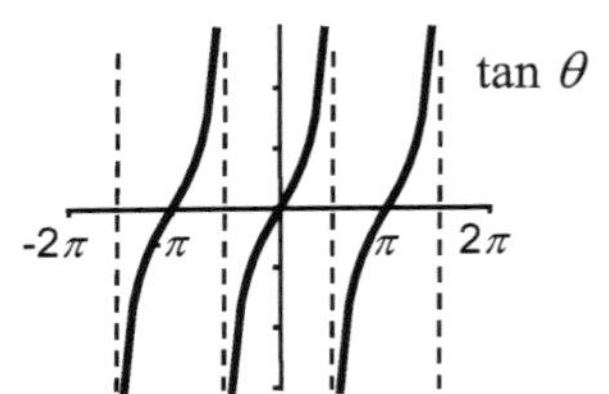

1.6.2 Derivatives & Integrals of Trigonometric Functions

The derivatives of trigonometric and **reciprocal trigonometric functions** are:

$$\frac{d}{dx}\sin x = \cos x$$

$$\frac{d}{dx}\cos x = -\sin x$$

$$\frac{d}{dx}\tan x = \sec^2 x$$

$$\frac{d}{dx}\csc x = -\csc x \cot x$$

$$\frac{d}{dx}\sec x = \sec x \tan x$$

$$\frac{d}{dx}\cot x = -\csc^2 x$$

The integrals of trigonometric and reciprocal trigonometric functions are:

$$\int \sin x dx = -\cos x + C$$

$$\int \cos x dx = \sin x + C$$

$$\int \sec^2 x dx = \tan x + C$$

$$\int \csc^2 x dx = -\cot x + C$$

$$\int \sec x \tan x dx = \sec x + C$$

$$\int \csc x \cot x dx = -\csc + C$$

$$\int \tan x dx = \ln(\sec x) + C = -\ln(\cos x) + C$$

$$\int \cot x dx = \ln(\sin x) + C$$

$$\int \sec x dx = \ln(\sec x + \tan x) + C$$

$$\int \csc x dx = \ln(\csc x - \cot x) + C$$

1.6.3 Inverse Trigonometric Functions

Consider the function $x = \sin y$.

The **inverse trigonometric function** is written $y = \sin^{-1} x$

It is important to note that the -1 as the index does not represent the reciprocal of $\sin x$, but means the **inverse function** where y is an angle whose sine is given by x. The inverse function in this case is a function if the value of x is restricted such that $-1 \le x \le 1$.

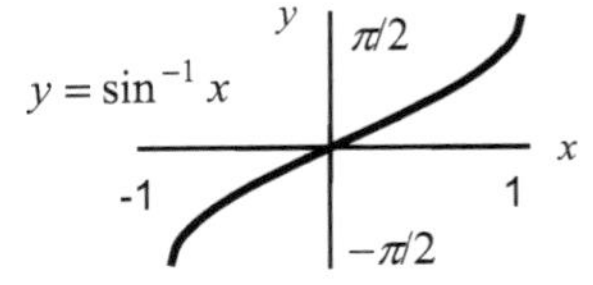

To avoid confusion with the reciprocal trigonometric function, the inverse trigonometric function is often referred to as the **arcsine** function.

$$y = \arcsin x$$

means that y is the angle whose sine is x.

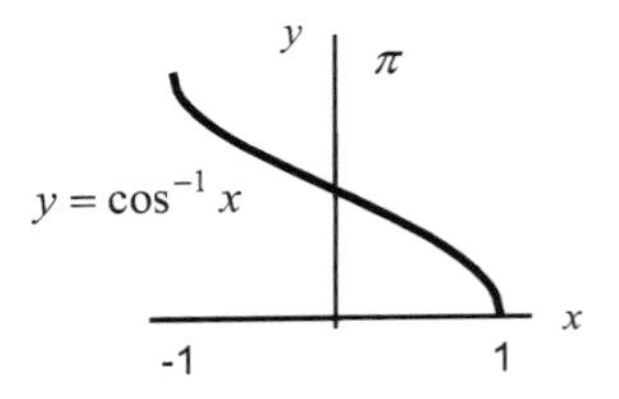

Similar inverse functions exist for cos (**arccos function**) and tan (**arctan function**) functions such that:

$$y = \cos^{-1} x$$

$$y = \tan^{-1} x$$

Inverse cosecant, secant and cotangent functions can also be expressed:

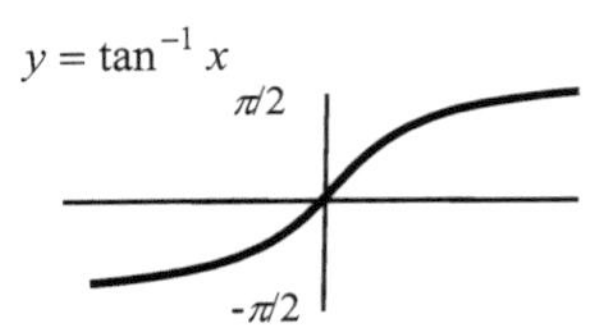

$$y = \csc^{-1} x$$

$$y = \sec^{-1} x$$

$$y = \cot^{-1} x$$

Derivatives of inverse trigonometric functions:

$$\frac{d}{dx}\sin^{-1} x = \frac{1}{\sqrt{1-x^2}} \qquad \frac{d}{dx}\csc^{-1} x = \frac{-1}{|x|\sqrt{x^2-1}}$$

$$\frac{d}{dx}\cos^{-1} x = \frac{-1}{\sqrt{1-x^2}} \qquad \frac{d}{dx}\sec^{-1} x = \frac{1}{|x|\sqrt{x^2-1}}$$

$$\frac{d}{dx}\tan^{-1} x = \frac{1}{1+x^2} \qquad \frac{d}{dx}\cot^{-1} x = \frac{-1}{1+x^2}$$

Integrals of inverse trigonometric functions can be found by integrating both sides of the above equations. For example: $\int \frac{1}{\sqrt{1-x^2}}\,dx = \sin^{-1} x + C$

1.6.4 Derivatives of Trigonometric Functions

Consider the function $y = \sin x$.

We wish to determine the derivative of y with respect to x. From the point of view of a limit, we have:

$$\frac{dy}{dx} = \lim_{\Delta x \to 0} \frac{\sin(x + \Delta x) - \sin x}{\Delta x}$$

$$\sin(x + \Delta x) - \sin x = 2\cos\frac{x + \Delta x + x}{2}\sin\frac{x + \Delta x - x}{2}$$

$$\frac{dy}{dx} = \lim_{\Delta x \to 0} \frac{2}{\Delta x}\cos\left(x + \frac{\Delta x}{2}\right)\sin\frac{\Delta x}{2}$$

$$= \lim_{\Delta x \to 0} \cos\left(x + \frac{\Delta x}{2}\right) \lim_{\Delta x \to 0} \frac{2}{\Delta x}\sin\frac{\Delta x}{2}$$

$$= \cos x$$

Similar treatments yield the derivatives of the other trigonometric functions. For the derivatives of the inverse trigonometric functions, consider the inverse sine:

$$y = \sin^{-1} x$$

$$x = \sin y$$

$$\frac{dy}{dx} = \frac{1}{dx/dy}$$

$$= \frac{1}{\cos y}$$

$$= \frac{1}{\sqrt{1 - \sin^2 y}}$$

but $x = \sin y$

thus $\frac{dy}{dx} = \frac{1}{\sqrt{1 - x^2}}$

For the inverse tangent:

$$y = \tan^{-1} x$$

$$x = \tan y$$

$$\frac{dy}{dx} = \frac{1}{\sec^2 y}$$

$$= \frac{1}{1 + \tan^2 y}$$

$$= \frac{1}{1 + x^2}$$

1.6.5 Hyperbolic Functions

Trigonometric, or circular, functions such as sine, cosine and tangent are related to the angle swept out by a point traveling on a circle of unit radius. There is a similar class of functions that are related to a **hyperbola** called **hyperbolic functions** and are defined as:

$$\sinh x = \frac{e^x - e^{-x}}{2}; \cosh x = \frac{e^x + e^{-x}}{2};$$

$$\tanh x = \frac{e^x - e^{-x}}{e^x + e^{-x}}$$

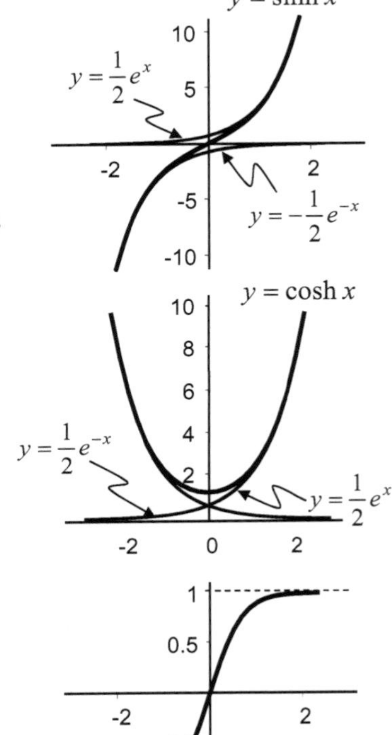

We might well ask, what is x? Is it an angle? Consider a point P with coordinates P(cosh t, sinh t) where t is a working (or dummy) variable which takes the place of θ for the case of circular functions. The equation for a hyperbola with the transverse axis parallel to the y axis with $a = b = 1$ is:

$$x^2 - y^2 = 1$$

Letting $x = \cosh t$ and $y = \sinh t$, then as t varies from $-\infty$ to $+\infty$, the point P traces out this hyperbola.

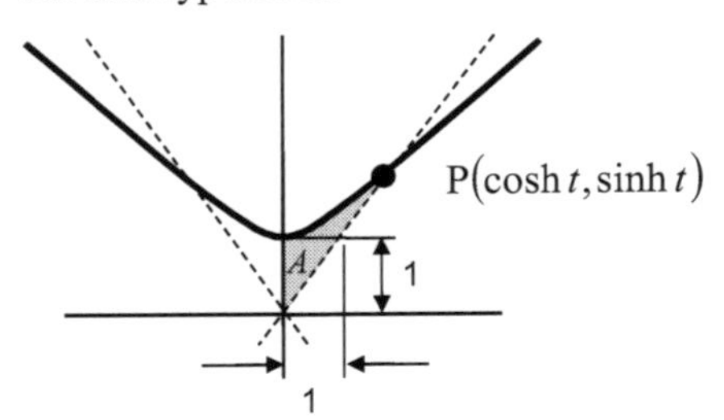

The variable t is *not* an angle, it is simply a variable. However, it can be shown that the area A swept out by the **hyperbolic sector** for a given value of t is $t/2$.

Analogous with trigonometric functions, we have **reciprocal hyperbolic functions**:

$$\operatorname{csch} x = \frac{1}{\sinh x}; \operatorname{sech} x = \frac{1}{\cosh x}; \coth x = \frac{\cosh x}{\sinh x}$$

1.6.6 Properties of Hyperbolic Functions

Many of the properties of the trigonometric functions apply to the hyperbolic functions (often with a change in sign).

$$\cosh^2 x - \sinh^2 x = 1$$
$$1 - \tanh^2 x = \text{sech}^2 x$$
$$\cosh x + \sinh x = e^x$$
$$\sinh(x+y) = \sinh x \cosh y + \sinh y \cosh x$$
$$\cosh(x+y) = \cosh x \cosh y + \sinh x \sinh y$$
$$\sinh 2x = 2\sinh x \cosh x$$
$$\tanh(x+y) = \frac{\tanh x + \tanh y}{1 + \tanh x \tanh y}$$
$$\cosh\frac{x}{2} = \sqrt{\frac{1+\cosh x}{2}}$$
$$\coth^2 x - 1 = \text{csch}^2 x$$
$$(\cosh x + \sinh x)^n = \cosh nx + \sinh nx$$

Example:

Verify that: $\sinh(x+y) = \sinh x \cosh y + \cosh x \sinh y$

Solution:

$$\sinh x \cosh y + \cosh x \sinh y$$
$$= \frac{e^x - e^{-x}}{2}\frac{e^y + e^{-y}}{2} + \frac{e^x + e^{-x}}{2}\frac{e^y - e^{-y}}{2}$$
$$= \frac{1}{4}\left[\left(e^x - e^{-x}\right)\left(e^y + e^{-y}\right) + \left(e^x + e^{-x}\right)\left(e^y - e^{-y}\right)\right]$$
$$= \frac{1}{4}\left[e^{x+y} + e^{x-y} - e^{y-x} - e^{-x-y} + e^{x+y} - e^{x-y} + e^{y-x} - e^{-x-y}\right]$$
$$= \frac{e^{x+y} - e^{-x-y}}{2}$$
$$= \sinh(x+y)$$

1.6.7 Derivative of Hyperbolic Functions

The exponential form of the hyperbolic functions makes it fairly straight-forward to determine expressions for the derivative of these functions.

$$y = \sinh x = \frac{e^x - e^{-x}}{2}$$

$$\frac{dy}{dx} = \frac{d}{dx}\frac{e^x}{2} - \frac{d}{dx}\frac{e^{-x}}{2}$$

$$= \frac{e^x + e^{-x}}{2}$$

$$\boxed{\frac{d}{dx}\sinh x = \cosh x}$$

$$y = \cosh x = \frac{e^x + e^{-x}}{2}$$

$$\frac{dy}{dx} = \frac{d}{dx}\frac{e^x}{2} + \frac{d}{dx}\frac{e^{-x}}{2}$$

$$= \frac{e^x - e^{-x}}{2}$$

$$\boxed{\frac{d}{dx}\cosh x = \sinh x}$$

$$y = \tanh x = \frac{\sinh x}{\cosh x}$$

$$\frac{dy}{dx} = \frac{\cosh x\left(\frac{d}{dx}\sinh x\right) - \sinh x\left(\frac{d}{dx}\cosh x\right)}{\cosh^2 x}$$

$$= \frac{\cosh^2 x - \sinh^2 x}{\cosh^2 x}$$

$$\boxed{\frac{d}{dx}\tanh x = \frac{1}{\cosh^2 x} = \operatorname{sech}^2 x}$$

The corresponding integrals are:

$$\int \sinh x dx = \cosh x + C$$

$$\int \cosh x dx = \sinh x + C$$

$$\int \tanh x dx = \ln \cosh x + C$$

$$\int \operatorname{sech}^2 dx = \tanh x + C$$

Derivatives and integrals of reciprocal hyperbolic functions:

$$\frac{d}{dx}\operatorname{csch} x = -\operatorname{csch} x \coth x$$

$$\frac{d}{dx}\operatorname{sech} x = -\operatorname{sech} x \tan x$$

$$\frac{d}{dx}\coth x = -\operatorname{csch}^2 x$$

$$\int \operatorname{csch} x \coth x dx = -\operatorname{csch} x + C$$

$$\int \operatorname{sech} x \tanh x dx = -\operatorname{sech} x + C$$

$$\int \operatorname{csch}^2 x dx = -\coth x + C$$

1.6.8 Inverse Hyperbolic Functions

Inverse hyperbolic functions are also defined and can be expressed:

If $y = \sinh^{-1} x$

then $x = \sinh y = \dfrac{e^y - e^{-y}}{2}$

$$2x - e^y + e^{-y} = 0$$

$$e^{2y} - 2xe^y - 1 = 0$$

by quadratic formula $\quad e^y = x + \sqrt{x^2 + 1}$

$$y = \ln\left(x + \sqrt{x^2 + 1}\right)$$

thus $\quad \sinh^{-1} x = \ln\left(x + \sqrt{x^2 + 1}\right)$

similarly $\cosh^{-1} x = \ln\left(x + \sqrt{x^2 - 1}\right) \qquad x \geq 1$

$$\tanh^{-1} x = \frac{1}{2} \ln \frac{1+x}{1-x} \qquad |x| < 1$$

$$\operatorname{sech}^{-1} x = \ln\left(\frac{1 + \sqrt{1 - x^2}}{x}\right) \qquad 0 < x \leq 1$$

Derivatives and integrals of inverse hyperbolic functions:

$$\frac{d}{dx} \sinh^{-1} x = \frac{1}{\sqrt{x^2 + 1}} \qquad \int \frac{1}{\sqrt{x^2 + a^2}} dx = \sinh^{-1} \frac{x}{a} + C$$

$$\frac{d}{dx} \cosh^{-1} x = \frac{1}{\sqrt{x^2 - 1}} \qquad \int \frac{1}{\sqrt{x^2 - a^2}} dx = \cosh^{-1} \frac{x}{a} + C$$

$$\frac{d}{dx} \tanh^{-1} x = \frac{1}{1 - x^2} \qquad \int \frac{1}{a^2 - x^2} dx = \frac{1}{a} \tanh^{-1} \frac{x}{a} + C$$

$$\frac{d}{dx} \operatorname{sech}^{-1} x = \frac{-1}{x\sqrt{1 - x^2}} \qquad \int \frac{1}{x\sqrt{a^2 - x^2}} dx = -\frac{1}{a} \operatorname{sech}^{-1} \frac{x}{a} + C$$

1.7 Methods of Integration

Summary

Integration by substitution

$$\int f(g(x))g'(x)dx = \int f(u)du \quad \text{where } u = g(x)$$
$$= F(u) + C$$
$$= F(g(x)) + C$$

$$\int udv = uv - \int vdu$$

Integration by parts

$$x = a\sin\theta$$
$$dx = a\cos\theta d\theta$$

Trigonometric substitutions

$$(ax+b)^n = \frac{A_1}{px+q} + \frac{A_2}{(px+q)^2} + \ldots + \frac{A_m}{(px+q)^m}$$

Partial fractions

$$\lim_{x\to a}\frac{f(x)}{g(x)} = \lim_{x\to a}\frac{f'(x)}{g'(x)}$$

L'Hôpital's rule

$$\int_a^\infty f(x)dx = \lim_{x\to\infty}\int_a^x f(x)dx$$

Improper integral

1.7.1 Integration by Substitution

One of the most common techniques in integration is a reverse application of the chain rule for differentiation: by **substitution** and **change of variable**:

$$\int f(g(x))g'(x)dx = \int f(u)du \text{ where } u = g(x)$$

$$= F(u) + C$$
$$= F(g(x)) + C$$

The method is particularly powerful, especially when one becomes accustomed to identifying the necessary functions and derivatives in the expression to be integrated. Consider the following simple example:

$$\int (2x-5)^5 dx$$

By a mental application of the reverse of the chain rule, we can see that the evaluation of this would be very much facilitated if we had a "2" in front of the expression in brackets, so, we add one (and put a factor of 0.5 in front to compensate):

More formally, we let $u = (2x-5)$ and so $du = 2dx$.

$$\text{Thus: } \int (2x-5)^5 dx = \frac{1}{2}\int (2x-5)^5 2dx$$

$$= \frac{1}{2}\int u^5 du$$

$$= \frac{1}{2}\frac{u^6}{6} + C$$

$$= \frac{1}{12}(2x-5)^6 + C$$

1.7.2 Integration by Parts

Integration by parts can often be used when the integrand is not in a standard form. The method works by treating the integrand as the product of a function with the differential of another function.

$$\frac{d}{dx}(f(x)g(x)) = f(x)g'(x) + g(x)f'(x)$$

$$f(x)g'(x) = \frac{d}{dx}(f(x)g(x)) - g(x)f'(x)$$

$$\int f(x)g'(x)dx = f(x)g(x) - \int g(x)f'(x)dx$$

Let $u = f(x)$ then $du = f'(x)dx$

$v = g(x)$ $dv = g'(x)dx$

$$\boxed{\int udv = uv - \int vdu}$$

The integral of the product of two functions is the first times the integral of the second minus the integral of the differential of the first times the integral of the second.

Example:

Integrate the following expression: $\int \sqrt{x}\ln x dx$

Solution:

Let: $u = \ln x$

and: $dv = x^{\frac{1}{2}}dx$

Thus: $\frac{du}{dx} = \frac{1}{x}$

$du = \frac{1}{x}dx$

and: $\frac{dv}{dx} = x^{\frac{1}{2}}$

$v = 2\frac{x^{\frac{3}{2}}}{3}$

Therefore:

$$\int \sqrt{x}\ln x dx = uv - \int vdu$$

$$(\ln x)\frac{2}{3}x^{\frac{3}{2}} - \frac{2}{3}\int x^{\frac{3}{2}}\frac{1}{x}dx$$

$$= \frac{2}{3}x^{\frac{3}{2}}\ln x - \frac{2}{3}\int x^{\frac{1}{2}}dx$$

$$= \frac{2}{3}x^{\frac{3}{2}}\left(\ln x - \frac{2}{3}\right) + C$$

$$= \frac{2}{9}x^{\frac{3}{2}}(3\ln x - 2) + C$$

1.7.3 Trigonometric Substitutions

When an integrand contains expressions of the type $a^2 - x^2, a^2 + x^2, x^2 - a^2$ and cannot be easily integrated directly, it is possible that a more convenient integrand can be obtained by making a **trigonometric substitution**. In this way, standard trigonometric identities can often be used to simply the expression. Suggested substitutions are:

Expression	Substitution
$a^2 - x^2$	$x = a\sin\theta$
$a^2 + x^2$	$x = a\tan\theta$
$x^2 - a^2$	$x = a\sec\theta$

After making the substitution, it is important to also include the differential $d\theta$ by expressing dx in terms of $d\theta$ in the integration. For example:

$$\text{If} \quad x = a\sin\theta$$
$$\text{then} \quad dx = a\cos\theta d\theta$$

Example:

Integrate the following expression: $\int \frac{x}{\sqrt{2^2 - x^2}} dx$

Solution:

Let: $x = 2\sin\theta$

Thus: $\frac{dx}{d\theta} = 2\cos\theta$

$$\sqrt{2^2 - x^2} = \sqrt{2^2 - 2^2 \sin\theta}$$
$$= \sqrt{2^2(1 - \sin^2\theta)}$$
$$= 2\cos\theta$$

Therefore:

$$\int \frac{x^2}{\sqrt{2^2 - x^2}} dx = \int \frac{2\sin\theta}{2\cos\theta} 2\cos\theta d\theta$$
$$= 2\int \sin\theta d\theta$$
$$= -2\cos\theta + C$$
$$= -\sqrt{4 - x^2} + C$$

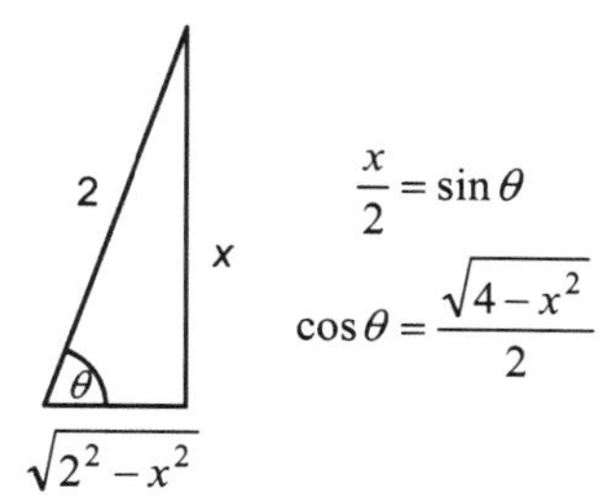

1.7.4 Integration by Partial Fractions

When the numerator and denominator of a fraction consist of polynomial expressions where the degree of the numerator is lower than that of the denominator, the fraction can be expressed as the sum of two or more **partial fractions**.

The decomposition of an expression into partial fractions is done by expressing the denominator (which may contain sums and differences) in terms of factors (or products) of expressions. The factors may take one or more of two general forms. For each factor of the form:

$(ax+b)^n$ where $n \geq 1$ we obtain partial fractions: $\rightarrow \dfrac{A_1}{px+q} + \dfrac{A_2}{(px+q)^2} + \ldots + \dfrac{A_m}{(px+q)^m}$

$(ax^2+bx+c)^m$ where $m \geq 1$ and no roots $\rightarrow \dfrac{A_1x+B_1}{ax^2+bx+c} + \dfrac{A_2x+B_1}{(ax^2+bx+c)^2} + \ldots + \dfrac{A_nx+B_n}{(ax^2+bx+c)^n}$

and $A_1, A_2, \ldots$ are constants.

Decomposition into partial fractions enables a complicated expression to be expressed as a sum of simpler expressions which may be each on their own more easily integrated.

Example:

Integrate the following expression: $\displaystyle\int \frac{x+16}{x^2+2x-8}dx$

Solution:

$$\frac{x+16}{x^2+2x-8} = \frac{x+16}{(x+4)(x-2)}$$

$$= \frac{A}{x+4} + \frac{B}{x-2}$$

$$x+16 = A(x-2) + B(x+4)$$

Let: $x = 2$

$$18 = 6B$$

$$B = 3$$

Let: $x = -4$

$$12 = A - 6$$

$$A = -2$$

Therefore:

$$\int \frac{x+16}{(x+4)(x-2)} = \int \frac{-2}{x+4}dx + \int \frac{3}{x-2}dx$$

$$= -2\ln(x+4) + 3\ln(x-2) + C$$

1.7.5 Quadratic Expressions

Any quadratic expression can be expressed as the sum or difference of two squares. The process of expressing a **quadratic polynomial** in this way is called **completing the square**. Such a process may be needed when evaluating an integral containing a quadratic expression that cannot be expressed as the product of two factors.

The general equation for a quadratic polynomial is:

$$y = ax^2 + bx + c$$

If the polynomial has no roots (lies completely above or below the x axis), then it is not possible to express this polynomial as a product of two factors. However, we can express the polynomial in terms of squared expressions as follows:

$$\begin{aligned} y &= ax^2 + bx + c \\ &= a\left(x^2 + \frac{b}{a}x\right) + c \\ &= a\left(x + \frac{b}{2a}\right)^2 + c - \frac{b^2}{4a} \end{aligned}$$

A substitution may then be made which results in an integrand of standard form.

Example:

Integrate the following expression: $\int \frac{1}{x^2 - 4x + 8} dx$

Solution:

$$\begin{aligned} &\int \frac{1}{x^2 - 4x + 8} dx \\ &= \int \frac{1}{\left(x^2 - 4x + 4\right) + 4} dx \\ &= \int \frac{1}{(x-2)^2 + 4} dx \end{aligned}$$

Let: $u = x - 2$

$$\frac{du}{dx} = 1$$

$$du = dx$$

Therefore:

$$\begin{aligned} \int \frac{1}{x^2 - 4x + 8} dx &= \int \frac{1}{u^2 + 4} du \\ &= \frac{1}{2}\tan^{-1}\frac{u}{2} + C \\ &= \frac{1}{2}\tan^{-1}\frac{x-2}{2} + C \end{aligned}$$

1.7.6 Indeterminate Forms

Consider the definition of the derivative of a function $f(x)$ expressed in terms of a limit:

$$\frac{dy}{dx} = \lim_{\Delta x \to 0} \frac{f(x+\Delta x) - f(x)}{\Delta x}$$

As written, the quotient has an **indeterminate form** 0/0 as Δx approaches 0. In some cases, the indeterminate form can be avoided by algebraic manipulation, but often this is not the case. Similar indeterminate forms exist when both denominator and numerator approach infinity.

To evaluate the limit of a quotient which has an indeterminate form 0/0 or ∞ / ∞ at x equal to some number a, **L'Hôpital's rule** may be used to find the limit as x approaches a.

$$\lim_{x \to a} \frac{f(x)}{g(x)} = \lim_{x \to a} \frac{f'(x)}{g'(x)}$$

It is important to note that L'Hôpital's rule can only be used for expressions which have an indeterminate form 0/0 or ∞ / ∞.

In cases where an expression has the form $f(x)g(x) = 0 \cdot \infty$ as $x \to a$, then L'Hôpital's rule can still be used:

$$\lim_{x \to a} f(x)g(x) = \lim_{x \to a} \frac{f(x)}{1/g(x)} = \frac{0}{0}$$

Similarly, for $f(x)g(x) = \infty \cdot 0$ as $x \to a$,

$$\lim_{x \to a} f(x)g(x) = \lim_{x \to a} \frac{f(x)}{1/g(x)} = \frac{\infty}{\infty}$$

For $f(x) - g(x) = \infty - \infty$ as $x \to a$,

$$\lim_{x \to a} f(x) - g(x) = \lim_{x \to a} \frac{(f(x) - g(x)) f(x) g(x)}{f(x)g(x)}$$

$$= \lim_{x \to a} \frac{(1/g(x) - 1/f(x))}{1/(f(x)g(x))}$$

$$= \frac{0}{0}$$

Note: L'Hôpital's rule may be applied several times to arrive at an expression in which the limit can be determined as long as the condition for an indeterminate form is satisfied at each application.

For expressions which have limiting values of the type $0^0; 1^\infty; \infty^0$ logarithms can be used to express the function in a form suitable for L'Hôpital's rule:

$$\lim_{x \to a} f(x)^{g(x)} = \exp\left(\lim_{x \to a} g(x) \ln f(x)\right)$$

$$= \exp\left(\lim_{x \to a} \frac{\ln f(x)}{1/g(x)}\right)$$

1.7.7 Improper Integrals

In many applications of integration it is necessary to integrate a function with an **infinite limit** of integration. This type of integral can be written in terms of a limit.

$$\int_a^{\infty} f(x)dx = \lim_{x\to\infty}\int_a^{x} f(x)dx$$

Limits approaching $-\infty$ are written in a similar way. The procedure for evaluating integrals of this type is to find the anti-derivative of the function in the normal way, and then evaluate the limit. If the limit exists, then the **improper integral** is said to **converge**.

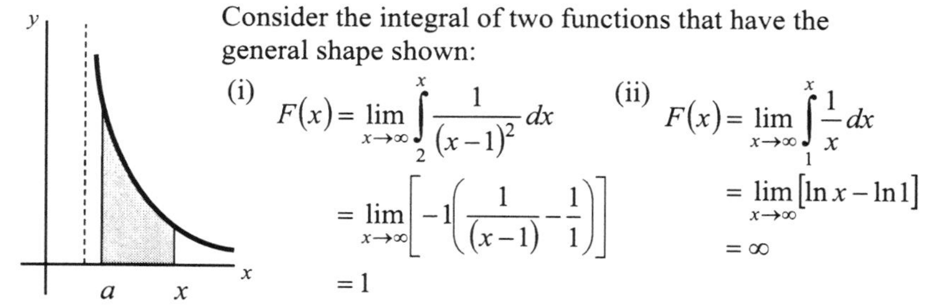

Consider the integral of two functions that have the general shape shown:

(i)
$$F(x) = \lim_{x\to\infty}\int_2^{x}\frac{1}{(x-1)^2}dx$$
$$= \lim_{x\to\infty}\left[-1\left(\frac{1}{(x-1)}-\frac{1}{1}\right)\right]$$
$$= 1$$

(ii)
$$F(x) = \lim_{x\to\infty}\int_1^{x}\frac{1}{x}dx$$
$$= \lim_{x\to\infty}\left[\ln x - \ln 1\right]$$
$$= \infty$$

One may intuitively think that all such integrals must diverge since as x gets large, we are surely adding more and more area to the summation under the curve. Actually, whether or not an improper integral is **convergent** or **divergent** is solely a question of the nature of the limit and can really only be established by mathematical techniques and not by inspection of the graph. One such test is called the ***p* test**.

$$\int_0^1 \frac{1}{x^p}dx \quad \text{Convergent if } p < 1 \qquad\qquad \int_1^{\infty}\frac{1}{x^p}dx \quad \text{Convergent if } p > 1$$

These tests show that, for functions like that shown above, whether or not the integral converges or diverges depends upon the rate at which the function approaches the x axis.

For cases where the function being integrated has a **discontinuity** within the limits of integration (say at c), the integral is expressed as the sum of two integrals with infinite limits:

$$\int_a^b f(x)dx = \int_a^c f(x)dx + \int_c^b f(x)dx \quad \text{where} \quad \begin{aligned}\int_a^c f(x)dx &= \lim_{x\to\infty}\int_a^x f(x)dx \\ \int_c^b f(x)dx &= \lim_{x\to\infty}\int_x^b f(x)dx\end{aligned}$$

1.8 Waves

Summary

$v = f\lambda$	Velocity of a wave
$y = A\sin(\omega t - kx + \phi)$	Displacement of a particle
$v_y = \omega A\cos(\omega t - kx + \phi)$	Velocity of a particle
$a_y = -\omega^2 A\sin(\omega t - kx + \phi)$	Acceleration of a particle
$\dfrac{\partial^2 y}{\partial x^2} = \dfrac{1}{v^2}\dfrac{\partial^2 y}{\partial t^2}$	General wave equation
$y(x,t) = Ae^{i(\omega t - kx + \phi)}$	Complex representation of a wave
$E = \dfrac{1}{2}\omega^2 A^2(\rho\lambda)$	Energy transmitted by one wavelength of stretched string
$P = \dfrac{1}{2}\sqrt{T\rho}\,\omega^2 A^2$	Power in one wavelength on stretched string

1.8.1 Simple Harmonic Motion

Consider the motion of a mass attached to a spring.

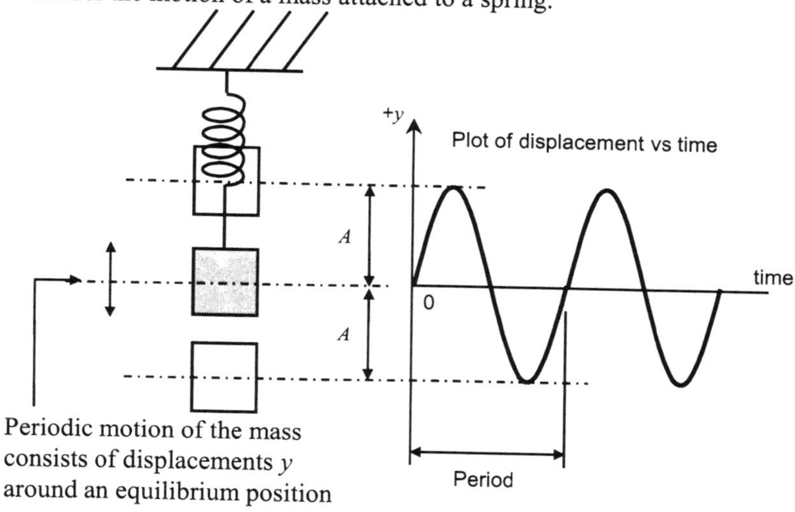

When y is plotted against time, we find the relationship is sinusoidal. The displacement y of the mass as a function of θ is

$$y = A\sin\theta$$

The motion of the mass would be the same as if it were attached to a rotating wheel and viewed edge on. That is, the angle θ (called the **phase**) would be the angular displacement of the wheel.

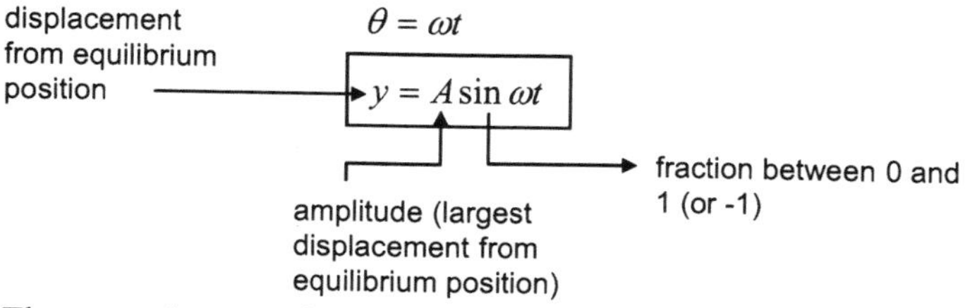

The general expression for position (i.e., displacement from the equilibrium position) for the mass is thus:

$$y = A\sin(\omega t + \phi)$$

where ϕ is the *initial* phase angle, or offset, that is added to the angle θ to account for the case when the time $t = 0$ does not correspond to $y = 0$. Regular motion of this kind is called **simple harmonic motion (SHM)**.

1.8.2 Waves

Consider a **transverse wave** in a string where the points on the string undergo SHM:

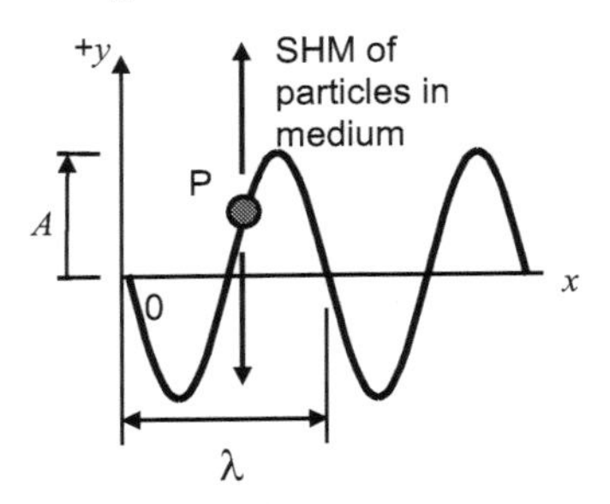

This is a "snapshot" of the disturbance, or wave, at some time "t".

- The shape of the wave is a repeating pattern.
- λ is called the **wavelength** and is the length of one complete cycle.

The wave travels with a velocity v. The time for one complete cycle is T. Thus, since:

$$v = \frac{d}{t}$$

then $v = \frac{\lambda}{T}$ since $f = \frac{1}{T}$ and since one complete wavelength passes a given point in a time T

and $\boxed{v = f\lambda}$

f in cycles per second

We wish to calculate the displacement y of any point P located at some distance x on the string as a function of time t.

Let's consider the motion of the point located at $x = 0$. If the points on the string are moving up and down with SHM, then:

$$y = A\sin(\omega t + \phi) \quad \text{if } y = 0 \text{ at } t = 0, \text{ then } \phi = 0$$

The disturbance, or wave, travels from left to right with velocity $v = x/t$. Thus, the disturbance travels from 0 to a point x in time x/v.

Now, let us consider the motion of a point P located at position x. The displacement of point P located at x at time t is the same as that of point P' located at $x = 0$ at time $(t - x/v)$.

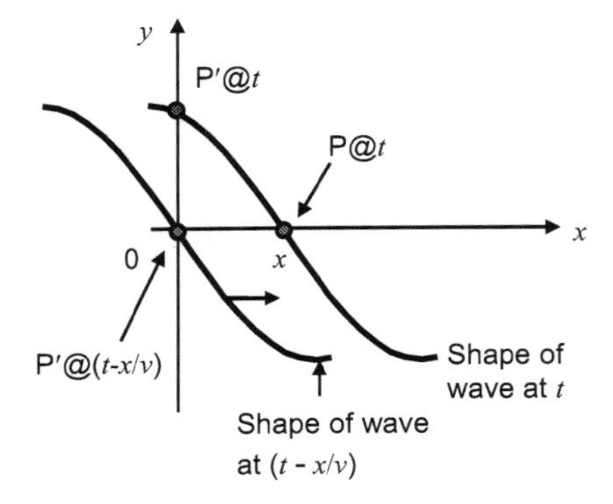

Thus, to get the displacement of the point P at (x, t) we use the same formula for the motion of point P' located at $x = 0$ but put in the time $t = (t - x/v)$:

$$y = A\sin(\omega t + \phi)$$

$$= A\sin\left(\omega\left(t - \frac{x}{v}\right) + \phi\right)$$

A wave that can be represented by either a sine or cosine function is called a sinusoidal wave, or a harmonic wave. The equation is known as the wave function.

Now, it is convenient to let: $k = \dfrac{2\pi}{\lambda}$

↓ Wave number

and since: $v = f\lambda$

then: $= f\dfrac{2\pi}{k}$

$$v = \frac{\omega}{k}$$

Thus:

$$y = A\sin\left(\omega\left(t - x\frac{k}{\omega}\right) + \phi\right)$$

$$\boxed{y = A\sin(\omega t - kx + \phi)}$$

Displacement of particle in the medium from equilibrium position as a function of x and t

The argument to the sine function $(\omega t - kx + \phi)$ is often called the **phase** of the wave – not to be confused with ϕ, which is the *initial* phase angle. The initial phase ϕ is an offset, or a constant which is there to start with and makes a constant contribution to the phase independent of x and t.

The velocity and acceleration *of the particles* in the medium are found by differentiating while holding x constant:

Displacement

$$\boxed{y = A\sin(\omega t - kx + \phi)}$$

Velocity = dy/dt

$$\boxed{v_y = \omega A\cos(\omega t - kx + \phi)}$$ (holding x constant)

Acceleration = dv/dt

$$\boxed{a_y = -\omega^2 A\sin(\omega t - kx + \phi)}$$ (holding x constant)

Note, these are the displacement, velocity and acceleration of the *particle*. The velocity of the *wave* is determined by the frequency and the wavelength of the wave.

1.8.3 Wave Equation

Let us now find *dy*/*dx* while holding *t* constant:

$$y = A\sin(\omega t - kx)$$

$$\frac{\partial y}{\partial x} = -kA\cos(\omega t - kx)$$

$$\frac{\partial^2 y}{\partial x^2} = -k^2 A\sin(\omega t - kx)$$

The symbol ∂ is used to remind us that we are taking the derivative with one (or more) of the variables in the equation held constant (i.e., in this case, t). Derivatives of this type are called **partial derivatives**.

but $\omega = vk$ v is the velocity of the *wave*

thus

$$\frac{\partial^2 y}{\partial x^2} = -\frac{\omega^2}{v^2} A\sin(\omega t - kx)$$

$$\boxed{\frac{\partial^2 y}{\partial x^2} = \frac{1}{v^2}\frac{\partial^2 y}{\partial t^2}}$$

a_y = acceleration of particle

velocity of wave

This is called the (one dimensional) **wave equation** and gives information about all aspects of the wave by tying together the motion of the particles and the wave.

The solution to this second order partial differential equation is a function. That is, the original **wave function**:

$$y = A\sin(\omega t - kx)$$

An important property of the wave equation is that it shows that two different solutions of the wave equation (having different values of the constant A) may be added to yield a third solution – this is the principle of **superposition**.

1.8.4 Sign Conventions

It has been shown that the wave function is written:

$$y(x,t) = A\sin(\omega t - kx + \phi)$$

A very important assumption in the above is that we were dealing with the transverse motion of a particle in a two dimensional situation. That is, the particle was moving up and down along the y axis, and the wave, or shape of the disturbance, moved towards the right along the x axis.

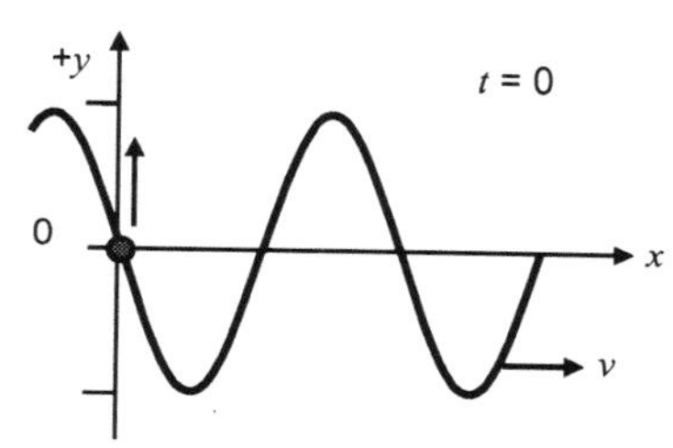

As the wave moves to right, the particle at $x = 0$ moves upwards.

However, in some physical situations, the wave itself may not represent the *motion* of a particle. For example, in quantum physics, the **matter wave** associated with a particle represents the probability of finding that particle at some place x at a time t. That is, the vertical axis for a matter wave when plotted against position x is a probability and has no relationship to a transverse spatial direction y.

In this case, it is convenient to reverse the signs in the phase terms such that:

$$y(x,t) = A\sin(kx - \omega t + \phi)$$

Note, both representations of the wave are correct, but not equivalent – their phases differ by a factor of π.

1.8.5 Complex Functions

Previously, we saw that for the case of $b^2 - 4ac < 0$, a quadratic equation has no roots. However, if we use complex numbers, we can use the quadratic formula to calculate complex roots of the equation.

For example, consider the quadratic equation:

$$m^2 - 2m + 2 = 0$$

By the quadratic formula, we have: $m = \dfrac{2 \pm \sqrt{4 - 4(2)}}{2}$

$$= \frac{2 \pm 2\sqrt{-1}}{2}$$

The roots of the equation are complex: $= 1 \pm i$

Now, consider a complex number $z = a + bi$. If a and b are themselves functions of a variable x (or more than one variable $a = f(x,y)$, etc.) then z is a **complex variable**. A **complex function** may involve the complex variable z in much the same way as real functions involve real variables.

For example, consider the power series expansion of e^z where now, z is a complex number given by $z = (a + ib)x$.

$$e^{(a+ib)x} = e^{ax} e^{ibx}$$

$$\downarrow$$

$$e^{ibx} = 1 + ibx + \frac{(ibx)^2}{2!} + \frac{(ibx)^3}{3!} \dots$$

$$= 1 + ibx + i^2 \frac{(bx)^2}{2!} + i^3 \frac{(bx)^3}{3!} \dots$$

$$= 1 + ibx - \frac{(bx)^2}{2!} - i\frac{(bx)^3}{3!} + \frac{(bx)^4}{4!} + i\frac{(bx)^5}{5!} + \dots$$

$$= \left(1 - \frac{(bx)^2}{2!} + \frac{(bx)^4}{4!} - \dots\right) + i\left(bx - \frac{(bx)^3}{3!} + \frac{(bx)^5}{5!} - \dots\right)$$

but: $\sin x = x - \dfrac{x^3}{3!} + \dfrac{x^5}{5!} - \dfrac{x^7}{7!} + \dots$ and $\cos x = 1 - \dfrac{x^2}{2!} + \dfrac{x^4}{4!} - \dfrac{x^6}{6!} + \dots$

Thus:

$$\boxed{e^{(a+bi)x} = e^{ax}(\cos bx + i \sin bx)}$$ **Euler's formula**

1.8.6 Complex Representation of a Wave

Representation of waves using sine and cosine functions can become unwieldy. An often more convenient method is to use complex numbers.

Imagine a point P that is represented by the complex number z = $a + bi$ on the complex plane. P is rotating about the origin. Viewed from side-on, P would be moving up and down with simple harmonic motion.

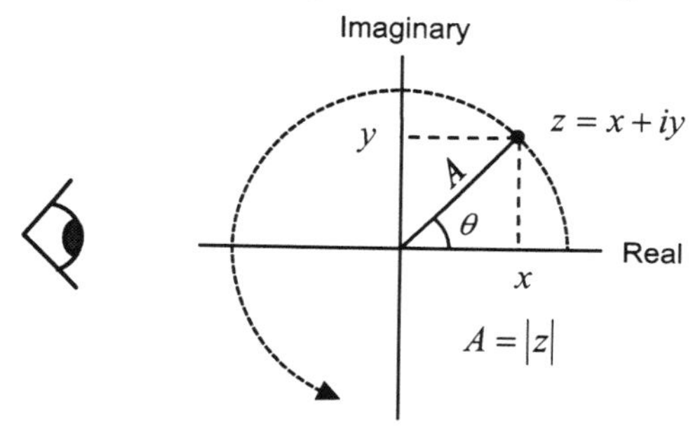

Now, in polar coordinates, we have:

$$x = A\cos\theta$$

$$y = A\sin\theta$$

Let $z = x + iy$

thus $z = A\cos\theta + iA\sin\theta$

$$= A(\cos\theta + i\sin\theta)$$

Euler's formula
$e^{i\theta} = \cos\theta + i\sin\theta$

Using Euler's formula, we can thus write:

$$z = Ae^{i\theta}$$

The amplitude A of the motion is given by the magnitude of z.

$$A = |z| = \sqrt{z^* z}$$

where z^* is the complex conjugate of (i.e., $z^* = x - iy$).

It can be seen from the above, that the cosine function is associated with the real part of the motion, and the sine function is associated with the imaginary part. Either cosine or sine can be used to represent a wave, but it is usual to use the real part. Thus, a travelling wave can be represented by the *real part* of:

$$y(x,t) = Ae^{i(\omega t - kx + \phi)}$$ or $$y(x,t) = Ae^{i(kx - \omega t + \phi)}$$

1.8.7 Superposition in Complex Form

The solution to certain types of differential equations often takes the form of a superposition of two waves, one travelling in the $+x$ direction and the other in the $-x$ direction, each with different amplitudes C_1 and C_2.

Beginning with Euler's formula:

$$e^{ikx} = \cos kx + i\sin kx \qquad \text{Wave travelling in } +x$$

$$\text{and} \quad e^{-ikx} = \cos(-kx) + i\sin(-kx) \qquad \text{Wave travelling in } -x$$

$$= \cos kx - i\sin kx$$

$$y(t) = C_1 e^{ikx} + C_2 e^{-ikx} \qquad \text{Superimposed waves}$$

$$= C_1(\cos kx + i\sin kx) + C_2(\cos kx - i\sin kx)$$

$$\text{let } C_1 = a + bi$$

$$C_2 = a - bi$$

$$C_1 + C_2 = 2a$$

$$C_1 - C_2 = 2bi$$

$$\text{thus } C_1 e^{ikx} + C_2 e^{-ikx} = (a + bi)(\cos kx + i\sin kx) + (a - bi)(\cos kx - i\sin kx)$$

$$= a\cos kx + ai\sin kx + bi\cos kx - b\sin kx + a\cos kx -$$

$$ai\sin kx - bi\cos kx - b\sin kx$$

$$= 2a\cos kx - 2b\sin kx$$

$$C_1 e^{ikx} + C_2 e^{-ikx} = A\cos kx - B\sin kx$$

$$\text{where } A = 2a$$

$$B = 2b$$

In this treatment, we have carefully chosen the values of C_1 and C_2 so that the resulting superposed wave is represented by a real function. This need not necessarily be the case. It depends on the physical situation being modelled. For example, for superimposed waves on a stretched string, the waves are real. For matter waves in quantum physics, the waves may be complex.

1.8.8 Energy in a Wave

A wave can be used to transfer energy between two locations.

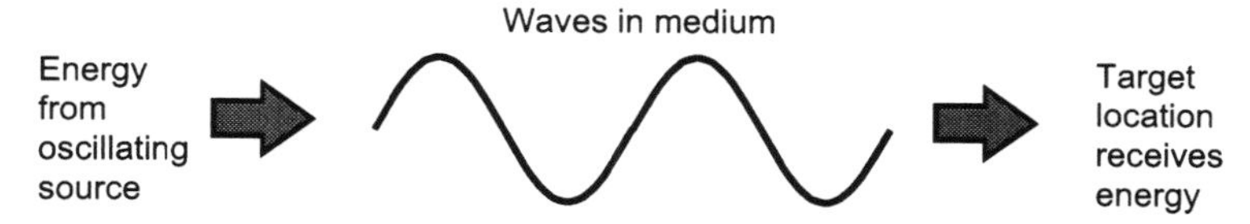

1. The external source performs work on the first particle in the string.
2. The particle moves with SHM. The energy of particle is converted from P.E. to K.E. etc. Total energy of the particle is unchanged.
3. The particle loses energy to the next particle at the same rate it receives energy from the external source. Total energy of the particle remains unchanged but energy from the source gets passed on from one particle to the next till it arrives at the target location.
4. Energy from the external source travels along the string with velocity v.
5. The total energy of each particle is $E = \frac{1}{2} m\omega^2 A^2$.
6. The total energy for all oscillating particles in a segment of string one wavelength long is: $\boxed{E = \frac{1}{2}\omega^2 A^2 (\rho\lambda)}$ since $m = \rho\lambda$ (mass per unit length)

In one time period T, the energy contained in one wavelength of string will have moved on to the next wavelength segment.

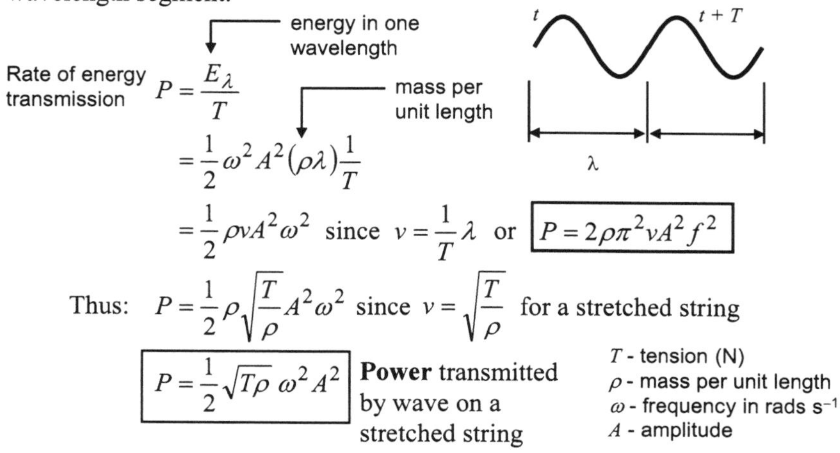

1.9 Infinite Series

Summary

$$f(n)=\left(1+\frac{1}{n}\right)^n$$ Sequence

$$L=\lim_{n\to\infty}\left(1+\frac{1}{n}\right)^n$$ Limit of a sequence

$$f(n)=a+(n-1)d$$ Arithmetic progression

$$f(n)=ar^{n-1}$$ Geometric progression

$$\sum_{n=1}^{\infty} f(n)$$ Infinite series

Power series

$$\sum_{n=0}^{\infty} a_n x^n = a_o + a_1 x + a_2 x^2 + a_3 x^3 \ldots + a_n x^n + \ldots$$

Taylor series

$$f(x)=f(c)+\frac{f'(c)}{1!}(x-c)+\frac{f''(c)}{2!}(x-c)^2+\ldots+\frac{f^n(c)}{n!}(x-c)^n+\ldots$$

Maclaurin series

$$f(x)=f(0)+\frac{f'(0)}{1!}x+\frac{f''(0)}{2!}x^2+\ldots+\frac{f^n(0)}{n!}x^n+\ldots$$

$$(1+x)^n = 1+nx+\frac{n(n-1)}{2!}x^2+\ldots$$ Binomial theorem

$$+\frac{n(n-1)\ldots(n-r+1)}{r!}x^r+\ldots+x^n$$

1.9.1 Sequences

An **infinite sequence** is a function where the independent variable is n, a positive integer. In a sequence, for each value of n, the function $f(n)$ gives the value of the nth term of the sequence. When $f(n)$ is written out term by term, the expression is also called a sequence.

For example, consider the sequence:

$$f(n) = \left(1 + \frac{1}{n}\right)^n$$

This sequence can be written: 2, 2, 25, 2.3704, 2.4414,...

If the sequence has a **limit** L as n approaches infinity, then:

$$L = \lim_{n\to\infty}\left(1 + \frac{1}{n}\right)^n$$

In the particular case above, the sequence does have a limit since it approaches a value close to 2.718 (in fact, the limit of this sequence is the irrational number **e** in this example).

Limits of infinite sequences have properties similar to limits in general. For example, if $f(n)$ and $g(n)$ are two infinite sequences, then:

$$\text{If } \lim_{n\to\infty} f(n) = L \text{ and } \lim_{n\to\infty} g(n) = M$$

$$\text{then } \lim_{n\to\infty}\left(f(n) \pm g(n)\right) = L \pm M$$

$$\lim_{n\to\infty}\left(f(n)g(n)\right) = LM$$

$$\lim_{n\to\infty}\left(\frac{f(n)}{g(n)}\right) = \frac{L}{M}$$

When a sequence is formed such that each term is found by adding a constant to the preceding term, then the sequence is called an **arithmetic sequence** or **arithmetic progression**:

$$a, a+d, a+2d, a+3d...$$

$$f(n) = a + (n-1)d$$

When a sequence is formed such that each term is found by multiplying the preceding term by a constant, then the sequence is called a **geometric sequence** or **geometric progression**:

$$a, ar, ar^2, ar^3 ... ar^{n-1} ...$$

$$f(n) = ar^{n-1}$$

1.9.2 Series

When the terms of a sequence are added together, the resulting summation is called a **series.**

$$\sum_{n=1}^{\infty} f(n)$$

A series may or may not converge to a number as n goes to infinity.

An infinite series of the form:

$$\sum_{n=1}^{\infty} ar^{n-1} = a + ar + ar^2 + ar^3 ... + ar^{n-1} ...$$

is called a **geometric series**. A geometric series with $|r| \geq 1$ is divergent. A geometric series with $|r| < 1$ is convergent and the limit is given by:

$$\lim_{n \to \infty} \sum_{n=1}^{\infty} ar^{n-1} = \frac{a}{1-r}$$

Whether or not a series is convergent can be tested by a number of standard tests (in a similar manner to the way in which improper integrals are handled). For example, the ***p* series** is:

$$\sum_{n=1}^{\infty} \frac{1}{n^p}$$

and converges for $p > 1$ and diverges for $p \leq 1$. The special case of $p = 1$ is called the **harmonic series**:

$$\sum_{n=1}^{\infty} \frac{1}{n} = 1 + \frac{1}{2} + \frac{1}{3} + \frac{1}{4} + ... + \frac{1}{n} ...$$

The harmonic series is divergent.

If the terms in a series are alternately positive and negative, then the series is an **alternating series.**

$$\sum_{n=1}^{\infty} (-1)^{n+1} a_n = a_1 - a_2 + a_3 - a_4 + ... + (-1)^{n+1} a_n + ...$$

An alternating series is convergent if each term is numerically less than or equal to the preceding term and if $\lim_{n \to \infty} a_n = 0$.

1.9.3 d'Alembert's Ratio Test

An infinite series is absolutely convergent if a series consisting of the absolute of the terms is convergent.

$$\sum_{n=1}^{\infty} |a_n| = |a_1| + |a_2| + |a_3| + ... |a_n|$$

The significance of this is that it forms the basis of a very important test for convergence, the **ratio test**. Consider an infinite series:

$$\boxed{\lim_{n\to\infty} \left| \frac{a_{n+1}}{a_n} \right| = L}$$

If $L < 1$ then the series is convergent.
If $L > 1$ then the series is divergent.
If $L = 1$ then the series may be convergent or divergent.

If all the terms in the series are positive, then the absolute signs need not be included.

As an example, consider the following series:

$$\sum_{n=1}^{\infty} nx^n$$

Applying the ratio test, we have:

$$\lim_{n\to\infty} \left| \frac{(n+1)x^{n+1}}{nx^n} \right| = \lim_{n\to\infty} \left| \frac{n+1}{n} x \right|$$
$$= |x|$$

This series is convergent if $-1 < x < 1$.

This series is divergent if $x < -1$ and $x > 1$.

In this example, the number 1, at the limits of the **interval of convergence**, is called the **radius of convergence**. Any series may diverge or converge at the radius of convergence.

1.9.4 Power Series

A series does not have to consist of constants. A series may consist of terms which are themselves functions of a variable. The **power series** is defined as:

$$\sum_{n=0}^{\infty} a_n x^n = a_o + a_1 x + a_2 x^2 + a_3 x^3 \ldots + a_n x^n + \ldots$$

When the variable x is given a value, we have a series of constant terms which may or may not be convergent. The range within the series if convergent is called the **interval of convergence**. Within the interval of convergence, a power series can be used to represent a function.

$$f(x) = \sum_{n=0}^{\infty} a_n x^n = a_o + a_1 x + a_2 x^2 + a_3 x^3 \ldots + a_n x^n + \ldots$$

The power series representation of a function allows the function to be differentiated and integrated in a different way from that normally used.

$$f'(x) = \sum_{n=0}^{\infty} \frac{d}{dx}\left(a_n x^n\right) = a_1 + 2a_2 x + 3a_3 x^2 \ldots + n a_n x^{n-1} + \ldots$$

$$\int_0^x f(x)dx = \sum_{n=0}^{\infty} \int_0^x \left(a_n x^n\right)dx = a_o x + \frac{1}{2}a_1 x^2 + \frac{1}{3}a_2 x^3 + \ldots + \frac{1}{n+1}a_n x^{n+1} + \ldots$$

A special case of the power series representation of a function is the **Taylor series** where:

$$f(x) = \sum_{n=0}^{\infty} a_n (x-c)^n = a_o + a_1(x-c) + a_2(x-c)^2 + \ldots$$

This is a series for $f(x)$ expanded around the number "c"

The **Taylor series** can be expressed:

$$f(x) = f(c) + \frac{f'(c)}{1!}(x-c) + \frac{f''(c)}{2!}(x-c)^2 + \ldots + \frac{f^n(c)}{n!}(x-c)^n + \ldots$$

A special case of the Taylor series occurs at $c = 0$ and is called the **Maclaurin series**:

$$f(x) = f(0) + \frac{f'(0)}{1!}x + \frac{f''(0)}{2!}x^2 + \ldots + \frac{f^n(0)}{n!}x^n + \ldots$$

Examples of Maclaurin series:

$$e^x = 1 + x + \frac{x^2}{2!} + \frac{x^3}{3!} \ldots + \frac{x^n}{n!} + \ldots$$

$$\sin x = x - \frac{x^3}{3!} + \frac{x^5}{5!} - \frac{x^7}{7!} \ldots + (-1)^n \frac{x^{2n+1}}{(2n+1)!} + \ldots$$

$$\cos x = 1 - \frac{x^2}{2!} + \frac{x^4}{4!} - \frac{x^6}{6!} \ldots + (-1)^n \frac{x^{2n}}{(2n)!} + \ldots$$

1.9.5 Binomial Series

The **binomial theorem** provides a series expansion of a function of the form $(a+b)^n$ where n is a positive integer. The term $(a+b)$ is a **binomial** as distinct from a **polynomial**, which usually has more terms. Historically, the coefficients of the expansion were determined using **Pascal's triangle**.

n	
0	1
1	1 1
2	1 2 1
3	1 3 3 1
4	1 4 6 4 1
5	1 5 10 10 5 1
6	1 6 15 20 15 6 1

For example,

$$(1+x)^5 = 1 + 5x + 10x^2 + 10x^3 + 5x^4 + 1x^5$$

Pascal's triangle can also be used for the more general expression:

$$(a+x)^n = a^n\left(1+\frac{x}{a}\right)^n$$

where the coefficients of this expansion are available directly from the triangle.

Pascal's triangle is inconvenient for expansions for n greater than about 8. A general expression for the expansion was given by Newton:

$$(1+x)^n = 1 + nx + \frac{n(n-1)}{2!}x^2 + \ldots + \frac{n(n-1)\ldots(n-r+1)}{r!}x^r + \ldots + x^n$$

where it can be seen that the expanded equation is a polynomial of degree n in x. The more general expression $(a+b)^n$ is written:

$$(a+b)^n = a^n + na^{n-1}b + \frac{n(n-1)}{2!}a^{n-2}b^2 + \ldots + \frac{n(n-1)(n-2)\ldots(n-r+1)}{r!}a^{n-r}b^r + \ldots + b^n$$

The **binomial theorem** has particular application to the theory of **probability** in random experiments which have two possible outcomes. If n is the number of trials, and a and b are the probabilities of each of the outcomes, then the binomial coefficients give the probability of 0, 1, 2 …n of one of the outcomes.

1.10 Probability

Summary

$\bar{x} = \dfrac{\sum x}{n}$	Sample mean
$s = \sqrt{\dfrac{\sum (x - \bar{x})^2}{n-1}}$	Sample standard deviation
${}^{n}p_{r} = \dfrac{n!}{(n-r)!}$	Permutation
${}^{n}c_{r} = \dfrac{n!}{r!(n-r)!}$	Combination
$p = \dfrac{s}{n}$	Probability
$\mu = \sum (xf(x))$	Expected value
$f(x) = {}^{n}c_{x} p^{x} (1-p)^{n-x}$	Binomial distribution
$f(x) = \dfrac{(np)^{x} e^{-np}}{x!}$	Poisson distribution
$\mu = np$	Mean of binomial distribution
$\sigma^2 = npq$	Variance of binomial distribution
$f(x) = \dfrac{1}{\sigma\sqrt{2\pi}} e^{\frac{-(x-\mu)^2}{2\sigma^2}}$	Normal distribution
$z = \dfrac{x - np}{\sqrt{npq}}$	Standard variable for normal distribution
$\sigma_{\bar{x}} = \dfrac{\sigma}{\sqrt{n}}$	Standard error of the mean

1.10.1 Mean, Median, Mode

The **mean**, or **average**, of a set of values is the sum of the values divided by the number of values:

$$\bar{x} = \frac{\sum x}{n}$$

Given the mean of several sets of data, the overall mean can be computed if the number of elements in each data set is known:

$$\bar{x} = \frac{n_1\bar{x}_1 + n_2\bar{x}_2 + n_3\bar{x}_3 + \ldots + n_k\bar{x}_k}{n_1 + n_2 + n_3 + \ldots + n_k} = \frac{\sum n\bar{x}}{\sum n}$$

The **median** of a set of values is the middle item (or the mean of the two middle items) when the data is arranged in an increasing or decreasing order of magnitude. The median is not so much affected by extreme values as is the mean, but is affected by the distribution of magnitudes within the data set.

The **mode** is the value which occurs with the highest frequency in a set of data.

The **range** is the difference between the largest and smallest element in the data set.

The **standard deviation** gives an indication of the mean distance of data from the mean, or the spread of data about the mean. The standard deviation is defined as:

$$s = \sqrt{\frac{\sum (x-\bar{x})^2}{n-1}}$$

$$= \sqrt{\frac{n\left(\sum x^2\right) - \left(\sum x\right)^2}{n(n-1)}}$$

It is important to distinguish between the mean $\bar{x}$ and standard deviation s of a sample of size n and the mean μ and standard deviation σ of the entire population of size N from which the sample is taken. Whether or not $x = \mu$ and $s = \sigma$ depends on the data. Often s is used in place of σ on the assumption that the two are equal (or nearly so).

The **sample variance** is defined as the standard deviation squared and is written s^2.

Chebyshev's theorem states that for any set of data, at least $1 - 1/k^2$ of the data must lie between k standard deviations on either side of the mean where $k > 1$.

The **coefficient of variation** gives a measure of the relative size of the standard deviation to the value of the mean and is given by:

$$v = \frac{\sigma}{\bar{x}}100$$

1.10.2 Permutations and Combinations

If a choice consists of k steps, and the first step can be made in n_1 ways, and for each of these, the second step can be made in n_2 ways, and for each of these, the kth step can be made in n_k ways, then the choice can be made in $n_1 n_2 \ldots n_k$ ways.

If r objects are selected from a set of n total objects (and not replaced after each object is selected), any particular ordered arrangement of these r objects is called a **permutation**. The total number of possible permutations of r objects selected from a total of n objects is given a special notation:

$$ {}^{n}p_r = \frac{n!}{(n-r)!} $$

Order is important.

The "!" symbol indicates **factorial** n which has the meaning:

$$ n! = n(n-1)(n-2)\ldots(3)(2)(1) $$

and $0! = 1$ by definition

Note: For large n, $n!$ can be approximated by **Stirling's formula**:

$$ n! \approx \left(\sqrt{2\pi n}\right) n^n e^{-n} $$

The number of n permutations of all the objects taken all together is:

$$ {}^{n}p_n = n! $$

If r objects within the n total objects are alike, and the others are all distinct, then the number of permutations of these n objects taken all together is:

$$ {}^{n}p_n = \frac{n!}{r!} $$

If r_1 objects within the n total objects are alike, and r_2 objects are alike, and the others are distinct, then the number of permutations of these n objects taken all together is:

$$ {}^{n}p_n = \frac{n!}{r_1! r_2!} $$

If r objects are selected from n total objects, and if the order of the r objects is not important, then each possible selection of r objects is called a **combination** and is calculated from:

$$ {}^{n}c_r = \frac{n!}{r!(n-r)!} $$

Order is not important.

1.10.3 Probabilities, Odds and Expectation

The classical definition of probability states that if there are n equally likely possibilities of which one must occur, and s are regarded as the total number of possible **successes** or **events**, then the probability p of a success is:

$$p = \frac{s}{n}$$

The probability of a **failure** q is:

$$q = 1 - p = \frac{n-s}{n}$$

If the probability of a **success** is p, the **odds** for its occurrence are a to b given by:

$$\frac{a}{b} = \frac{p}{1-p} = \frac{p}{q}$$

Note: It is customary to express odds as the ratio of two positive integers having no common factor.

A second way of defining probability is to say that if after n trials an event is observed to occur s times, then the probability of an event (or a success) occurring is:

$$p = \frac{s}{n}$$

If individual events are mutually exclusive, then the probability that one or the other of them will occur is given by the sum of the individual probabilities.

Example:

Calculate the probability that two cards withdrawn from a deck of cards will both be black.

Solution:

Since 2 cards are drawn from 52 cards, the number of possibilities is given by:

$$n = {}^{n}c_r = {}^{52}c_2 = \frac{52!}{2!(52-2)!} = 1326$$

Since half the cards are black, the number of possible successes is given by:

$$n = {}^{n}c_r = {}^{26}c_2 = \frac{26!}{2!(26-2)!} = 325$$

The probability of a success is thus: $p = \frac{325}{1326} = 0.245$

Or, we can say that the probabilities of the first and second cards being black are:

$$p_1 = \frac{26}{52}; p_2 = \frac{25}{52}$$

$$p = p_1 p_2 = 0.245$$

1.10.4 Probability Distribution

Let X be a random variable which can take on the values x_1, x_2, x_3,... The probability of X being equal to some particular value x is p_1, p_2, p_3...that is, the probability: $P(X = x) = f(x)$

Graphically, this can be represented by a **histogram** which shows the distribution in probabilities for each value of X. For example, X might be the number of heads obtained in n tosses of a coin and so $P(1)$ represents the probability of obtaining one head, $P(2)$, the probability of obtaining 2 heads.

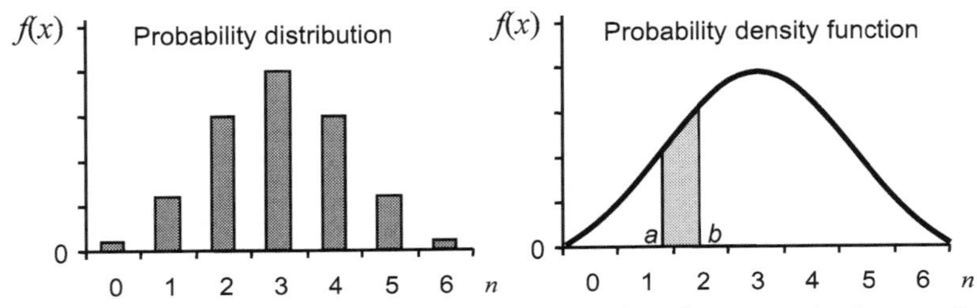

Now, if the variable X can take on any value rather than a particular number, it is called a **continuous random variable** and the probability that X takes on a particular value x is zero. However, the probability that X lies between two different values, say a and b, is by definition given by:

$$P(a < X < b) = \int_a^b f(x)\, dx$$

$$\text{where } f(x) \geq 0 \text{ and } \int_{-\infty}^{\infty} f(x) = 1$$

Note that it is the area under the curve of $f(x)$ that gives the probability. For the continuous case, the value of $f(x)$ at any point is not a probability. Rather, $f(x)$ in this case is called the **probability density function**.

A **cumulative probability distribution function** for the number of successes gives the probability that X takes on some value less than or equal to x. The cumulative probability function can be obtained from the probability function by adding the probabilities for all values of X less than x.

$$P(X \leq x) = F(x) = \int_{-\infty}^{x} f(x) dx$$

F(X)
1
Distribution function
X

1.10.5 Expected Value

If the probabilities of obtaining the amounts $x_1, x_2, \ldots x_k$ are $p_1, p_2, \ldots p_k$, then the **expected value** is given by:

$$E = x_1 p_1 + x_2 p_2 + \ldots + x_k p_k$$

This is the **expected value** of the random variable x and gives the value of x that would occur if the same set of trials were to be repeated many times. The expected value is the **mean** value of the probability distribution.

The **mean** of a discrete probability distribution is given by:

$$\mu = \sum (xf(x))$$

For X being a continuous random variable, the mean is given by:

$$\mu = \int_{-\infty}^{\infty} xf(x)dx$$

The **variance** σ^2 of a probability distribution is:

$$\sigma^2 = \sum (x-\mu)^2 f(x) \quad \text{or} \quad \sigma^2 = \int_{-\infty}^{\infty} (x-\mu)^2 f(x)dx$$

The **variance** gives a measure of the dispersion of probabilities about the mean value. The square root of the variance is the **standard deviation** (hence the notation σ^2 for variance).

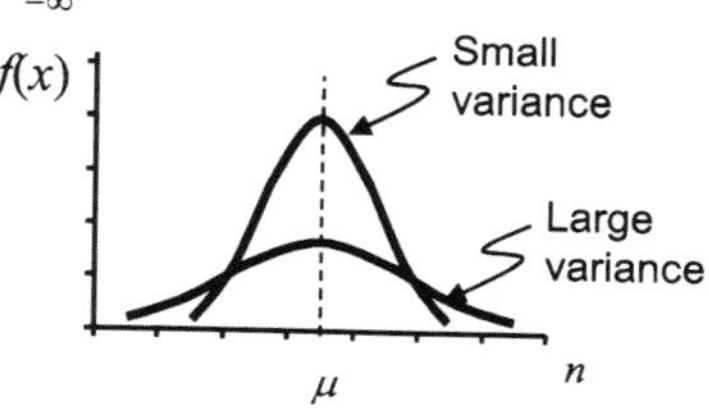

Example:

Determine the expected value for the number of heads when a coin is tossed 4 times if the probability distribution is given:

Solution:

$$E = x_1 p_1 + x_2 p_2 + \ldots + x_k p_k$$

x	p	xp
0	1/16	0
1	4/16	4/16
2	6/16	12/16
3	4/16	12/16
4	1/16	4/16

Expected value: 32/16 = 2

1.10.6 Binomial Distribution

Consider a series of n trials where there are only two possible outcomes, one of which represents a success and the other a failure and where the probability p of a success remains constant throughout the n trials. The probability of obtaining x successes is a function of x and is called the **binomial distribution**:

$$f(x) = {}^{n}c_{x} p^{x}(1-p)^{n-x}$$

Note: The two possible outcomes might be the either heads or tails on the toss of a coin, either a black or not a black card drawn from a deck. Note, to maintain a constant probability the card would need to be replaced before drawing another card.

If n is large and p is small, then the binomial distribution is approximated by the **Poisson distribution**:

$$f(x) = \frac{(np)^{x} e^{-np}}{x!}$$

where np represents the average number of successes.

The **mean** of a **binomial distribution** is given by:

$$\mu = np$$

The **variance** is:

$$\sigma^{2} = npq$$

where $q = 1 - p$ and is the probability of obtaining x failures in the n trials.

Example:

Determine the probability distribution for obtaining x heads from 4 tosses of a coin.

Solution:

There are 2 possible outcomes for each toss, so the total number of possibilities is $2^4 = 16$. The number of possible successes is equal to the number of ways that x heads can be had from 4 tosses. Thus $s = {}^{4}c_{x}$. For 0 heads, we have $s = 1$. The probability of obtaining 0 heads is thus 1/16.

For 1 head, we have $s = {}^{4}c_{1} = 4$ and the probability of obtaining 1 head is thus: $p_1 = \frac{4}{16}$

For 2 heads, we have $s = {}^{4}c_{2} = 6$ and the probability of obtaining 2 heads is thus: $p_2 = \frac{6}{16}$

For 3 heads, we have $s = {}^{4}c_{3} = 4$ and the probability of obtaining 3 heads is thus: $p_3 = \frac{4}{16}$

For 4 heads, we have $s = {}^{4}c_{4} = 1$ and the probability of obtaining 4 heads is thus: $p_4 = \frac{1}{16}$

Tabulating these results, we obtain:

No. heads	p
0	1/16
1	4/16
2	6/16
3	4/16
4	1/16

In the case of 4 tosses of a coin, $n = 4$ and $p = 0.5$ so the probability of obtaining x heads in 4 tosses is:

$$f(x) = {}^{4}c_{x} \frac{1}{2}^{2}\left(1 - \frac{1}{p}\right)^{4-x} = {}^{4}c_{x} \frac{1}{16}$$

Note: This distribution of probabilities is called the binomial distribution and the numerators correspond to coefficients given by the **binomial theorem**.

1.10.7 Normal Distribution

The **normal distribution**, or **Gaussian distribution**, is a continuous probability distribution. If μ is the mean and σ^2 the variance of the distribution, then the probability density function is defined as:

$$f(x) = \frac{1}{\sigma\sqrt{2\pi}} e^{\frac{-(x-\mu)^2}{2\sigma^2}}$$

The **cumulative distribution function** is given by:

$$F(x) = P(X \leq x) = \frac{1}{\sigma\sqrt{2\pi}} \int_{-\infty}^{x} e^{\frac{-(x-\mu)^2}{2\sigma^2}} dx$$

If the mean μ of a probability distribution is equal to 0 and the standard deviation $\sigma = 1$, then this gives the **standard normal distribution** and can be drawn:

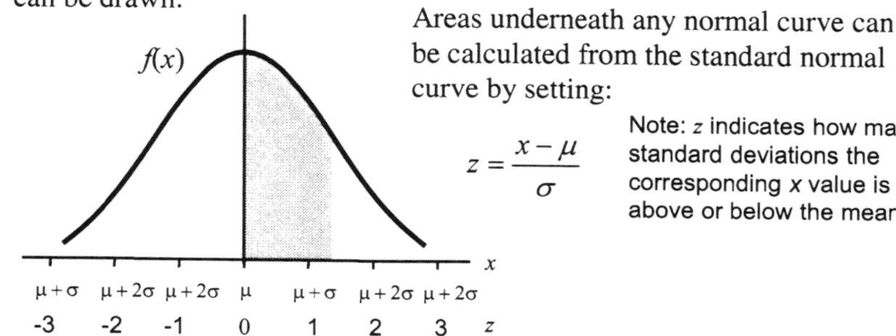

Areas underneath any normal curve can be calculated from the standard normal curve by setting:

$$z = \frac{x - \mu}{\sigma}$$

Note: z indicates how many standard deviations the corresponding x value is above or below the mean.

For large n and p and q ($= 1 - p$) not too close to zero, the **binomial distribution** is approximately equal to a normal distribution with a standard variable z given by:

$$z = \frac{x - np}{\sqrt{npq}}$$

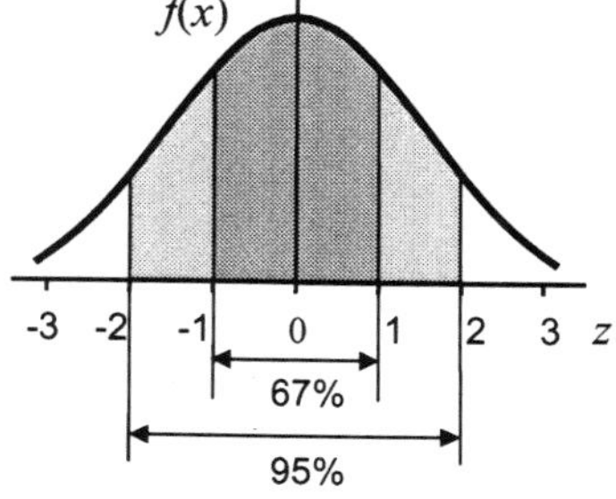

The area under the curve between any two points represents the probability of the outcome of a trial having a value between those two points. The total area under the curve is 1.

The probability that a value for x lies between $\pm\sigma$ is 67%. The probability that a value for x lies between $\pm 2\sigma$ is 95%.

Example:

The ideal value of a geometrical constant G for a pyramidal indenter used for hardness testing of materials is 24.5. Measurements on a large number of specimens yield a mean value of 24.52 with a standard deviation of 0.852. What is the probability of selecting an indenter from the population which has a value of the constant within 2% of the ideal value?

Solution:

The question we need to answer is "What is the probability of the value of G for an indenter chosen at random from the entire population of indenters having a value between 24.5 −2% to 24.5 +2% (i.e., 24.01 to 24.99)?" We shall assume that the sample standard deviation is representative of the population standard deviation and that the data is normally distributed about the mean. Thus:

$$z_1 = \frac{x_1 - \mu}{\sigma} = \frac{24.01 - 24.52}{0.852} = -0.60$$

$$z_2 = \frac{x_2 - \mu}{\sigma} = \frac{24.99 - 24.52}{0.852} = 0.55$$

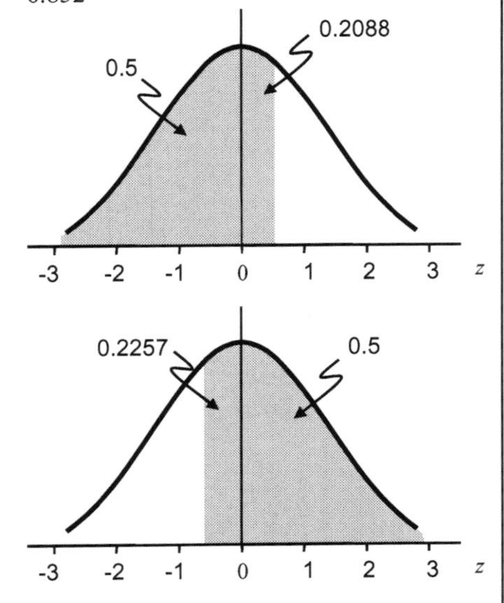

The standard table of normal distributions gives the area under the curve from 0 to z. For z = 0.55, we have from a table of normal curve areas, P = 0.2088. The total probability of choosing an indenter with a value of $G < 24.99$ is 0.5+0.2088 = 0.7088.

For z = 0.60, the table gives P = 0.2257. The probability of choosing an indenter with a value of $G > 24.01$ is thus 0.5–0.2257 = 0.2743.

The probability we are looking for is the difference between the two:

$$P = 0.7088 - 0.2743$$
$$= 43.4\%$$

1.10.8 Sampling

A population has a **population mean** μ and standard deviation σ. A random sample of size n taken from this population has itself a mean which we call the **sample mean** $\bar{x}$ and a **sample standard deviation** s. As the number of samples taken approaches all possible samples, then the mean of all the sample means approaches the mean of the population.

$$\mu_{\bar{x}} = \mu$$

The standard deviation of the sample means is given by:

$$\sigma_{\bar{x}} = \frac{\sigma}{\sqrt{n}}$$ **standard error of the mean** of the sample means

If the sample size n is large (usually > 30) and several samples are taken, then the distribution of the means of each sample is approximately described by a **normal distribution**. This is called the **central limit theorem**.

Often, it is required to estimate the population mean from the sample mean on the basis of just one sample of size n. Using the central limit theorem, it is possible to estimate the error or degree of confidence in which the mean of the sample approximates the mean of the population.

Let $z_{\alpha/2}$ be the value of z for which the area under the normal curve to its right is equal to $\alpha/2$.

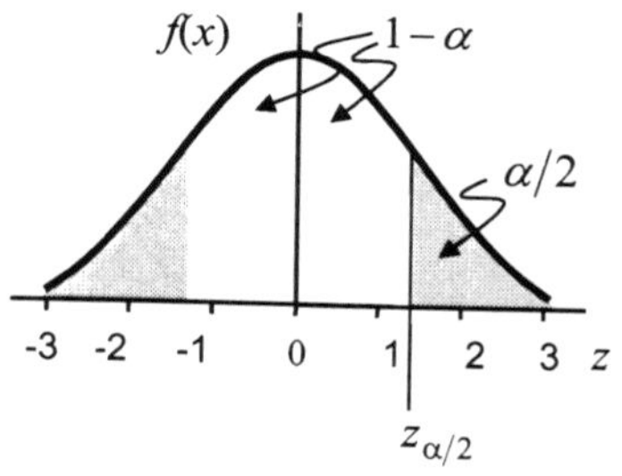

A particular sample mean will differ from the population mean by less than $z_{\alpha/2}$ standard errors of the mean.

We can thus say that the probability is $1 - \alpha$ that $\bar{x} - \mu$ is less than:

$$z_{\alpha/2}\left(\sigma_{\bar{x}}\right) = z_{\alpha/2}\left(\frac{\sigma}{n}\right)$$

Note: if σ (population) is not known, then s (of the sample) is used instead.

The quantity $1 - \alpha$ is called the **degree of confidence**. We say that with probability $1 - \alpha$, the population mean is somewhere within the confidence interval.

$$\bar{x} - z_{\alpha/2}\left(\frac{\sigma}{\sqrt{n}}\right) \le \mu \le \bar{x} + z_{\alpha/2}\left(\frac{\sigma}{\sqrt{n}}\right)$$

Typically, α is chosen to be 0.05 or 0.01 (95% or 99% confidence level). From a table of the standard normal distribution, this corresponds to:

$$z_{\alpha/2} = z_{0.025} = 1.96$$

$$z_{\alpha/2} = z_{0.005} = 2.58$$

1.10.9 *t* Distribution

If the sample size n is small, and if the population shows a normal distribution, then the statistical quantity $\boldsymbol{t}$ is defined such that:

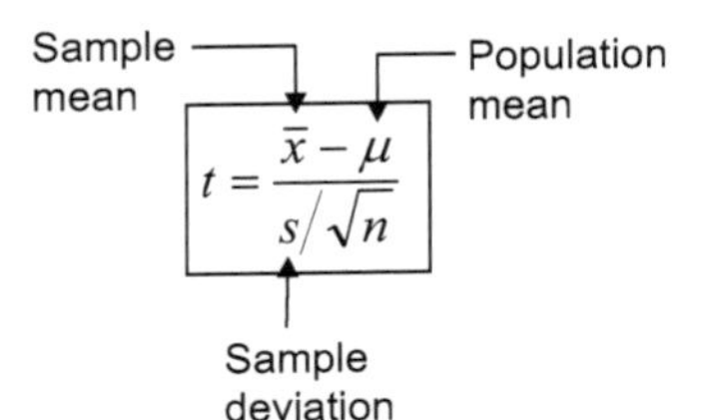

Note: Here we use the symbol s to denote the standard deviation of the sample as distinct from the standard deviation of the population.

Standard tables give the area under the t distribution curve such that:

$$\bar{x} - t_{\alpha/2}\left(\frac{s}{\sqrt{n}}\right) \leq \mu \leq \bar{x} + t_{\alpha/2}\left(\frac{s}{\sqrt{n}}\right)$$

The shape of the t distribution depends upon the number of **degrees of freedom** $\nu = n - 2$ and so the tables that provide the areas are constructed usually showing the most commonly used values of $t_{\alpha/2} = 0.1, 0.05, 0.025, 0.01, 0.005$. The t distribution is similar in shape to the normal distribution.

Example:

10 measurements were performed where the mean value was determined to be 106.3 with a variance of 8.51. Determine the range around this mean which, at the 90% confidence level, would include the mean of the population using the t distribution.

Solution:

$$n = 10$$
$$\bar{x} = 106.3$$
$$s^2 = 8.51$$

At 90% confidence level:

$$\alpha = \frac{1 - 0.9}{2} = 0.05$$
$$\nu = 10 - 2 = 8$$

$t = 1.860$ from table of t distribution

Thus:

$$t = \frac{\bar{x} - \mu}{\frac{s}{\sqrt{n}}}$$

$$1.86 = \frac{106.3 - \mu}{\sqrt{\frac{8.51}{10}}}$$

$$\mu = 104.6$$

The difference is: $\delta x = 106.3 - 104.6 = 1.7$

$$\bar{x} = 106.3 \pm 1.7$$

At 90% confidence level the mean of the population is between 106.3–1.7 and 106.3+1.7.

1.10.10 Chi-Squared Distribution

The central limit theorem and the t distribution are used to estimate the confidence in which we can estimate the mean of a population from the mean of a single sample taken from the population. We also may wish to estimate the standard deviation of a population. This is based upon the standard deviation of the samples taken from the population. If the population has a normal distribution, then the statistical quantity χ^2 (**Chi-squared**) is defined as:

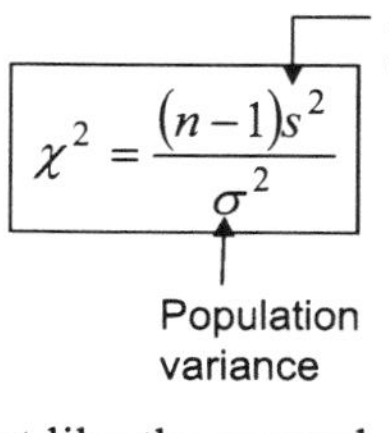

$$\chi^2 = \frac{(n-1)s^2}{\sigma^2}$$

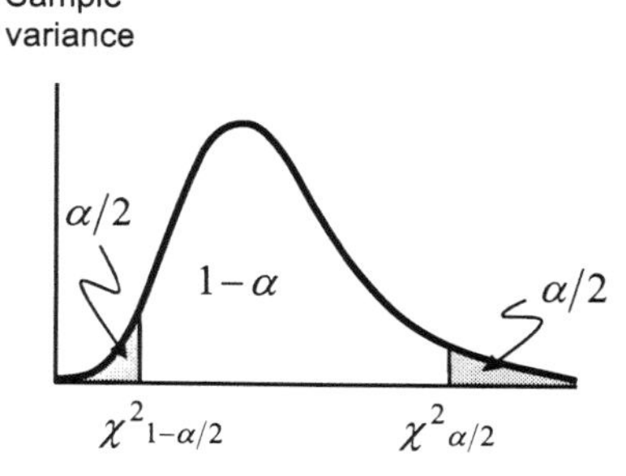

The χ^2 distribution is not like the normal distribution it is defined only for $\chi^2 \geq 0$ and is not symmetric.

Standard tables give the area under the Chi-square distribution curve such that:

$$\frac{(n-1)s^2}{\chi^2_{\alpha/2}} \leq \sigma^2 \leq \frac{(n-1)s^2}{\chi^2_{1-\alpha/2}}$$

Note: since the distribution is not symmetric, these confidence limits are different (unlike the normal and t distributions).

The shape of the χ^2 distribution depends upon the number of **degrees of freedom** $v = n - 1$ and so the tables that provide the areas are constructed usually showing the most commonly used values $\chi^2_{\alpha/2} = 0.1, 0.05, 0.025, 0.01, 0.005$.

Example:

10 measurements were performed where the mean value was determined to be 106.3 with a variance of 8.51 (s = 2.91). Determine the 90% confidence level for the standard deviation of the population based upon these measurements.

Solution:

$n = 10$

$\bar{x} = 106.3$

$s^2 = 8.51$

At 90% confidence level:

$$\alpha = \frac{1-0.9}{2} = 0.05$$

$$v = 10 - 1 = 9$$

$\chi^2_{0.975} = 2.700$, $\chi^2_{0.025} = 19.023$ from standard table

Thus:

$$\sigma^2_{0.975} = \frac{(10-1)8.51}{2.7} = 28.366$$

$$\sigma^2_{0.025} = \frac{(10-1)8.51}{19.023} = 4.026$$

The confidence limits for estimating σ are thus:

$$\sqrt{4.026} < \sigma < \sqrt{28.366}$$

$$2.00 < \sigma < 5.32$$

@90% confidence level

1.11 Matrices

Summary

$c_{jk} = \sum_{i=1}^{n} a_{ji} b_{ik}$	Product of two matrices
$\mathbf{A} = (a_{jk}), \mathbf{A}^{\mathrm{T}} = (a_{kj})$	Transpose of a matrix
$\lvert \mathbf{A} \rvert = \sum_{k=1}^{n} a_{jk} A_{jk}$	Determinant
$\mathbf{X} = \mathbf{A}^{-1}\mathbf{Y}$	Inverse of a matrix
$\lvert \mathbf{A} - m\mathbf{I} \rvert = 0$	Characteristic equation
$\mathbf{A}^{n} = p\mathbf{A} + q\mathbf{I}$	Matrix raised to a power

1.11.1 Matrices

A matrix is a rectangular array of numbers with m rows and n columns. When $m = n$, the matrix is square. Numbers in the matrix are called elements and are written a_{jk} where the subscripts refer to the row and column, respectively.

$$\mathbf{A} = \begin{bmatrix} a_{11} & a_{12} & a_{13} \\ a_{21} & a_{22} & a_{23} \\ a_{31} & a_{32} & a_{33} \end{bmatrix}$$

Addition and subtraction of matrices: $\mathbf{A} + \mathbf{B} = (a_{jk} + b_{jk})$

Multiplication by a number: $k\mathbf{A} = (ka_{jk})$

Multiplication of two matrices:

If **A** is an $m \times n$ matrix, and **B** is an $n \times p$ matrix, then the product **C** of **A** and **B** is given by:

$$c_{jk} = \sum_{i=1}^{n} a_{ji} b_{ik}$$

The number of columns in **A** must be the same as the number of rows in **B**. The result matrix **C** has dimensions $m \times p$.

For example:

$$\begin{bmatrix} \cdot & \cdot & \cdot \\ a_{21} & a_{22} & a_{23} & a_{24} \\ \cdot & \cdot & \cdot \end{bmatrix} \begin{bmatrix} \cdot & \cdot & b_{13} & \cdot & \cdot \\ \cdot & \cdot & b_{23} & \cdot & \cdot \\ \cdot & \cdot & b_{33} & \cdot & \cdot \\ \cdot & \cdot & b_{43} & \cdot & \cdot \end{bmatrix} = \begin{bmatrix} \cdot & \cdot & \cdot & \cdot & \cdot \\ \cdot & \cdot & c_{23} & \cdot & \cdot \\ \cdot & \cdot & \cdot & \cdot & \cdot \end{bmatrix}$$

The **transpose** of a matrix is when the rows and columns are interchanged: $\mathbf{A} = (a_{jk}), \mathbf{A}^{\mathrm{T}} = (a_{kj})$

The diagonal elements of a square matrix are called the principal or main diagonal and the sum of these is called the **trace** of the matrix.

A square matrix in which the main diagonal elements are all 1 and all other elements are 0 is called a **unit matrix**.

For a square matrix **A**, if there exists a matrix **B** such that **AB** = **I**, then **B** is the **inverse** of **A** and **I** written $\mathbf{B}^{-1}$. Not every square matrix has an inverse. If a matrix has an inverse, then there is only one inverse. If the matrix is a **singular matrix**, then the inverse does not exist.

A matrix **A** is orthogonal if $\mathbf{A}^{\mathrm{T}}\mathbf{A} = \mathbf{I}$

Two column matrices, or vectors, are **orthogonal** if $\mathbf{A}^{\mathrm{T}}\mathbf{B} = 0$.

1.11.2 Determinants

The **determinant** of a square matrix is denoted $\det \mathbf{A} = \Delta\mathbf{A} = |\mathbf{A}|$

The determinant of a second order matrix is given by:

$$\begin{vmatrix} a_{11} & a_{12} \\ a_{21} & a_{22} \end{vmatrix} = a_{11}a_{22} - a_{12}a_{21}$$

Determinants are only defined for square matrices.

The determinant of a third order matrix is given by:

$$\begin{vmatrix} a_{11} & a_{12} & a_{13} \\ a_{21} & a_{22} & a_{23} \\ a_{31} & a_{32} & a_{33} \end{vmatrix} = a_{11}\begin{vmatrix} a_{22} & a_{23} \\ a_{32} & a_{33} \end{vmatrix} - a_{12}\begin{vmatrix} a_{21} & a_{23} \\ a_{31} & a_{33} \end{vmatrix} + a_{13}\begin{vmatrix} a_{21} & a_{22} \\ a_{31} & a_{32} \end{vmatrix}$$

$$= a_{11}(a_{22}a_{33} - a_{23}a_{32}) - a_{12}(a_{21}a_{33} - a_{23}a_{31}) + a_{13}(a_{21}a_{32} - a_{22}a_{31})$$

Consider a square matrix **A**. If one element a_{jk} is selected, and the elements in the *j*th row and *k*th column are removed, then the determinant of the remaining elements (which will be of order $n-1$) is called the **minor** of **A**. For example:

$$\mathbf{A} = \begin{bmatrix} a_{11} & a_{12} & a_{13} \\ a_{21} & a_{22} & a_{23} \\ a_{31} & a_{32} & a_{33} \end{bmatrix} \qquad \mathbf{A} = \begin{bmatrix} a_{11} & a_{12} & a_{13} \\ a_{21} & a_{22} & a_{23} \\ a_{31} & a_{32} & a_{33} \end{bmatrix}$$

$$|M_{11}| = \begin{vmatrix} a_{22} & a_{23} \\ a_{32} & a_{33} \end{vmatrix} \qquad |M_{22}| = \begin{vmatrix} a_{11} & a_{13} \\ a_{31} & a_{33} \end{vmatrix}$$

The **cofactor** A_{jk} of an element a_{jk} is found when the minor is multiplied by the factor: $(-1)^{j+k}$. The determinant can be found by summing the products if the elements of any row or column by their cofactors.

$$|\mathbf{A}| = \sum_{k=1}^{n} a_{jk}A_{jk}$$

This is called the **Laplace expansion** of a determinant.

Properties of determinants: $|\mathbf{AB}| = |\mathbf{A}||\mathbf{B}|$

$$|\mathbf{A}| = |\mathbf{A}^T|$$

If all the elements in a row (or a column) are multiplied by a number and then added to the corresponding elements of another row (or column) then the value of the determinant is unchanged.

The determinant of a **singular matrix** is zero.

1.11.3 Systems of Equations

A **linear system of equations** is an ordered set of coefficients and unknowns.

$$a_{11}x_1 + a_{12}x_2 + \dots a_{1n}x_n = y_1$$
$$a_{21}x_1 + a_{22}x_2 + \dots a_{2n}x_n = y_2$$
$$\vdots$$
$$a_{m1}x_1 + a_{m2}x_2 + \dots a_{mn}x_n = y_n$$

The linear system can be written in matrix form:

$$\begin{bmatrix} a_{11} & a_{12} & & a_{1n} \\ a_{21} & a_{22} & & a_{2n} \\ & & & \\ a_{m1} & a_{m2} & & a_{mn} \end{bmatrix} \begin{bmatrix} x_1 \\ x_2 \\ \\ x_n \end{bmatrix} = \begin{bmatrix} y_1 \\ y_2 \\ \\ y_n \end{bmatrix}$$

$$\mathbf{AX} = \mathbf{Y}$$

If the matrix $\mathbf{A}$ is non-singular, then there exists an inverse $\mathbf{A}^{-1}$. The unknowns are the values of x and can be found from:

$$\mathbf{X} = \mathbf{A}^{-1}\mathbf{Y}$$

The unknown matrix $\mathbf{X}$ can also be determined by **elementary row transformations** in which the **augmented matrix** is operated upon to yield a unit matrix and a solution column.

$$\begin{bmatrix} a_{11} & a_{12} & & a_{1n} \mid y_1 \\ a_{21} & a_{22} & & a_{2n} \mid y_2 \\ & & & \mid \\ a_{m1} & a_{m2} & & a_{mn} \mid y_n \end{bmatrix} \to \begin{bmatrix} 1 & 0 & & 0 \mid x_1 \\ 0 & 1 & & 0 \mid x_2 \\ & & & \mid \\ 0 & 0 & & 1 \mid x_n \end{bmatrix}$$

The unknown matrix $\mathbf{X}$ can also be determined from **Cramer's rule** where the determinant of the matrix $\mathbf{A}$:

$$x_1 = \frac{|\mathbf{A}_1|}{|\mathbf{A}|}$$

$$x_2 = \frac{|\mathbf{A}_2|}{|\mathbf{A}|}$$

$$x_n = \frac{|\mathbf{A}_n|}{|\mathbf{A}|}$$

where $|\mathbf{A}_n|$ is the determinant of a new matrix formed by replacing the nth column with the solution column y. For example:

$$\mathbf{A}_2 = \begin{bmatrix} a_{11} & y_1 & & a_{1n} \\ a_{21} & y_2 & & a_{2n} \\ & & & \\ a_{m1} & y_n & & a_{mn} \end{bmatrix}$$

1.11.4 Eigenvalues and Eigenvectors

The **characteristic equation** of a matrix is given by the determinant:

$$|\mathbf{A} - m\mathbf{I}| = 0$$

This determinant leads to an equation in m of degree n if **A** is a square matrix of order n. The roots of the characteristic equation are called the **eigenvalues** of **A**.

A non-zero vector **C** which is a solution of the equation:

$$|\mathbf{A} - m\mathbf{I}|\mathbf{C} = \mathbf{0}$$

for a particular eigenvalue is called an **eigenvector** of the matrix **A** corresponding to the eigenvalue m. If **C** is an eigenvector, then so is any multiple of **C**. Every real symmetric matrix **A** is orthogonally similar to a diagonal matrix whose diagonal elements are the eigenvalues of **A**.

Example:

Find an orthogonal matrix **C** such that: $\mathbf{C}^{-1}\mathbf{AC} = \mathbf{D}$

where **D** is a diagonal matrix with diagonal elements equal to the eigenvalues of **A** where **A** is given by:

$$\mathbf{A} = \begin{bmatrix} 7 & -2 & 1 \\ -2 & 10 & -2 \\ 1 & -2 & 7 \end{bmatrix}$$

Solution:

The characteristic equation is:

$$\begin{vmatrix} m-7 & 2 & -1 \\ 2 & m-10 & 2 \\ -1 & 2 & m-7 \end{vmatrix} = 0$$

$$m^3 - 24m^2 + 180m - 432 = 0$$

$m = 6,6,12$ eigenvalues

For m = 6

$$(\mathbf{A} - 6\mathbf{I})\mathbf{C} = 0$$

$$\begin{bmatrix} -1 & 2 & -1 \\ 2 & -4 & 2 \\ -1 & 2 & -1 \end{bmatrix}\begin{bmatrix} x_1 \\ x_2 \\ x_3 \end{bmatrix} = [0]$$

$$x_1 - 2x_2 + x_3 = 0$$

$$\mathbf{X_1} = \begin{bmatrix} 1 \\ 0 \\ -1 \end{bmatrix}; \quad \mathbf{X_2} = \begin{bmatrix} 1 \\ 1 \\ 1 \end{bmatrix}$$

two eigenvectors corresponding to this eigenvalue

For m = 12

$$\mathbf{X_3} = \begin{bmatrix} 1 \\ -2 \\ 1 \end{bmatrix}$$

We then form the unit matrix of the eigenvalues:

$$\mathbf{C} = \begin{bmatrix} 1/\sqrt{2} & 1/\sqrt{3} & 1/\sqrt{6} \\ 0 & 1/\sqrt{3} & -2/\sqrt{6} \\ -1/\sqrt{2} & 1/\sqrt{3} & -1/\sqrt{6} \end{bmatrix}$$

It is then easy to show that: $\mathbf{C}^{-1}\mathbf{AC} = \mathbf{D}$

$$= \begin{bmatrix} 6 & 0 & 0 \\ 0 & 6 & 0 \\ 0 & 0 & 12 \end{bmatrix}$$

where we say that we have diagonalised the matrix **A**.

1.11.5 Cayley–Hamilton Theorem

The **Cayley–Hamilton theorem** states that every square matrix satisfies its own characteristic equation.

Let: $\mathbf{A} = \begin{bmatrix} a & c \\ b & d \end{bmatrix}$ The characteristic equation is thus:

$$\begin{vmatrix} a-m & b \\ c & d-m \end{vmatrix} = 0$$

$$m^2 - (a+d)m + (ad - bc) = 0$$

By the theorem, the following matrix equation is true:

$$\mathbf{A}^2 - (a+d)\mathbf{A} + (ad-bc)\mathbf{I} = 0$$

Now: $$\mathbf{A}^2 = \begin{bmatrix} a & c \\ b & d \end{bmatrix}\begin{bmatrix} a & c \\ b & d \end{bmatrix}$$

$$= \begin{bmatrix} a^2 + bc & b(a+d) \\ ac + cd & bc + d^2 \end{bmatrix}$$

$$(a+d)\mathbf{A} = \begin{bmatrix} a(a+d) & b(a+d) \\ c(a+d) & d(a+d) \end{bmatrix}$$

$$\mathbf{A}^2 - (a+d)\mathbf{A} = \begin{bmatrix} a^2 + bc - a^2 - ad & 0 \\ 0 & bc - ad \end{bmatrix}$$

$$= -(ad - bc)\mathbf{I}$$

Thus: $$\mathbf{A}^2 - (a+d)\mathbf{A} + (ad-bc)\mathbf{I} = 0$$

$$\mathbf{A}^2 = (a+d)\mathbf{A} - (ad-bc)\mathbf{I}$$

$$\mathbf{A}^3 = (a+d)\mathbf{A}^2 - (ad-bc)\mathbf{I}\mathbf{A}$$

$$= (a+d)\mathbf{A}^2 - (ad-bc)\mathbf{A}$$

$$= (a+d)((a+d)\mathbf{A} - (ad-bc)\mathbf{I}) - (ad-bc)\mathbf{A}$$

$$= ((a+d)^2 - (ad-bc))\mathbf{A} - (a+d)(ad-bc)\mathbf{I}$$

In general, for a 2×2 matrix,

$\mathbf{A}^n = p\mathbf{A} + q\mathbf{I}$ which provides a way of determining a matrix raised to a power.

Example:

Solve the following system of equations:

$$\frac{d^2x}{dt^2}+3x-2y=0$$

$$\frac{d^2y}{dt^2}+3y-2x=0$$

Solution:

We begin by writing the equations in matrix form: $\begin{bmatrix} x'' \\ y'' \end{bmatrix} = \begin{bmatrix} -3 & 2 \\ 2 & -3 \end{bmatrix}\begin{bmatrix} x \\ y \end{bmatrix}$

Let: $\begin{bmatrix} x \\ y \end{bmatrix} = \mathbf{C}\begin{bmatrix} X \\ Y \end{bmatrix}$

Thus: $\mathbf{C}\begin{bmatrix} X'' \\ Y'' \end{bmatrix} = \begin{bmatrix} -3 & 2 \\ 2 & -3 \end{bmatrix}\mathbf{C}\begin{bmatrix} X \\ Y \end{bmatrix}$

$$= \mathbf{C}^{-1}\begin{bmatrix} -3 & 2 \\ 2 & -3 \end{bmatrix}\mathbf{C}\begin{bmatrix} X \\ Y \end{bmatrix}$$

Diagonal matrix **D**

Characteristic equation:

$$\begin{vmatrix} -3-m & 2 \\ 2 & -3-m \end{vmatrix} = 0$$

$$(3+m)(3+m)-4=0$$

$$m^2+6m+5=0$$

$$(m+5)(m+1)=0$$

$$m=-5,-1$$

Thus:

$$\mathbf{D} = \begin{bmatrix} -5 & 0 \\ 0 & -1 \end{bmatrix} = \mathbf{C}^{-1}\mathbf{A}\mathbf{C}$$

$$\begin{bmatrix} X'' \\ Y'' \end{bmatrix} = \begin{bmatrix} -5 & 0 \\ 0 & -1 \end{bmatrix}\begin{bmatrix} X \\ Y \end{bmatrix}$$

$X''=-5X$
$Y''=-Y$ Second order differential equations

$$X = A\sin\left(\sqrt{5}t+\alpha\right)$$

$$Y = B\sin(t+\alpha)$$

m = 5:

$$\begin{bmatrix} 2 & 2 \\ 2 & 2 \end{bmatrix}\begin{bmatrix} x_1 \\ y_1 \end{bmatrix} = 0$$

$$2x_1+2y_1=0$$

m = 1:

$$\begin{bmatrix} -2 & 2 \\ 2 & -2 \end{bmatrix}\begin{bmatrix} x_2 \\ y_2 \end{bmatrix} = 0$$

$$-2x_2+2y_2=0$$

$$\mathbf{C} = \frac{1}{\sqrt{2}}\begin{bmatrix} 1 & 1 \\ -1 & 1 \end{bmatrix}$$

$$\begin{bmatrix} x \\ y \end{bmatrix} = \mathbf{C}\begin{bmatrix} X \\ Y \end{bmatrix} = \frac{1}{\sqrt{2}}\begin{bmatrix} 1 & 1 \\ -1 & 1 \end{bmatrix}\begin{bmatrix} X \\ Y \end{bmatrix} = \frac{1}{\sqrt{2}}\begin{bmatrix} X+Y \\ -X+Y \end{bmatrix}$$

$$x = \frac{1}{\sqrt{2}}(X+Y) = \frac{1}{\sqrt{2}}\left(A\sin\left(\sqrt{5}t+\alpha\right)+B\sin(t+\alpha)\right)$$

$$y = \frac{1}{\sqrt{2}}(Y-X) = \frac{1}{\sqrt{2}}\left(B\sin(t+\alpha)-A\sin\left(\sqrt{5}t+\alpha\right)\right)$$

Solution

Part 2

Advanced Mathematics

2.1 Ordinary Differential Equations
2.2 Laplace Transforms
2.3 Vector Analysis
2.4 Partial Derivatives
2.5 Multiple Integrals
2.6 Fourier Series
2.7 Partial Differential Equations
2.8 Numerical Methods

2.1 Ordinary Differential Equations

Summary

$$M(x)dx + N(y)dy = 0$$ First order differential equation

$$\int M(x)dx + \int N(y)dy = 0$$ Separation of variables

$$y = f(x)$$

$$f(kx, ky) = k^n f(x, y)$$ Homogeneous equation

$$\frac{\partial M}{\partial y} = \frac{\partial N}{\partial x}$$ Exact equation

Linear equation

$$A_o(x)\frac{d^n y}{dx^n} + A_1(x)\frac{d^{n-1}y}{dx^{n-1}} \ldots A_{n-1}(x)\frac{dy}{dx} + A_n(x)y = F(x)$$

$$[X'] = [A][X]$$ System of equations

$$\|[A] - m[I]\| = 0$$

System with complex eigenvalues

$$X = e^{at}\begin{bmatrix} D_1([C]_{re}\cos bt - [C]_{im}\sin bt) + \\ iD_2([C]_{im}\cos bt + [C]_{re}\sin bt) \end{bmatrix}$$

$$y = a_o + a_1 x + a_2 x^2 + a_3 x^3 + a_4 x^4 \ldots$$ Power series solution

2.1.1 Ordinary Differential Equations

A **differential equation** is any equation that contains derivatives. Some examples of differential equations are:

(i) $\frac{dy}{dx} = x + 8$

(ii) $\frac{d^2y}{dx^2} + 3\frac{dy}{dx} + 5y = 0$

(iii) $\frac{d^3y}{dx^3} + 2\frac{d^2y}{dx^2} + 7\frac{dy}{dx} = \cos x$

(iv) $\left(\frac{d^2y}{dx^2}\right)^2 + \left(\frac{dy}{dx}\right)^3 + 4y = x^2$

Ordinary differential equations have one independent variable, e.g., $y = f(x)$.

(v) $\frac{\partial^2 z}{\partial x^2} + \frac{\partial z}{\partial y} = x^2 + y$

Partial differential equations have more than one independent variable, e.g., $z = g(x,y)$.

The **order** of the equation is that of the highest derivative. The **degree** of the equation is the index of the highest derivative. Thus, in example (iv) above, the equation has an order 2 and degree 2.

It is often necessary to find a solution to a differential equation. This means finding an equation involving y as a function of x that does not contain differentials or derivatives. The **general solution** to a differential equation is an equation of the form $y = f(x)$. The function $f(x)$ will have a number of **arbitrary constants** equal to the order of the equation. A **particular solution** to the differential equation is found from the general solution for particular values of the constants. The constants may be determined from the **initial conditions** or **boundary conditions**.

First order differential equations can be generally written in the form:

$$M(x,y)dx + N(x,y)dy = 0$$

Special methods can be employed to find the solutions to differential equations. The skill is often choosing the most suitable method that suits the equation under consideration.

2.1.2 Separation of Variables

Ordinary differential equations may be solved, in some circumstances, using a method of **separation of variables**. This involves collecting terms involving like variables and integrating. If the differential equation can be expressed in the form:

$$M(x)dx + N(y)dy = 0 \qquad \text{(first order equation)}$$

then the solution is found by integrating to give:

$$\int M(x)dx + \int N(y)dy = 0$$

$$y = f(x)$$

Example:

Find the general solution to the differential equations shown:

(i)

$$\frac{dy}{dx} = x + 5$$

$$dy = (x+5)dx$$

$$\int dy = \int x + 5dx$$

$$y = \frac{x^2}{2} + 5x + C$$

(ii)

$$\frac{d^2x}{dt^2} = t^2$$

$$\frac{d}{dt}\left(\frac{dx}{dt}\right) = t^2$$

$$\int \frac{d}{dt}\left(\frac{dx}{dt}\right)dt = \int t^2 dt$$

$$\frac{dx}{dt} = \frac{t^3}{3} + A$$

$$\int dx = \int \frac{t^3}{3} + Adt$$

$$x = \frac{t^4}{12} + At + B$$

These equations are general solutions to the differential equation. A particular solution to the differential equation is found when the constants are given some values (i.e., when the limits of the integration are specified or initial conditions given).

2.1.3 Homogeneous Equations

A function $f(x,y)$ is said to be **homogeneous** and of degree n if for some value of k:

$$f(kx,ky)=k^n f(x,y)$$

For example:

(i) $f(x,y)=x^2+xy+y^2$

$$f(kx,ky)=(kx)^2+kxky+(ky)^2$$
$$=k^2(x^2+xy+y^2)$$
$$=k^2 f(x,y)$$ Homogeneous of degree 2

(ii) $f(x,y)=e^{\frac{y}{x}}$

$$f(kx,ky)=e^{\frac{kx}{ky}}$$
$$=e^{\frac{y}{x}}$$
$$=k^0 f(x,y)$$ Homogeneous of degree 0

Consider the first order differential equation below where $M(x,y)$ and $N(x,y)$ are both homogeneous and of the same degree:

$$M(x,y)dx-N(x,y)dy=0$$
$$Mdx=Ndy$$
$$\frac{dy}{dx}=\frac{M}{N}$$ Homogeneous of degree 0

Now, $\frac{dy}{dx}=f(x,y)$

$$=f(kx,ky)$$ because homogeneous

$$=f\left(1,\frac{ky}{kx}\right)$$

$$=f\left(\frac{y}{x}\right)$$ The ratio of the two functions M and N is a function of y/x.

The first equation can thus be expressed in the form:

$$\frac{dy}{dx}+g\left(\frac{y}{x}\right)=0$$

Putting $y=vx$, where v is a new variable, we have an equation in which the variables are separable.

$$x\frac{dv}{dx}+v+g(v)=0$$

Example:

Find the general solution to the differential equation $(x^2 + y^2)dx - xydy = 0$.

Solution:

(i) Determine if the equation is homogeneous. To do this, we recognise from inspection that the coefficients are each homogeneous and of degree two. Thus, we can proceed with the substitution $y = vx$.

$$M = (x^2 + y^2)$$
$$N = xy$$

(ii) Let $y = vx$

so that $dy = vdx + xdv$

Thus: $(x^2 + v^2x^2)dx - (x^2v)(vdx + xdv) = 0$

$$(1 + v^2)dx - v(vdx + xdv) = 0$$
$$(1 + v^2)dx - v^2dx - xvdv = 0$$
$$(1 + v^2)dx = v^2dx + xvdv$$
$$(1 + v^2)dx - v^2dx = xvdv$$
$$(1 + v^2 - v^2)dx = xvdv$$
$$dx = xvdv$$
$$vdv = \frac{1}{x}dx$$

Variables have been separated.

$$\int vdv = \int \frac{1}{x}dx$$
$$\frac{1}{2}v^2 = \ln x + C_1$$
$$\frac{y^2}{x^2} = 2(\ln x + C_2)$$
$$y^2 = 2x^2 \ln \frac{x}{C_3}$$

General solution

2.1.4 Exact Equations

Consider the general form of a first order differential equation:

$$M(x,y)dx + N(x,y)dy = 0$$

If $\frac{\partial M}{\partial y} = \frac{\partial N}{\partial x}$ then the original equation is called an **exact equation**.

If the differential equation is exact, then there exists a function $F(x,y)$ such that:

$$dF = M(x,y)dx + N(x,y)dy = 0 \text{ and } \frac{\partial F}{\partial x} = M(x,y); \frac{\partial F}{\partial y} = N(x,y)$$

The solution of an exact differential equation has the form $F(x,y) = C$. In some cases, an equation that is not exact can be made so by the multiplication of each side by an **integrating factor**.

Example:

Solve the differential equation $(6x + y^2)dx + (2xy - 3y^2)dy = 0$.

Solution:

$$M = (6x + y^2); \frac{\partial M}{\partial y} = 2y$$

$$N = (2xy - 3y^2); \frac{\partial N}{\partial x} = 2y$$

Therefore exact

$$M = \frac{\partial F}{\partial x} = (6x + y^2)$$

$$F = \int (6x + y^2)\partial x$$

$$= \frac{6x^2}{2} + xy^2 + T(y)$$

$$= 3x^2 + xy^2 + T(y)$$

$$\frac{\partial F}{\partial y} = 2xy + T'(y)$$

$$N = \frac{\partial F}{\partial y} = (2xy - 3y^2)$$

$$= 2xy + T'(y)$$

$$T'(y) = -3y^2$$

$$T(y) = -\int 3y^2 dy$$

$$= -y^3$$

Thus: $F = 3x^2 + xy^2 - y^3$

$$dF = Mdx + Ndy = 0$$

$$\therefore F(x,y) = C$$

$$C = 3x^2 + xy^2 - y^3$$

2.1.5 Linear Equations

A **linear differential equation** has the form:

$$A_0(x)\frac{d^n y}{dx^n}+A_1(x)\frac{d^{n-1}y}{dx^{n-1}}\ldots A_{n-1}(x)\frac{dy}{dx}+A_n(x)y=F(x)$$

This equation is linear in terms of the dependent variable y and also in terms of the derivatives of y.

A first order linear differential equation can be thus written:

$$\frac{dy}{dx}+P(x)y=g(x)$$

where the equation is of order 1 in terms of the dependent variable y. If a first order linear equation is not exact, then it can be made so by the incorporation of a factor $u(x)$.

$$u(x)\left(\frac{dy}{dx}+P(x)y\right)=u(x)g(x)$$

Being a first order equation, this can be put into the standard form:

$$M(x,y)dx+N(x,y)dy=0$$

and so:

$$[u(x)P(x)y-u(x)g(x)]dx+[u(x)]dy=0$$

Examples of linear equations

(i) $$\frac{d^2y}{dx^2}+\frac{dy}{dx}-8y=0$$

(ii) $$x^2\frac{d^2y}{dx^2}+x\frac{dy}{dx}+xy=8x^2$$

(iii) $$\left(x+2y^3\right)\frac{dy}{dx}=y$$

Not linear in y and its derivative but linear in x $$\frac{dx}{dy}-\frac{x}{y}=\frac{1}{2y^2}$$

Examples of non-linear equations

(i) $$\frac{d^2y}{dx^2}+\left(\frac{dy}{dx}\right)^2-y=0$$

(ii) $$\frac{d^2y}{dx^2}+\frac{dy}{dx}+xy^2=x^3$$

(iii) $$\frac{1}{x}\frac{dy}{dx}+xy^3=0$$

Our aim is to determine the nature of $u(x)$ that would make the equation an **exact equation** and thus amenable to solution. We therefore assume that the new equation is exact and proceed accordingly:

$$\frac{\partial}{\partial y}[u(x)P(x)y-u(x)g(x)]=\frac{\partial}{\partial x}[u(x)]$$

by definition of exactness. $\frac{\partial M}{\partial y}=\frac{\partial N}{\partial x}$

$$u(x)P(x)=\frac{du}{dx}$$

$$P(x)dx=\frac{1}{u}du$$

$$\ln u=\int P(x)dx$$

$$u=e^{\int P(x)dx}$$

If $u(x)$ has this form and is applied to the original equation, then the equation becomes exact. $u(x)$ is called an **integrating factor**.

Example:

Find the general solution to the differential equation $(x^4 + 2y)dx - xdy = 0$.

Solution:

This equation is not exact but it is linear in y.

$$\frac{\partial M}{\partial y} = 2; \frac{\partial N}{\partial x} = -1$$

To find the integrating factor, we must first put the equation into standard form:

$$\frac{dy}{dx} + P(x)y = g(x)$$

$$(x^4 + 2y)dx - xdy = 0$$
$$(x^4 + 2y)dx = xdy$$
$$\frac{(x^4 + 2y)}{x}dx = dy$$
$$\frac{(x^4 + 2y)}{x} = \frac{dy}{dx}$$
$$x^3 + \frac{2y}{x} = \frac{dy}{dx}$$
$$\frac{dy}{dx} - \frac{2y}{x} = x^3 \quad \text{Standard form}$$

The integrating factor is thus given by: $u(x) = e^{\int -\frac{2}{x}dx}$

$$= e^{-2\ln x}$$
$$= x^{-2}$$

An exact equation is thus formed:

$$x^{-2}\frac{dy}{dx} - \frac{2y}{x^3} = x$$
$$\frac{d}{dx}(x^{-2}y) = x$$
$$x^{-2}y = \int xdx$$
$$x^{-2}y = \frac{x^2}{2} + C$$
$$2y = x^4 + Ax^2 \quad \text{General solution}$$

2.1.6 Linear Equations with Constant Coefficients

A **linear differential equation** is described as being **homogeneous** when it has the form:

$$A_0(x)\frac{d^n y}{dx^n} + A_1(x)\frac{d^{n-1}y}{dx^{n-1}} \ldots A_{n-1}(x)\frac{dy}{dx} + A_n(x)y = 0$$

The term homogeneous as used here has a different meaning from that used previously.

If the coefficients are constants, rather than being functions of x, then we have:

$$A_0\frac{d^n y}{dx^n} + A_1\frac{d^{n-1}y}{dx^{n-1}} \ldots A_{n-1}\frac{dy}{dx} + A_n y = 0$$

In operator notation, where $D = d/dx$, we write:

$$A_0 D^n y + A_1 D^{n-1} y \ldots A_{n-1} Dy + A_n y = 0$$

$$\left(A_0 D^n + A_1 D^{n-1} \ldots A_{n-1} D + A_n\right)y = 0$$

The bracketed term above can be regarded as a polynomial of degree n in D. It can also be considered a **differential operator** L since if it is applied to the variable y, it yields the original differential equation. That is:

$$Ly = \left(A_0 D^n + A_1 D^{n-1} \ldots A_{n-1} D + A_n\right)y = 0$$

Thus, we may write the operator L in terms of a polynomial function $f(D)$:

$$L = f(D)$$

Now, the operator D, if acting on the function e^{mx}, can be written:

$$D^n e^{mx} = m^n e^{mx}$$

Consider now the effect of the operator L on the function e^{mx}.

$$Le^{mx} = \left(A_o m^n e^{mx} + A_1 m^{n-1} e^{mx} \ldots A_{n-1} m e^{mx} + A_n e^{mx}\right)$$

compare

That is, the solution to the original differential equation has the form $y = e^{mx}$. The bracketed term is now a polynomial in m, $f(m)$, and further, since $e^{mx} \neq 0$, then:

$$f(m) = \left(A_o m^n + A_1 m^{n-1} \ldots A_{n-1} m + A_n\right) = 0$$

This is called the **auxiliary equation**. The solution to the original differential equation involves finding the roots of the auxiliary equation which is a polynomial in m of degree n. There are n roots of the polynomial leading to n different solutions of the original differential equation. If the roots are all real and distinct, then the general solution is the superposition of the solutions:

$$y(x) = C_1 e^{m_1 x} + C_2 e^{m_2 x} \ldots + C_n e^{m_n x}$$

Example:

Find the general solution to the differential equation $2\frac{d^2y}{dx^2}+\frac{dy}{dx}-y=0$.

Solution:

The equation can be written in operator notation such that:

$$\left(2D^2+D-1\right)y=0$$

The auxiliary equation is thus written:

$$2m^2+m-1=0$$
$$(2m-1)(m+1)=0$$

thus $m=+\frac{1}{2},-1$ are the roots of the auxiliary equation.

But, the solutions have the form $y = e^{mx}$, thus:

$$y_1=e^{\frac{1}{2}x}$$
$$y_2=e^{-x}$$

The general solution is thus written $y(x)=C_1e^{\frac{1}{2}x}+C_2e^{-x}$.

Example:

Find the general solution to the differential equation: $\left(D^2-2D+2\right)y=0$

Solution:

The auxiliary equation is written $m^2-2m+2=0$

And by the quadratic formula, we have: $m=\frac{2\pm\sqrt{4-4(2)}}{2}$

$$=\frac{2\pm2\sqrt{-1}}{2}$$

$=1\pm i$ are roots.

The general solution is thus written: $y=C_1e^{(1+i)x}+C_2e^{(1-i)x}$

$=e^x\left(A\cos x+iB\sin x\right)$ by use of Euler's formula where

$$A=C_1+C_2$$
$$B=C_1-C_2$$

The imaginary part of this expression can be omitted to provide a real solution if required by physical considerations since we only need either the real *or* the imaginary part of the expression to represent a wave.

2.1.7 Method of Undetermined Coefficients

Consider a **nonhomogeneous linear differential equation** with constant coefficients:

$$A_0 \frac{d^n y}{dx^n} + A_1 \frac{d^{n-1} y}{dx^{n-1}} \ldots A_{n-1} \frac{dy}{dx} + A_n y = f(D)y = Ly = F(x)$$

If $F(x) = 0$, then we have a homogenous equation. The **auxiliary equation** for this homogeneous equation has roots m. If y_c is the solution to this homogeneous equation, and y_p is a particular solution of the nonhomogeneous equation, then we can show that $y = y_c + y_p$ is a solution of the original non-homogeneous equation:

$$\begin{aligned} f(D)(y_c + y_p) &= F(x) \\ &= f(D)y_c + f(D)y_p \\ &= 0 + f(D)y_p \end{aligned}$$

y_c is called the **complementary function** for the original equation. To find y_p, we can proceed in the following manner. If $y_q = F(x)$ is in itself a particular solution of some other homogenous linear equation with constant coefficients whose auxiliary equation has roots m', then this other homogenous equation can be written:

$$g(D)y_q = g(D)F(x) = 0$$

Since the original equation can be written:

$$Ly = f(D)y = F(x)$$

Then:

$$g(D)f(D)y = g(D)F(x) = 0$$

i.e., $g(D)$ applied to both sides of the original non-homogeneous equation transforms results in a new homogenous equation.

represents a new homogeneous equation which contains within it the original non-homogeneous equation. Since the new homogeneous equation contains within it the original non-homogeneous equation, the solution y must be common to both. The general solution of the new homogeneous equation therefore has the form:

$$y = y_c + y_q$$

where y_q is a particular solution to the new homogeneous equation. The desired general solution of the original non-homogeneous equation is $y = y_c + y_p$. But, y is also a general solution of the new homogeneous equation so there must be a particular solution y_p which equals y_q. That is, the as yet **undetermined coefficients** of y_q can be found such that $y_q = y_p$ and so the general solution of the original equation can be found.

Example:

Find the general solution to the differential equation $\left(D^2 - 3D - 4\right)y = 6e^x$.

Solution:

(i) Find the complementary function y_c from: $\left(D^2 - 3D - 4\right)y = 0$

$m^2 - 3m - 4 = 0$

$$m = \frac{3 \pm \sqrt{9 - 4(-4)}}{2} = 4, -1$$

$$y_c = C_1 e^{4x} + C_2 e^{-x}$$

(ii) Let $y_q = F(x)$ be a particular solution of a certain homogeneous linear differential equation $g(D)$.

$$F(x) = 6e^x$$

We can say immediately that $m' = 1$ and so $g(D) = D - 1$ and the homogeneous equation is written:

$$(D - 1)y_q = 0$$

(iii) We apply $g(D)$ to both sides of the original equation so that:

$$\underbrace{(D - 1)\left(D^2 - 3D - 4\right)}y = (D - 1)6e^x = 0$$

Here we have a homogenous equation in which the roots of the auxiliary function can be easily written down:

$$m' = 1, 4, -1$$

The general solution of this homogeneous equation is:

$$y = \underbrace{C_3 e^x}_{y_q} + \underbrace{C_1 e^{4x} + C_2 e^{-x}}_{y_c}$$

This general solution can be seen to be comprised of $y_q + y_c$. But this general solution is also a general solution of the original non-homogeneous equation, so we need to find coefficients of y_p such that $y_p = y_q$. Now, it is evident y_p has the form of y_q, i.e.:

$y_p = Ae^x$ where A is the as yet undetermined coefficient.

Inserting into the original equation, we find:

$$\left(D^2 - 3D - 4\right)y = 6e^x$$

$$Ae^x - 3Ae^x - 4Ae^x = 6e^x$$

$$A - 3A - 4A = 6$$

$$A = -1$$

And so the general solution to the original non-homogeneous equation is:

$$y = C_1 e^{4x} + C_2 e^{-x} - e^x$$

2.1.8 Systems of Equations

Consider the first order homogeneous **linear system** of equations with constant coefficients:

$$\frac{dx}{dt} = 0x + y$$

$$\frac{dy}{dt} = -2x + 3y$$

where x and y are both functions of the variable t.

This system can be written in matrix form:

$$[\mathbf{X'}] = [\mathbf{A}][\mathbf{X}]$$

$$\begin{bmatrix} dx/dt \\ dy/dt \end{bmatrix} = \begin{bmatrix} 0 & 1 \\ -2 & 3 \end{bmatrix} \begin{bmatrix} x \\ y \end{bmatrix}$$

Let the solutions be of the form: $\left.\begin{array}{l} x = C_1 e^{mt} \\ y = C_2 e^{mt} \end{array}\right\} \quad [\mathbf{X}] = [\mathbf{C}]e^{mt}$

Required is the matrix $[\mathbf{C}]$ and the index m.

Substituting back we have:

$$\text{Original equation: } [\mathbf{X'}] = [\mathbf{A}][\mathbf{X}]$$

$$\text{Solution equation: } [\mathbf{X}] = [\mathbf{C}]e^{mt}$$

$$[\mathbf{X'}] = [\mathbf{C}]me^{mt}$$

$$[\mathbf{C}]me^{mt} = [\mathbf{A}][\mathbf{C}]e^{mt}$$

$$[\mathbf{A}][\mathbf{C}]e^{mt} - m[\mathbf{C}]e^{mt} = [0]$$

$$\text{Now, } [\mathbf{C}] = [\mathbf{I}][\mathbf{C}]$$

$$\text{Thus, } ([\mathbf{A}] - m[\mathbf{I}])[\mathbf{C}]e^{mt} = [0]$$

$$\text{Since, } e^{mt} > 0$$

$$\text{Then } ([\mathbf{A}] - m[\mathbf{I}])[\mathbf{C}] = [0]$$

The matrix $[\mathbf{C}]$ contains only constants, so the derivative of $[\mathbf{X}]$ with respect to t is just me^{mt}. Also, we can see that

$$m[\mathbf{C}] = [\mathbf{A}][\mathbf{C}]$$

However, it can be shown that a non-trivial result for $[\mathbf{C}]$ can only be obtained if $([\mathbf{A}] - m[\mathbf{I}])$ is a singular matrix. A singular matrix is one where the determinant = 0; thus:

$$\|[\mathbf{A}] - m[\mathbf{I}]\| = 0$$

leads to a polynomial equation in m of degree n (for an $n \times n$ matrix). This equation is called the **characteristic equation** of the matrix $[\mathbf{A}]$. The roots of the characteristic equation are called the **eigenvalues** of $[\mathbf{A}]$. A non-zero matrix $[\mathbf{C}]$ for a particular eigenvalue is called an **eigenvector** of the matrix $[\mathbf{A}]$.

The first step in solving the homogeneous system is to find the **eigenvalues** and the corresponding **eigenvectors** of the matrix $[\mathbf{A}]$.

Consider the example from the previous page: $[\mathbf{X'}]=[\mathbf{A}][\mathbf{X}]$

$$[\mathbf{A}]=\begin{bmatrix}0 & 1\\ -2 & 3\end{bmatrix}$$

The characteristic equation is written:

$$\left|\begin{bmatrix}0 & 1\\ -2 & 3\end{bmatrix}-m\begin{bmatrix}1 & 0\\ 0 & 1\end{bmatrix}\right|=\begin{vmatrix}-m & 1\\ -2 & (3-m)\end{vmatrix}=0$$

$$-m(3-m)-(1)(-2)=0$$

$$-3m+m^2+2=0$$

The eigenvalues are: $m_1=1; m_2=2$

For m_1, the equation becomes:

$$([\mathbf{A}]-m[\mathbf{I}])[\mathbf{C}]=[0]$$

$$\begin{bmatrix}-1 & 1\\ -2 & 2\end{bmatrix}\begin{bmatrix}C_1\\ C_2\end{bmatrix}=\begin{bmatrix}0\\ 0\end{bmatrix}$$

$$-C_1+C_2=0$$

$$C_1=C_2$$

Eigenvector corresponding to m_1 $[\mathbf{C}]=\begin{bmatrix}C_1\\ C_2\end{bmatrix}=\begin{bmatrix}C_1\\ C_1\end{bmatrix}=C_1\begin{bmatrix}1\\ 1\end{bmatrix}$

For m_2, the equation becomes:

$$([\mathbf{A}]-m[\mathbf{I}])[\mathbf{C}]=[0]$$

$$\begin{bmatrix}-2 & 1\\ -2 & 1\end{bmatrix}\begin{bmatrix}C_1\\ C_2\end{bmatrix}=\begin{bmatrix}0\\ 0\end{bmatrix}$$

$$-2C_1+C_2=0$$

$$C_2=2C_1$$

Eigenvector corresponding to m_2 $[\mathbf{C}]=\begin{bmatrix}C_1\\ C_2\end{bmatrix}=\begin{bmatrix}C_1\\ 2C_1\end{bmatrix}=C_1\begin{bmatrix}1\\ 2\end{bmatrix}$

There are thus two solutions for the system:

$$X_1=k_1\begin{bmatrix}1\\ 1\end{bmatrix}e^{1t}; X_2=k_2\begin{bmatrix}1\\ 2\end{bmatrix}e^{2t}$$

where k_1 and k_2 are arbitrary constants.

But, $[\mathbf{X}]=\begin{bmatrix}x\\ y\end{bmatrix}$

Thus, the solutions to the system are written:

$$\begin{cases} x=k_1e^t+k_2e^{2t}\\ y=k_1e^t+2k_2e^{2t}\end{cases}$$

Example:

Find the general solution of the system: $[\mathbf{X}'] = [\mathbf{A}][\mathbf{X}]$

where: $[\mathbf{A}] = \begin{bmatrix} 8 & -3 \\ 16 & -8 \end{bmatrix}$

Solution:

$$\left|[\mathbf{A}] - m[\mathbf{I}]\right| = 0$$

$$\left| \begin{bmatrix} 8 & -3 \\ 16 & 8 \end{bmatrix} - \begin{bmatrix} m & 0 \\ 0 & m \end{bmatrix} \right| = 0$$

$$\left.\begin{vmatrix} 8-m & -3 \\ 16 & -8-m \end{vmatrix} = 0 \right\} \quad -(8-m)(8+m)+3(16)=0$$

$$m_1 = 4$$

$$m_2 = -4$$

$$([\mathbf{A}] - m[\mathbf{I}])[\mathbf{C}] = 0$$

$$\begin{bmatrix} 8-m & -3 \\ 16 & -8-m \end{bmatrix} \begin{bmatrix} C_1 \\ C_2 \end{bmatrix} = \begin{bmatrix} 0 \\ 0 \end{bmatrix}$$

For m_1

$$\begin{bmatrix} 4 & -3 \\ 16 & -12 \end{bmatrix} \begin{bmatrix} C_1 \\ C_2 \end{bmatrix} = \begin{bmatrix} 0 \\ 0 \end{bmatrix}$$

$$4C_1 - 3C_2 = 0$$

$$4C_1 = 3C_2$$

$$16C_1 + -12C_2 = 0$$

$$16C_1 = 12C_2$$

$$C_1 = 0.75C_2$$

$$[\mathbf{C}] = \begin{bmatrix} 0.75C_2 \\ C_2 \end{bmatrix}$$

$$= C_2 \begin{bmatrix} 0.75 \\ 1 \end{bmatrix}$$

$$= C_2 \begin{bmatrix} 3 \\ 4 \end{bmatrix}$$

$$X_1 = e^{4t} k_1 \begin{bmatrix} 3 \\ 4 \end{bmatrix}$$

For m_2

$$\begin{bmatrix} 12 & -3 \\ 16 & -4 \end{bmatrix} \begin{bmatrix} C_1 \\ C_2 \end{bmatrix} = \begin{bmatrix} 0 \\ 0 \end{bmatrix}$$

$$12C_1 = 3C_2$$

$$4C_1 = C_2$$

$$[\mathbf{C}] = \begin{bmatrix} C_1 \\ 4C_1 \end{bmatrix}$$

$$= C_1 \begin{bmatrix} 1 \\ 4 \end{bmatrix}$$

$$X_2 = e^{-4t} k_2 \begin{bmatrix} 1 \\ 4 \end{bmatrix}$$

2.1.9 Complex Eigenvalues

The solutions to some systems of equations are complex. **Complex eigenvalues** always occur in pairs as complex conjugates. For example, consider the solution to $[\mathbf{X}'] = [\mathbf{A}][\mathbf{X}]$

$$[\mathbf{A}] = \begin{bmatrix} 2 & -5 \\ 2 & -4 \end{bmatrix}$$

The characteristic equation is: $\begin{vmatrix} 2-m & -5 \\ 2 & -4-m \end{vmatrix} = 0$ and so: $m_1 = -1+i$, $m_2 = -1-i$

$$m^2 + 2m + 2 = 0$$

For m_1

$$\begin{bmatrix} 3-i & -5 \\ 2 & -3-i \end{bmatrix} \begin{bmatrix} C_1 \\ C_2 \end{bmatrix} = \begin{bmatrix} 0 \\ 0 \end{bmatrix}$$

$$(3-i)C_1 - 5C_2 = 0$$

$$\frac{(3-i)}{5} C_1 = C_2$$

Let $C_1 = 5; C_2 = (3-i)$

$$[\mathbf{C}] = k_1 \begin{bmatrix} 5 \\ 3-i \end{bmatrix}$$

$$X_1 = k_1 \begin{bmatrix} 5 \\ 3-i \end{bmatrix} e^{(-1+i)t}$$

For m_2

$$\begin{bmatrix} 3+i & -5 \\ 2 & -3+i \end{bmatrix} \begin{bmatrix} C_1 \\ C_2 \end{bmatrix} = \begin{bmatrix} 0 \\ 0 \end{bmatrix}$$

$$[\mathbf{C}] = k_2 \begin{bmatrix} 5 \\ 3+i \end{bmatrix}$$

$$\downarrow$$

$$X_2 = k_2 \begin{bmatrix} 5 \\ 3+i \end{bmatrix} e^{(-1-i)t}$$

$$X = k_1 \begin{bmatrix} 5+0i \\ 3-i \end{bmatrix} e^{(-1+i)t} + k_2 \begin{bmatrix} 5+0i \\ 3+i \end{bmatrix} e^{(-1-i)t}$$ is the general solution

But, by using **Euler's formula**, this becomes:

$$X = e^{-t}\left[D_1\left(\begin{bmatrix} 5 \\ 3 \end{bmatrix} \cos t - \begin{bmatrix} 0 \\ -1 \end{bmatrix} \sin t \right) + D_2\left(\begin{bmatrix} 0 \\ -1 \end{bmatrix} \cos t + \begin{bmatrix} 5 \\ 3 \end{bmatrix} \sin t \right) \right]$$

Since m_1 is the complex conjugate of m_2, when $m_1 = a + bi$, the general solution takes the form:

$$\boxed{X = e^{at}\left[D_1\left([C]_{re} \cos bt - [C]_{im} \sin bt \right) + D_2\left([C]_{im} \cos bt + [C]_{re} \sin bt \right) \right]}$$

2.1.10 Power Series

Differential equations with either constant or variable coefficients can be solved by assuming that the solution to the equation is of the form of a **power series**.

For example, consider the differential equation $y''+y=0$.

Let the solution be in the form of a power series:

$y=a_o+a_1x+a_2x^2+a_3x^3+a_4x^4\ldots$ where a_n are arbitrary constants.

Thus: $y'=a_1+2a_2x+3a_3x^2+4a_4x^3\ldots$

$y''=2a_2+6a_3x+12a_4x^2$

Substituting into the original equation, we obtain:

$$2a_2+6a_3x+12a_4x^2\ldots.+a_o+a_1x+a_2x^2+a_3x^3\ldots=0$$

$$(2a_2+a_o)+(6a_3+a_1)x+(12a_4+a_2)x^2\ldots=0$$

Therefore:

$$\left.\begin{aligned}(2a_2+a_o)&=0\\(6a_3+a_1)&=0\\(12a_4+a_2)&=0\end{aligned}\right\}$$

$$a_2=-\frac{1}{2}a_o=-\frac{1}{2!}a_o$$

$$a_3=-\frac{1}{6}a_1=-\frac{1}{3!}a_1$$

$$a_4=-\frac{a_2}{4(3)}=\frac{a_o}{4!}$$

$$a_5=\frac{a_1}{5!}$$

It can be thus shown that all the even a's can be expressed in terms of a_0 and all the odd a's can be expressed in terms of a_1.

Thus:

$$y=a_o+a_1x+\frac{-a_o}{2!}x^2+\frac{-a_1}{3!}x^3+\frac{-a_ox^4}{4!}\ldots$$

$$=a_o\left(1+\frac{-x^2}{2!}+\frac{x^4}{4!}+\ldots\right)+a_1\left(x+\frac{-x^3}{3!}+\frac{x^5}{5!}+\ldots\right)$$

Maclaurin series for $\cos x$ and $\sin x$

$y=a_o\cos x+a_1\sin x$ is the required solution.

Note, in this example, we can see that (using the methods described previously) the auxiliary equation can be written: $m^2+1=0$

$$m=\pm i$$

and so $y=A\cos x+B\sin x$

Example:

Use the power series method to solve $y''-4y=0.$

Solution:

Auxiliary equation: $m^2-4=0$

$$m=\pm 2 \quad \text{thus} \quad y=C_1e^{2x}+C_2e^{-2x}$$

Let the solution be in the form of a power series:

$$y=a_0+a_1x+a_2x^2+a_3x^3+a_4x^4\ldots \qquad \text{or} \qquad y=\sum_{n=1}^{\infty}a_nx^n$$

Thus: $y'=\sum_{n=1}^{\infty}na_nx^{n-1}$ and: $y''=\sum_{n=2}^{\infty}n(n-1)a_nx^{n-2}$

Substituting into the original equation, we obtain:

$$\sum_{n=2}^{\infty}n(n-1)a_nx^{n-2}-4\sum_{n=1}^{\infty}a_nx^n=0$$

let $t=n-2$

when $n=2, t=0$

and $n=\infty, t=\infty$

We need to make the limits of the summations the same so that we can combine expressions.

$$\sum_{n=2}^{\infty}n(n-1)a_nx^{n-2}=\sum_{t=0}^{\infty}(t+2)(t+1)a_{t+2}x^t$$

$$=\sum_{n=0}^{\infty}(n+2)(n+1)a_{n+2}x^n$$

$$\sum_{n=0}^{\infty}(n+2)(n+1)a_{n+2}x^n-4\sum_{n=0}^{\infty}a_nx^n=0$$

$$\sum_{n=0}^{\infty}[(n+2)(n+1)a_{n+2}-4a_n]x^n=0$$

$$(n+2)(n+1)a_{n+2}-4a_n=0$$

$$a_{n+2}=\frac{4a_n}{(n+2)(n+1)}$$

$$\begin{cases} a_n=\dfrac{4^{n-1}a_o}{n!} & n \text{ 0, even} \\ a_n=\dfrac{4^{n-2}a_1}{n!} & n \text{ odd} \end{cases}$$

Thus $y=a_o+a_1x+a_2x^2+a_3x^3+a_4x^4\ldots$

$$=a_o+a_1x+\frac{4a_o}{2!}x^2+\frac{4a_1}{3!}x^3+\frac{4^2a_o}{4!}x^4+\frac{4^2a_1}{5!}x^5+\ldots$$

$$=a_o\left(1+\frac{4x^2}{2!}+\frac{4^2x^4}{4!}+\ldots\right)+a_1\left(x+\frac{4x^3}{3!}+\frac{4^2x^5}{5!}+\ldots\right)$$

Using a Maclaurin series for e^{2x} and e^{-2x}, we obtain:

$$y=2a_o\left(e^{2x}+e^{-2x}\right)+4a_1\left(e^{2x}-e^{-2x}\right)$$

$$=Ae^{2x}+Be^{-2x}$$

2.2 Laplace Transforms

Summary

$$L(f(t)) = \int_0^\infty e^{-st} f(t)dt = F(s)$$

Laplace transform

$$L\left(f^n(t)\right) = s^n F(s) - \sum_{k=0}^{n-1} s^{n-1-k} f^k(0)$$

Laplace transform of a derivative

$$L(U(t-C)f(t-C)) = e^{-sC} L(f(T))$$

Step function

2.2.1 Laplace Transform

The **Laplace transform** $L(f(t))$ of a function $f(t)$ is defined as:

$$L(f(t)) = \int_0^\infty e^{-st} f(t)\,dt = F(s)$$

$F(s)$ is an algebraic equation in s. s is called the Laplace variable.

In operator notation, we can say that the **Laplace operator** L acts on the function $f(t)$ to provide the Laplace transform of the function. The Laplace transform is an example of an **integral transform**. The Laplace operator is a linear operator and so:

$$L(c_1 f_1(t) + c_2 f_2(t)) = c_1 L(f_1(t)) + c_2 L(f_2(t))$$

The **inverse Laplace transform** $L^{-1}(F(s))$ is given by:

$$L^{-1}(F(s)) = f(t)$$

Standard results

$L(1) = \frac{1}{s}$	$s > 0$
$L(t^n) = \frac{n!}{s^{n+1}}$	$s > 0$
$L(e^{at}) = \frac{1}{s-a}$	$s > a$
$L(\sin \omega t) = \frac{\omega}{s^2+\omega^2}$	$s > 0$
$L(\cos \omega t) = \frac{s}{s^2+\omega^2}$	$s > 0$
$L\left(\frac{t^n}{n!}\right) = \frac{1}{s^{n+1}}$	$s > 0$
$L(\sinh \omega t) = \frac{\omega}{s^2-\omega^2}$	$s > \|a\|$
$L(\cosh \omega t) = \frac{s}{s^2-\omega^2}$	$s > \|a\|$
$L(A+B) = L(A) + L(B)$	$s > 0$
$L(0) = 0$	

Theorems for the Laplace transform

$$L(e^{at} f(t)) = F(s-a)$$

$$L(U(t-a) f(t-a)) = e^{-as} F(s)$$

(Step function: U)

$$L(f(at)) = \frac{1}{a} F\left(\frac{s}{a}\right)$$

Function with period T:

$$L(f(t)) = \frac{\int_0^T e^{-st} f(t)\,dt}{1 - e^{-st}}$$

$$L\left(\int_0^t f(u)\,du\right) = \frac{F(s)}{s}$$

$$L\left(\frac{f(t)}{t}\right) = \int_s^\infty F(u)\,du$$

$$L\left(\int_0^t f(u) g(t-u)\,du\right) = F(s)G(s) = f * g$$

Convolution

2.2.2 Laplace Transform of Derivatives

Consider the **Laplace transform** of the first derivative of a function of t:

$$L(f'(t)) = \int_0^\infty e^{-st} f'(t)dt$$

Let $u = e^{-st}; dv = f'(t)dt$

$du = -se^{-st}dt; v = f(t)$

$$L(f'(t)) = \left[e^{-st} f(t)\right]_0^\infty + s\int_0^\infty f(t)e^{-st}dt$$

$$= \left[e^{-st} f(t)\right]_0^\infty + sL(f(t))$$

$$= -f(0) + sL(f(t))$$

But $L(f(t)) = F(s)$

Thus $L(f'(t)) = -f(0) + sF(s)$

$$= sF(s) - f(0)$$

$$L(f'(t)) = sL(f(t)) - f(0)$$

Similarly $L(f''(t)) = s^2 L(f(t)) - sf(0) - f'(0)$

In general: $$L\left(f^n(t)\right) = s^n F(s) - \sum_{k=0}^{n-1} s^{n-1-k} f^k(0)$$

The Laplace transform is given by:

$$L(f(t)) = \int_0^\infty e^{-st} f(t)dt = F(s)$$

Taking the derivative with respect to s gives:

$$\frac{d}{ds}F(s) = \int_0^\infty \frac{\partial}{\partial s}\left(e^{-st} f(t)\right)dt$$

This is the same as the Laplace transform of the product $-t\,F(t)$.

$$= \int_0^\infty -te^{-st} f(t)dt$$

$$= L(-tf(t))$$

$$= -L(tf(t))$$

$$L(tf(t)) = -\frac{d}{ds}F(s)$$

$$= -F'(s)$$

In general, the derivative of a Laplace transform is given by:

$$L\left(t^n f(t)\right) = (-1)^n F^n(s)$$

2.2.3 Step Functions

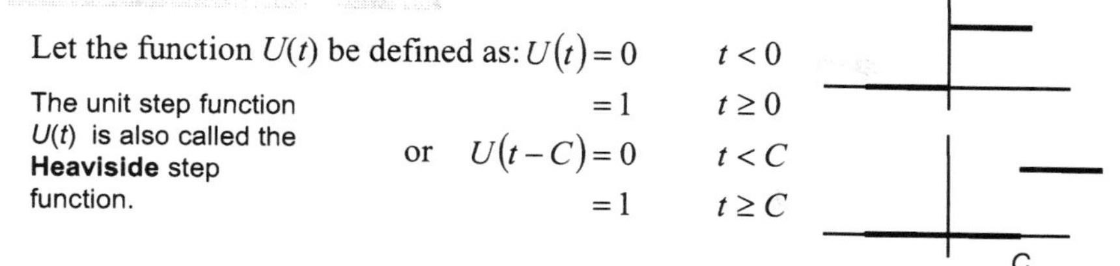

Let the function $U(t)$ be defined as:

$$U(t) = 0 \qquad t < 0$$
$$= 1 \qquad t \geq 0$$
$$\text{or} \quad U(t-C) = 0 \qquad t < C$$
$$= 1 \qquad t \geq C$$

The unit step function $U(t)$ is also called the **Heaviside** step function.

The Laplace transform is:

$$L(U(t-C)) = \int_0^\infty e^{-st} U(t-C)dt$$
$$= \int_0^C e^{-st}(0)dt + \int_C^\infty e^{-st}(1)dt$$
$$= \frac{e^{-sc}}{s}$$

Now, consider some other function $f(t)$ where:

$$U(t)f(t) = 0 \qquad t < 0$$
$$= f(t) \qquad t \geq 0$$
$$U(t-C)f(t-C) = 0 \qquad t < C$$
$$= f(t-C) \qquad t \geq C$$

The Laplace transform of this product of functions is given by:

$$L(U(t-C)f(t-C)) = \int_0^\infty e^{-st} U(t-C) f(t-C)dt$$
$$= \int_C^\infty e^{-st}(1) f(t-C)dt$$
$$= \int_0^\infty e^{-s(T+C)} f(T)dT$$

Let $T = (t-C)$
Thus $dT = dt$
When $t = C$
$T = 0$
When $t = \infty$
$T = \infty$

$$\boxed{L(U(t-C)f(t-C)) = e^{-sC} L(f(T))}$$

2.2.4 Laplace Transforms and Differential Equations

Laplace transforms are suitable for solving ordinary differential equations with constant coefficients. The **Laplace operator** transforms the initial differential equation into an algebraic expression which is easy to solve. Once the algebraic expression is solved, the inverse transform gives a solution to the original differential equation.

Consider the motion of a pendulum. The equation of motion (for small values of θ) is:

$$\frac{d^2\theta}{dt^2}+\omega^2\theta=0 \text{ where } \omega=\sqrt{\frac{g}{L}}$$

The initial conditions at $t = 0$ are: $\left.\dfrac{d\theta}{dt}\right|_{t=0}=0$

$$\theta(0)=A$$

where A is the amplitude of the swing.

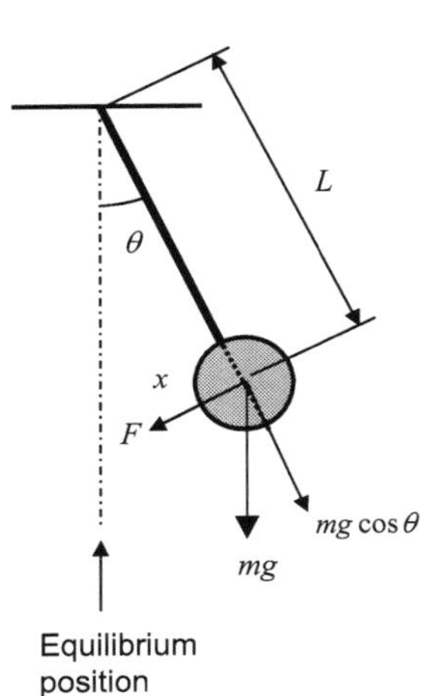

The differential equation can be solved by taking the transform of both sides:

$$L\left(\frac{d^2\theta}{dt^2}\right)+L\left(\omega^2\theta\right)=0$$

$$s^2L(\theta(t))-s\theta(0)-\left.\frac{d\theta}{dt}\right|_{t=0}+\omega^2L(\theta(t))=0$$

$$s^2L(\theta(t))-sA+\omega^2L(\theta(t))=0$$

This is now an algebraic equation in s which may be solved for $\theta(t)$:

$$s^2L(\theta(t))-sA+\omega^2L(\theta(t))=0$$

$$\left(s^2-\omega^2\right)L(\theta(t))=sA$$

$$L(\theta(t))=A\left(\frac{s}{s^2-\omega^2}\right)$$

$$\theta(t)=AL^{-1}\left(\frac{s}{s^2-\omega^2}\right)$$

The inverse transform can be obtained from standard results.

Thus: $\theta(t)=A\cos\omega t$

is a solution to the differential equation.

2.2.5 Laplace Transforms and Partial Fractions

The solution of differential equations by Laplace transforms often involves finding the **inverse transform** of an expression containing polynomial functions of s:

$$L^{-1}\left(\frac{P(s)}{Q(s)}\right)$$

The degree of polynomial of Q is greater than that of P. The expression to be simplified (P/Q) is expanded into a sum of terms for each factor in the denominator. For example, for the following expression, there are constants A, B and C such that:

$$\frac{2}{s(s+2)(s-3)}=\frac{A}{s}+\frac{B}{s+2}+\frac{C}{s-3}$$

We multiply each **partial fraction** by the denominator on the left hand side, giving:

$$2=A(s+2)(s-3)+Bs(s-3)+Cs(s+2)$$

$$\text{Letting } s=0 \quad 2=-6A$$

$$s=3 \quad 2=15C$$

$$s=-2 \quad 2=10B$$

$$A=-\frac{1}{3};B=\frac{1}{5};C=\frac{2}{15}$$

The Laplace transform is then more easily determined as the individual transforms of the partial fractions, which are more likely to be in a standard form.

For more complicated expressions where the denominator contains repeated linear factors of the form $(as+b)^n$ with $n \geq 1$, we write:

$$\frac{A_1}{as+b}+\frac{A_2}{(as+b)^2}+\ldots\frac{A_n}{(as+b)^n}$$

For example:

$$\frac{3}{(s+2)(s-2)^3}=\frac{A}{s+2}+\frac{B}{s-2}+\frac{C}{(s-2)^2}+\frac{D}{(s-2)^3}$$

Example:

Find the Laplace transform of $\sinh kt$.

Solution:

We start by expressing sinh *kt* in terms of exponential functions using Euler's equations:

$$\sinh kt = \frac{1}{2}\left[e^{jkt} - e^{-kt}\right]$$

$$\text{Thus: } L(\sinh kt) = \frac{1}{2}\int_0^\infty e^{-st}\left(e^{kt} - e^{-kt}\right)dt$$

$$= \frac{1}{2}\int_0^\infty e^{-(s-k)t} - e^{-(s+k)t}\,dt$$

$$= \frac{1}{2}\left[\frac{-1}{s-k}e^{-(s-k)t} - \frac{-1}{s+k}e^{-(s+k)t}\right]_0^\infty$$

$$= -\frac{1}{2}\left(\frac{-1}{s-k} + \frac{1}{s+k}\right)$$

$$= -\frac{1}{2}\left(\frac{-(s+k)+(s-k)}{(s-k)(s+k)}\right)$$

$$= -\frac{1}{2}\left(\frac{-2k}{s^2 + sk - sk - k^2}\right)$$

$$L(\sinh kt) = \frac{k}{s^2 - k^2}$$

Example:

Solve the differential equation: $f''(t)-2f'(t)=-4$

subject to the initial conditions:

$$f(0)=0$$
$$f'(0)=4$$

Solution:

$$L(f''(t))-2L(f'(t))=-4L(1)$$
$$s^2L(f(t))-sf(0)-f'(0)-2(sL(f(t))-f(0))=-4L(1)$$
$$s^2L(f(t))-4-2sL(f(t))=-4L(1)$$
$$(s^2-2s)L(f(t))-4=-\frac{4}{s}$$
$$L(f(t))=\frac{4}{s(s-2)}-\frac{4}{s^2(s-2)}$$

$$\frac{4}{s(s-2)}=\frac{A}{s}+\frac{B}{s-2}$$
$$4=A(s-2)+Bs$$
$$=As-2A+Bs$$
$$=(A+B)s-2A$$
$$-2A=4$$
$$A=-2$$
$$B=2$$

$$\frac{-4}{s^2(s-2)}=\frac{A}{s}+\frac{B}{s^2}+\frac{C}{s-2}$$
$$-4=As(s-2)+B(s-2)+Cs^2$$
$$=As^2-2As+Bs-2B+Cs^2$$
$$=(A+C)s^2+(B-2A)s-2B$$
$$-2B=-4$$
$$B=2$$
$$2=2A$$
$$A=1$$
$$A=-C$$
$$C=-1$$

$$L(f(t))=-\frac{2}{s}+\frac{2}{s-2}+\frac{1}{s}+\frac{2}{s^2}-\frac{1}{s-2}$$
$$=-\frac{1}{s}+\frac{1}{s-2}+\frac{2}{s^2}$$

Thus: $f(t)=-1+e^{2t}+2t$

2.3 Vector Analysis

Summary

$$\mathbf{r} = x\mathbf{i} + y\mathbf{j} + z\mathbf{k}$$

Position vector

$$|\mathbf{r}| = \sqrt{x^2 + y^2 + z^2}$$

$$\mathbf{u} = l\mathbf{i} + m\mathbf{j} + n\mathbf{k}$$

Unit vector
(l, m, n are direction cosines)

$$\mathbf{a} \cdot \mathbf{b} = a_1b_1 + a_2b_2 + a_3b_3 = |\mathbf{a}||\mathbf{b}|\cos\theta$$

Dot product

$$\mathbf{a} \times \mathbf{b} = \begin{vmatrix} \mathbf{i} & \mathbf{j} & \mathbf{k} \\ a_1 & a_2 & a_3 \\ b_1 & b_2 & b_3 \end{vmatrix} = (a_2b_3 - a_3b_2)\mathbf{i} - (a_1b_3 - a_3b_1)\mathbf{j} + (a_1b_2 - a_2b_1)\mathbf{k}$$

Cross product

$$\frac{d^2\mathbf{r}}{dt^2} = \frac{dv}{dt}\mathbf{T} + \frac{v^2}{R}\mathbf{N}$$

Motion of a particle

$$\frac{d}{dt}(\mathbf{a} + \mathbf{b}) = \frac{d}{dt}\mathbf{a} + \frac{d}{dt}\mathbf{b}$$

Vector differentiation

$$\frac{d}{dt}k\mathbf{a} = k\frac{d\mathbf{a}}{dt}$$

$$\frac{d}{dt}(\phi\mathbf{a}) = \frac{d\phi}{dt}\mathbf{a} + \phi\frac{d\mathbf{a}}{dt}$$

$$\frac{d}{dt}(\mathbf{a} \cdot \mathbf{b}) = \frac{d\mathbf{a}}{dt} \cdot \mathbf{b} + \mathbf{a} \cdot \frac{d\mathbf{b}}{dt}$$

$$\frac{d}{dt}(\mathbf{a} \times \mathbf{b}) = \frac{d\mathbf{a}}{dt} \times \mathbf{b} + \mathbf{a} \times \frac{d\mathbf{b}}{dt}$$

$$\frac{d\mathbf{a}}{dt} = \frac{d\mathbf{a}}{du}\frac{du}{dt}$$

2.3.1 Vectors

If a vector extends from point P_o to point P_1, then we write this as: $\overrightarrow{P_oP_1}$ or simply **a**.

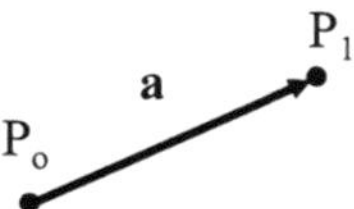

The magnitude of the vector is written: $\left|\overrightarrow{P_oP_1}\right|$ or $|\mathbf{a}|$

The components of the vector with respect to the *x,y,z* coordinate axes are written:

$$\begin{aligned}\overrightarrow{P_oP_1} &= \mathbf{a} = \langle a_1, a_2, a_3 \rangle \\ &= \langle x_1 - x_o, y_1 - y_o, z_1 - z_o \rangle\end{aligned}$$

The special vectors: $\mathbf{i} = \langle 1,0,0 \rangle$, $\mathbf{j} = \langle 0,1,0 \rangle$, $\mathbf{k} = \langle 0,0,1 \rangle$ are **unit vectors** in the direction of the *x*, *y* and *z* coordinate axes.

Thus:

$$\begin{aligned}\mathbf{a} &= \langle a_1, a_2, a_3 \rangle \\ &= a_1\mathbf{i} + a_2\mathbf{j} + a_3\mathbf{k}\end{aligned}$$

A **unit vector** in the direction of **a** is given by $\mathbf{u} = \dfrac{\mathbf{a}}{|\mathbf{a}|} = \left\langle \dfrac{a_1}{|\mathbf{a}|}, \dfrac{a_2}{|\mathbf{a}|}, \dfrac{a_3}{|\mathbf{a}|} \right\rangle$.

A vector drawn from the origin to a point *P(x,y,z)* is called the **position vector r** of that point.

$$\mathbf{r} = x\mathbf{i} + y\mathbf{j} + z\mathbf{k}$$

$$|\mathbf{r}| = \sqrt{x^2 + y^2 + z^2}$$

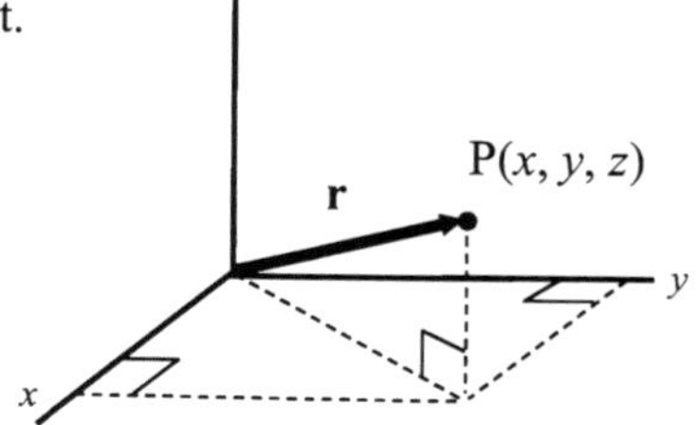

A **vector function** or **vector field** describes a vector in terms of components whose magnitudes are functions of *x*, *y* and *z*. That is, for each point P(*x,y,z*) there is a corresponding vector **F**:

$$\begin{aligned}\mathbf{F} &= f(x,y,z)\mathbf{i} + g(x,y,z)\mathbf{j} + h(x,y,z)\mathbf{k} \\ &= F_1\mathbf{i} + F_2\mathbf{j} + F_3\mathbf{k}\end{aligned}$$

2.3.2 Direction Cosines

Consider the vector **a** in a three dimensional coordinate system.

The **direction angles** of the vector **a** are the angles α, β and γ between the vectors **i**, **j**, and **k**, respectively.

The **direction cosines** l, m and n of the vector **a** are the cosines of these angles:

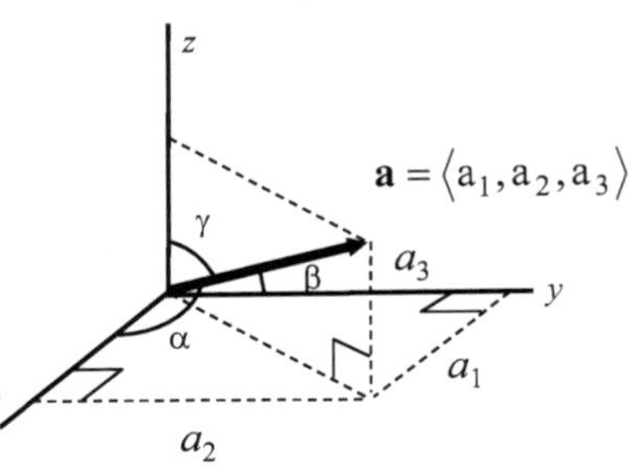

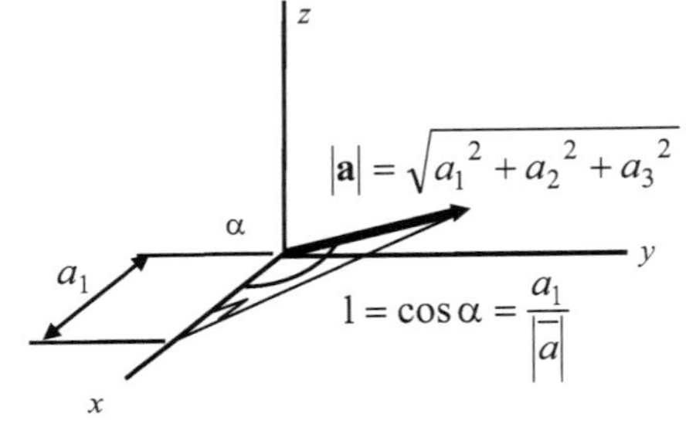

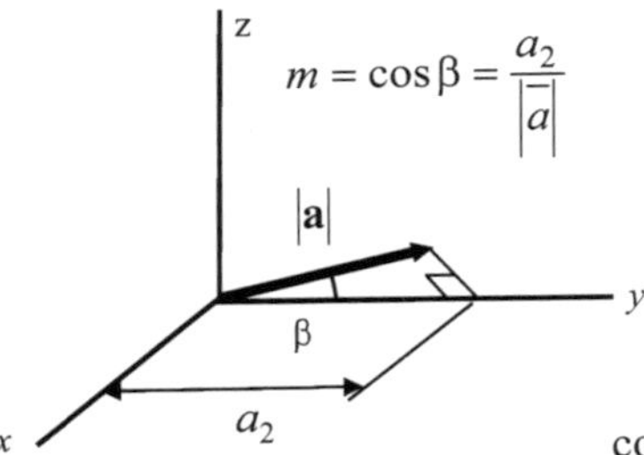

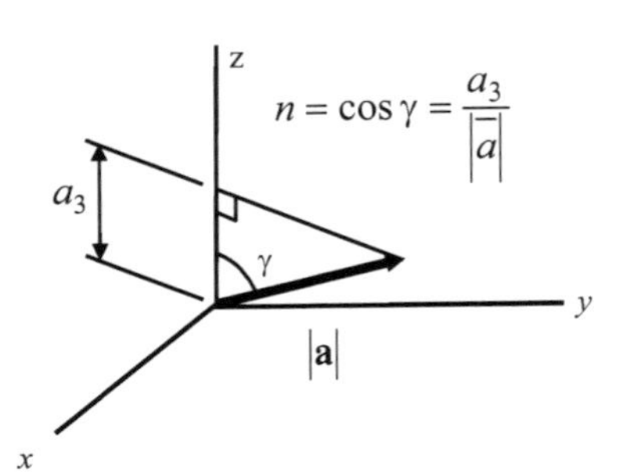

$$
\begin{aligned}
a_1 &= |\mathbf{a}|\cos\alpha \\
a_2 &= |\mathbf{a}|\cos\beta \\
a_3 &= |\mathbf{a}|\cos\gamma \\
\mathbf{a} &= \langle |\mathbf{a}|\cos\alpha, |\mathbf{a}|\cos\beta, |\mathbf{a}|\cos\gamma \rangle \\
&= |\mathbf{a}|\langle \cos\alpha, \cos\beta, \cos\gamma \rangle
\end{aligned}
$$

$$
\begin{aligned}
\cos^2\alpha + \cos^2\beta + \cos^2\gamma &= \frac{a_1^2}{|\mathbf{a}|^2} + \frac{a_2^2}{|\mathbf{a}|^2} + \frac{a_3^2}{|\mathbf{a}|^2} \\
&= \frac{a_1^2 + a_2^2 + a_3^2}{a_1^2 + a_2^2 + a_3^2} \\
&= 1 \\
&= l^2 + m^2 + n^2
\end{aligned}
$$

The direction cosines of **a** are the components of the unit vector: $\mathbf{u} = l\mathbf{i} + m\mathbf{j} + n\mathbf{k}$

2.3.4 Vector Dot Product

If $\mathbf{a} = \langle a_1, a_2, a_3 \rangle$

$\mathbf{b} = \langle b_1, b_2, b_3 \rangle$

then the **dot product** is defined as:

$$\mathbf{a} \cdot \mathbf{b} = a_1 b_1 + a_2 b_2 + a_3 b_3$$
$$= |\mathbf{a}||\mathbf{b}| \cos\theta$$

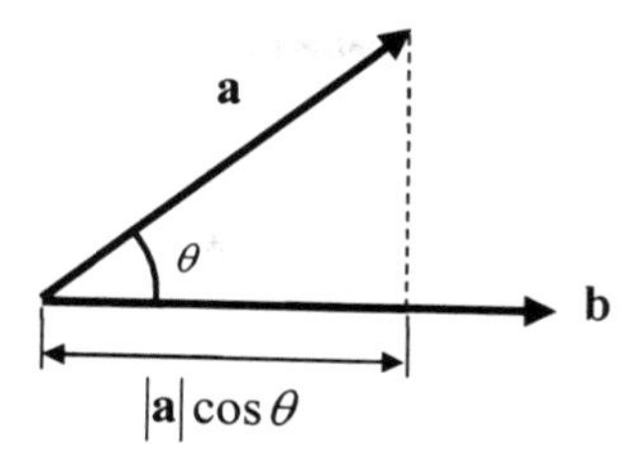

But, the quantity $|\mathbf{a}|\cos\theta$ is the component of **a** in the direction of **b**. The dot product is a scalar and represents the magnitude of the component of **a** in the direction of **b** multiplied by the magnitude of **b**.

To find the component of **a** in **b**'s direction, we form the dot product of **a** with a unit vector in the direction of **b**.

$$|\mathbf{a}|\cos\theta = \mathbf{a} \cdot \frac{\mathbf{b}}{|\mathbf{b}|}$$
$$= \mathbf{a} \cdot \mathbf{u}$$

↓

Unit vector in the direction of **b**

The magnitude of a vector is given by the square root of the dot product of it and itself.

$$|\mathbf{a}| = \sqrt{\mathbf{a} \cdot \mathbf{a}}$$
$$= \sqrt{a_1{}^2 + a_2{}^2 + a_3{}^2}$$

Properties of the dot product:

$$\mathbf{a} \cdot \mathbf{a} = |\mathbf{a}|^2$$
$$\mathbf{a} \cdot \mathbf{b} = \mathbf{b} \cdot \mathbf{a}$$
$$\mathbf{a} \cdot (\mathbf{b} + \mathbf{c}) = \mathbf{a} \cdot \mathbf{b} + \mathbf{a} \cdot \mathbf{c}$$
$$(\mathrm{k}\mathbf{a}) \cdot \mathbf{b} = \mathrm{k}(\mathbf{a} \cdot \mathbf{b})$$
$$\mathbf{a} \cdot 0 = 0$$
$$\mathbf{u} \cdot \mathbf{u} = 1$$

2.3.5 Equation of a Line in Space

It is desirable to be able to describe a **line** in space using an equation. If we know the coordinates of two points P_o and P_1 on the line, then we can draw a vector **a** between the two points:

$$\overrightarrow{P_oP_1} = \mathbf{a} = \langle a_1, a_2, a_3 \rangle = a_1\mathbf{i} + a_2\mathbf{j} + a_3\mathbf{k}$$
$$= \langle x_1 - x_o, y_1 - y_o, z_1 - z_o \rangle$$

We now wish to express the coordinates of any point P on this line in terms of the known values of a_1, a_2 and a_3.

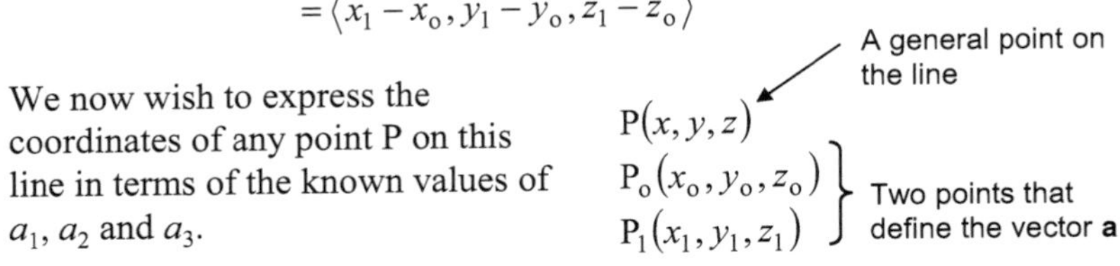

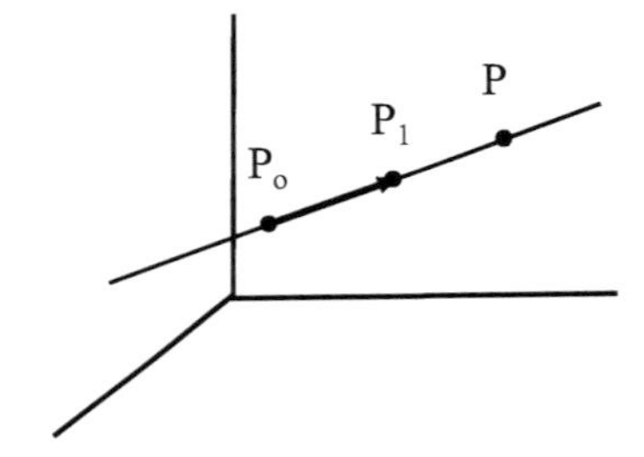

The equation of a line is given as a series of equations in terms of a parameter, say, t. Such equations are called **parametric equations**.

$$\overrightarrow{P_oP} = t\overrightarrow{P_oP_1}$$
$$\langle x - x_o, y - y_o, z - z_o \rangle = t\langle x_1 - x_o, y_1 - y_o, z_1 - z_o \rangle$$
$$= t\langle a_1, a_2, a_3 \rangle$$

$$\boxed{\begin{aligned} x &= x_o + a_1 t \\ y &= y_o + a_2 t \\ z &= z_o + a_3 t \end{aligned}} \quad \text{or} \quad \frac{x - x_o}{a_1} = \frac{y - y_o}{a_2} = \frac{z - z_o}{a_3} = t$$

If a point with coordinates x, y and z falls upon the line, then those values of x, y and z will satisfy the parametric equations for that line for a particular value of t.

2.3.6 Equation of a Plane

It is desirable to be able to describe a **plane** in space using an equation. If we know the coordinates of two points P_o and P_1 on a line which is normal to the plane, then we can draw a vector **n** between these two points:

$$\overrightarrow{P_oP_1} = \mathbf{n} = \langle a,b,c \rangle$$

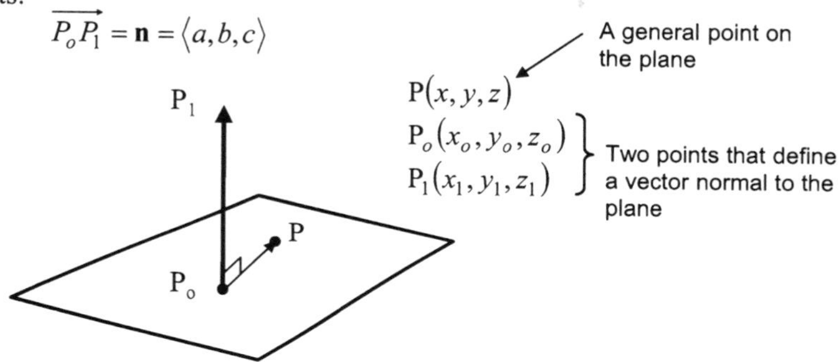

$$\text{Now, } \overrightarrow{P_oP_1} \cdot \overrightarrow{P_oP} = 0$$

Therefore, the **equation of a plane** is: $\langle a,b,c \rangle \cdot \langle x-x_o, y-y_o, z-z_o \rangle = 0$

$$a(x-x_o)+b(y-y_o)+c(z-z_o)=0$$

But the sum of the products $ax_o\ by_o\ cz_o$ is a constant (since these terms are the same no matter what the value of x,y,z).

$$d = -ax_o - by_o - cz_o$$

So we can write the equation of a plane as:

$$ax + by + cz + d = 0$$

That is, a plane is completely described by a vector **n** which is normal to the plane and a constant d.

2.3.7 Distance from a Point to a Plane

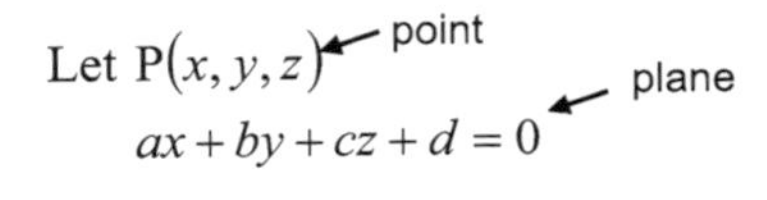

Let $\mathrm{P}(x, y, z)$ ← point

$ax + by + cz + d = 0$ ← plane

The point $\mathrm{P_o}(x_o, y_o, z_o)$ is on the plane.

The vector **a** joins the points P and $\mathrm{P_o}$.

$$\mathbf{a} = \langle x - x_o, y - y_o, z - z_o \rangle$$

Thus:

$$\mathbf{a} \cdot \mathbf{n} = |\mathbf{a}||\mathbf{n}| \cos\theta$$

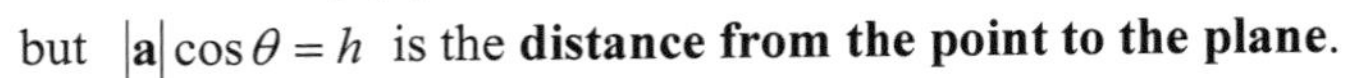

but $|\mathbf{a}|\cos\theta = h$ is the **distance from the point to the plane**.

thus $h = \mathbf{a} \cdot \dfrac{\mathbf{n}}{|\mathbf{n}|}$

$$= \langle x - x_o, y - y_o, z - z_o \rangle \cdot \frac{\langle a, b, c \rangle}{\sqrt{a^2 + b^2 + c^2}}$$

$$= \left(a(x - x_o) + b(y - y_o) + c(z - z_o)\right) \frac{1}{\sqrt{a^2 + b^2 + c^2}}$$

$$= \frac{ax + bx + cx + d}{\sqrt{a^2 + b^2 + c^2}}$$

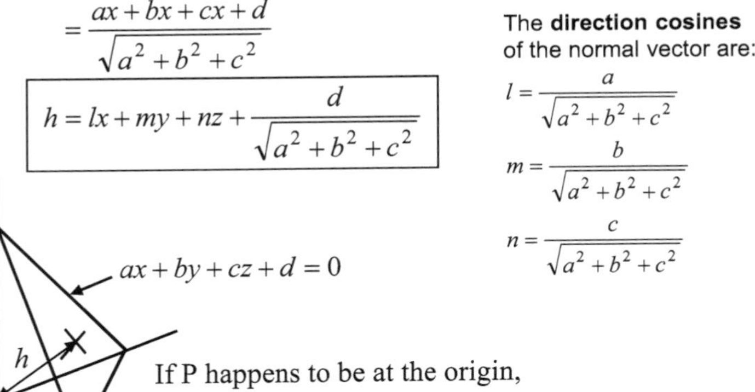

$$\boxed{h = lx + my + nz + \frac{d}{\sqrt{a^2 + b^2 + c^2}}}$$

The **direction cosines** of the normal vector are:

$$l = \frac{a}{\sqrt{a^2 + b^2 + c^2}}$$

$$m = \frac{b}{\sqrt{a^2 + b^2 + c^2}}$$

$$n = \frac{c}{\sqrt{a^2 + b^2 + c^2}}$$

If P happens to be at the origin, then the formula gives the distance between the plane and the origin. $h = \dfrac{d}{\sqrt{a^2 + b^2 + c^2}}$

2.3.8 Vector Cross Product

Consider two vectors **a** and **b** that lie in a plane.

$$\mathbf{a} = \langle a_1, a_2, a_3 \rangle$$
$$\mathbf{b} = \langle b_1, b_2, b_3 \rangle$$

The direction of **R** is given by the right hand rule. If the fingers curl in the direction from **a** to **b**, then the direction of the resulting vector **a** × **b** is given by the thumb. Thus, **a** × **b** is opposite in direction to **b** × **a**.

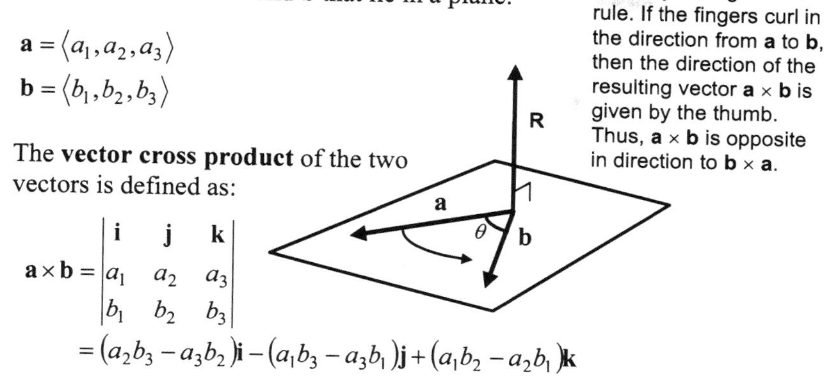

The **vector cross product** of the two vectors is defined as:

$$\mathbf{a}\times\mathbf{b} = \begin{vmatrix} \mathbf{i} & \mathbf{j} & \mathbf{k} \\ a_1 & a_2 & a_3 \\ b_1 & b_2 & b_3 \end{vmatrix}$$
$$= (a_2b_3 - a_3b_2)\mathbf{i} - (a_1b_3 - a_3b_1)\mathbf{j} + (a_1b_2 - a_2b_1)\mathbf{k}$$

The resultant vector **R** is perpendicular to the plane of the two vectors **a** and **b**. The magnitude of the resultant vector is:

$$|\mathbf{R}| = |\mathbf{a}||\mathbf{b}|\sin\theta$$

But, the quantity $|\mathbf{b}|\sin\theta$ is the height of a parallelogram formed by the two vectors.

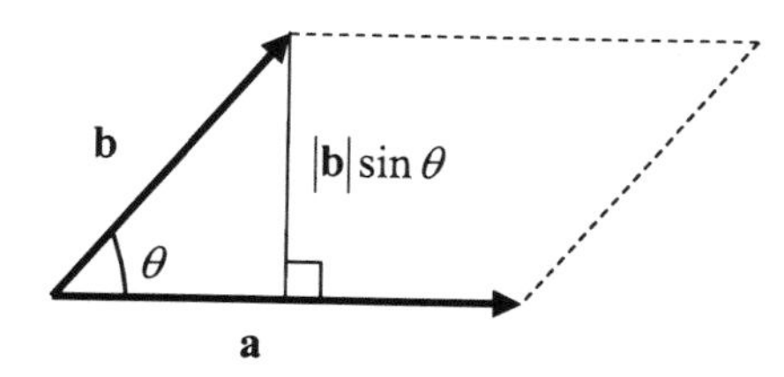

The magnitude of the resultant **R** is therefore the area of the parallelogram formed by the two vectors **a** and **b**.

Properties of the cross product:

$$\mathbf{a}\times\mathbf{b} = -\mathbf{b}\times\mathbf{a}$$
$$k\mathbf{a}\times\mathbf{b} = k(\mathbf{a}\times\mathbf{b})$$
$$\mathbf{a}\times(\mathbf{b}+\mathbf{c}) = \mathbf{a}\times\mathbf{b} + \mathbf{a}\times\mathbf{c}$$
$$(\mathbf{a}\times\mathbf{b})\cdot\mathbf{c} = \mathbf{a}\cdot(\mathbf{b}\times\mathbf{c})$$
$$\mathbf{a}\times(\mathbf{b}\times\mathbf{c}) = (\mathbf{a}\cdot\mathbf{c})\mathbf{b} - (\mathbf{a}\cdot\mathbf{b})\mathbf{c}$$
$$|\mathbf{a}\times\mathbf{b}| = \sqrt{(\mathbf{a}\cdot\mathbf{a})(\mathbf{b}\cdot\mathbf{b}) - (\mathbf{a}\cdot\mathbf{b})^2}$$

2.3.9 Distance from a Point to a Line

The **vector cross product** has some significance in determining the **distance from a point to a line**.

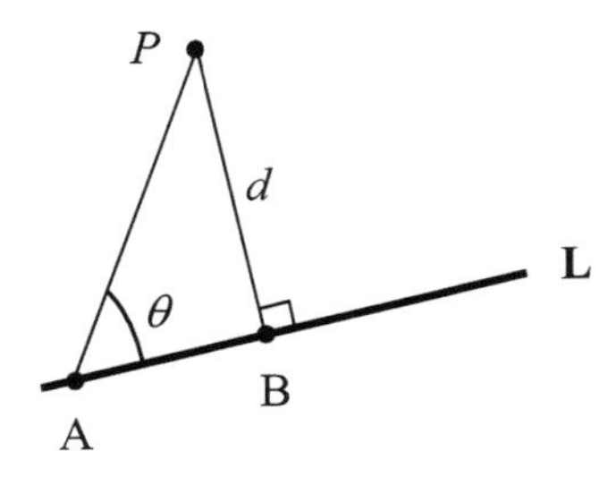

Let A and B be two points on the line **L**. The distance d is thus:

$$\left|\overrightarrow{AP}\times\overrightarrow{AB}\right| = \left|\overrightarrow{AP}\right|\left|\overrightarrow{AB}\right|\sin\theta$$

$$\left|\overrightarrow{AP}\right|\sin\theta = \frac{\left|\overrightarrow{AP}\times\overrightarrow{AB}\right|}{\left|\overrightarrow{AB}\right|}$$

$$= d$$

Example:

Find the distance of the point P(1,2,–1) from the line: $\dfrac{x-2}{3} = \dfrac{y+2}{-1} = \dfrac{z+1}{2}$

Solution:

Vector representation of the line is $\langle 3,-1,2\rangle$

A point on the line is (by inspection) Q(2,–2,-1)

Thus, a vector joining P to the line at Q is $\langle -1,4,0\rangle$

P(1,2,–1)

d

L$\langle 3,-1,2\rangle$

θ

Q(2,–2,–1)

$$d = \left|\overrightarrow{QP}\right|\sin\theta$$

$$= \frac{\left|\overrightarrow{QP}\times\mathbf{L}\right|}{\left|\mathbf{L}\right|}$$

$$\overrightarrow{QP}\times\mathbf{L} = \begin{vmatrix} \mathbf{i} & \mathbf{j} & \mathbf{k} \\ -1 & 4 & 0 \\ 3 & -1 & 2 \end{vmatrix}$$

$$= 8\mathbf{i} + 2\mathbf{j} - 11\mathbf{k}$$

$$\left|\overrightarrow{QP}\times\mathbf{L}\right| = \sqrt{189}$$

$$\left|\mathbf{L}\right| = \sqrt{3^2+1^2+2^2}$$

$$= \sqrt{14}$$

$$d = \frac{\sqrt{189}}{\sqrt{14}}$$

2.3.10 Distance between Two Skew Lines

Consider now two **skew lines in space**. We wish to find the distance between any two points on the lines.

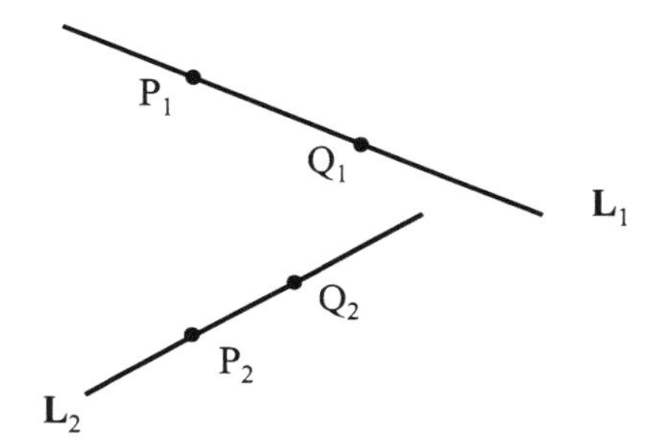

We first choose any two points P and Q on the lines as shown and draw vectors between them along the lines. The vector cross product

$$\overrightarrow{P_1Q_1} \times \overrightarrow{P_2Q_2}$$

is a vector normal to both lines.

If T_1 and T_2 are parallel planes that contain $\mathbf{L}_1$ and $\mathbf{L}_2$, then the distance between these planes is measured along a common normal. This distance is also the shortest distance between the two lines.

Example:

Find the perpendicular distance from the point (1,-1,2) to the plane

$$2x + 2y - z + 4 = 0$$

Solution:

A vector perpendicular to the plane is $<2,2,-1>$ with magnitude $|\mathbf{n}| = \sqrt{4+4+1} = 3$

The distance from the point to the plane is given by: $h = \dfrac{ax + bx + cx + d}{\sqrt{a^2 + b^2 + c^2}}$

$$= \frac{2(1) + 2(-1) + -1(2) + 4}{3}$$

$$= \frac{2}{3}$$

2.3.11 Vector Differentiation

The **position vector** of a point in space is given by: $\mathbf{r} = x\mathbf{i} + y\mathbf{j} + z\mathbf{k}$

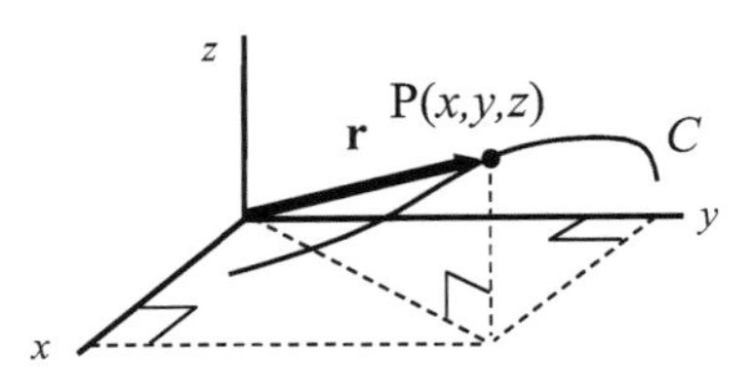

A curve C in space may be represented by a series of points, each with its own position vector.

For any two points on the curve, the length of the chord joining P_o and P_1 is $\Delta\mathbf{r}$. It is of interest to determine the rate of change of the vector $\mathbf{r}$ with respect to the arc length s.

$$\frac{d\mathbf{r}}{ds} = \lim_{\Delta t \to 0} \frac{\mathbf{r}(s_o + \Delta s) - \mathbf{r}(s_o)}{\Delta s}$$

$$= \lim_{\Delta s \to 0} \frac{\Delta\mathbf{r}}{\Delta s}$$

As $\Delta s \to 0$, $|\Delta\mathbf{r}|$ becomes equal to Δs and so: $\dfrac{d\mathbf{r}}{ds} = 1$.

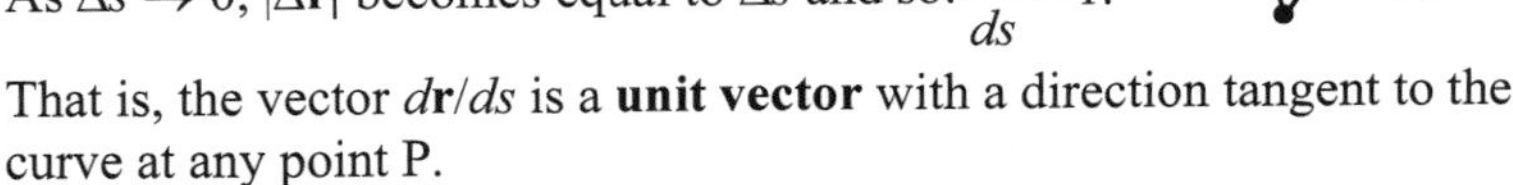

That is, the vector $d\mathbf{r}/ds$ is a **unit vector** with a direction tangent to the curve at any point P.

The position vectors may also be expressed in terms of time (t) such that:

$$\mathbf{r} = x(t)\mathbf{i} + y(t)\mathbf{j} + z(t)\mathbf{k}$$

Differentiating with respect to t gives:

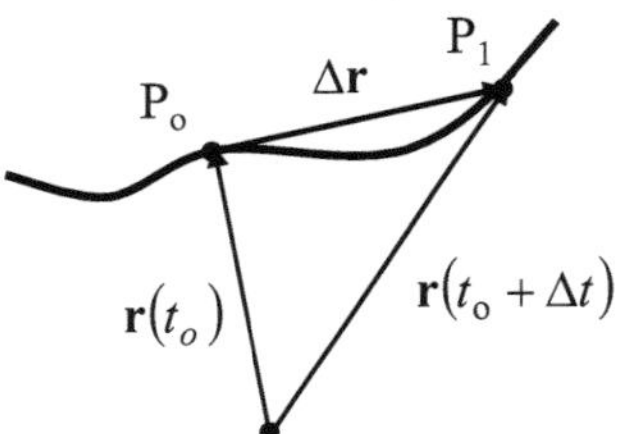

$$\frac{d\mathbf{r}}{dt} = \lim_{\Delta t \to 0} \frac{\mathbf{r}(t_o + \Delta t) - \mathbf{r}(t_o)}{\Delta t}$$

$$= \lim_{\Delta t \to 0} \frac{\Delta\mathbf{r}}{\Delta t}$$

As $\Delta t \to 0$, $\dfrac{d\mathbf{r}}{dt}$ evaluated at $t = t_o$ is a vector whose direction is tangent to the curve at P and represents the instantaneous velocity of a body which may be at point P.

In general, a vector function is said to be differentiable if

$$\lim_{\Delta t \to 0} \frac{\mathbf{r}(t_o + \Delta t) - \mathbf{r}(t_o)}{\Delta t} \text{ exists.}$$

Whereupon:

$$\frac{d\mathbf{r}}{dt} = \frac{dx}{dt}\mathbf{i} + \frac{dy}{dt}\mathbf{j} + \frac{dz}{dt}\mathbf{k} \quad \text{and} \quad \left|\frac{d\mathbf{r}}{dt}\right| = \sqrt{\frac{d\mathbf{r}}{dt} \cdot \frac{d\mathbf{r}}{dt}} = \sqrt{\left(\frac{dx}{dt}\right)^2 + \left(\frac{dy}{dt}\right)^2 + \left(\frac{dz}{dt}\right)^2}$$

2.3.12 Motion of a Body

The vector $d\mathbf{r}/dt$ was found tangent to the curve and represented the velocity of a body at a point P.

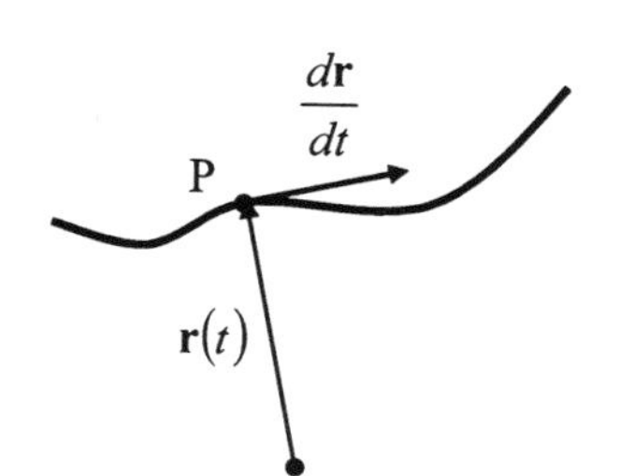

As shown previously, a unit vector $\mathbf{T}$ tangent to the curve is given by $d\mathbf{r}/ds$ and so:

$$\mathbf{T} = \frac{d\mathbf{r}}{ds}$$

$$\frac{d\mathbf{r}}{dt} = \frac{d\mathbf{r}}{ds}\frac{ds}{dt}$$

$$= \mathbf{T}\frac{ds}{dt}$$

which expresses the tangential velocity in terms of a unit tangent vector times the speed.

The **acceleration** is given by differentiating the velocity with respect to time thus:

$$\frac{d^2\mathbf{r}}{dt^2} = \frac{d}{dt}\left(\mathbf{T}\frac{ds}{dt}\right)$$

$$= \mathbf{T}\frac{d^2s}{dt^2} + \frac{ds}{dt}\frac{d\mathbf{T}}{dt}$$

$$= \mathbf{T}\frac{d^2s}{dt^2} + \frac{ds}{dt}\frac{d\mathbf{T}}{ds}\frac{ds}{dt}$$

$$= \mathbf{T}\frac{d^2s}{dt^2} + \left(\frac{ds}{dt}\right)^2\frac{d\mathbf{T}}{ds}$$

$$|\mathbf{T}| = 1$$

$$\mathbf{T}\cdot\mathbf{T} = 1$$

$$\frac{d}{ds}(\mathbf{T}\cdot\mathbf{T}) = \frac{d\mathbf{T}}{ds}\cdot\mathbf{T} + \mathbf{T}\cdot\frac{d\mathbf{T}}{ds}$$

$$= 0 \quad \text{since } \mathbf{T}\cdot\mathbf{T} = 1\text{, a constant}$$

Let $v = \dfrac{ds}{dt}$

$$\frac{d^2\mathbf{r}}{dt^2} = \mathbf{T}\frac{dv}{dt} + v^2\frac{d\mathbf{T}}{ds} \longrightarrow$$

Thus: $0 = 2\mathbf{T}\cdot\dfrac{d\mathbf{T}}{ds}$ which says that dT/ds is perpendicular to $\mathbf{T}$.

$$\frac{d\mathbf{T}}{ds} = \frac{1}{R}\mathbf{N}$$

$\mathbf{N}$ → unit normal vector to the curve

$1/R$ is the magnitude of $d\mathbf{T}/ds$ and is equal to the radius of curvature of the curve.

$$\boxed{\frac{d^2\mathbf{r}}{dt^2} = \frac{dv}{dt}\mathbf{T} + \frac{v^2}{R}\mathbf{N}}$$

The acceleration of the body is the vector addition of the tangential component and a normal (inwards or **centripetal**) component.

2.4 Partial Derivatives

Summary

$$\frac{\partial^2 z}{\partial x \partial y} = \frac{\partial}{\partial x}\left(\frac{\partial z}{\partial y}\right)$$

$$\frac{\partial^2 z}{\partial y \partial x} = \frac{\partial}{\partial y}\left(\frac{\partial z}{\partial x}\right)$$

$$\frac{\partial^2 z}{\partial x^2} = \frac{\partial}{\partial x}\left(\frac{\partial z}{\partial x}\right)$$

$$\frac{\partial^2 z}{\partial y^2} = \frac{\partial}{\partial y}\left(\frac{\partial z}{\partial y}\right)$$

Partial derivatives

$$dz = \frac{\partial z}{\partial x} dx + \frac{\partial z}{\partial y} dy$$

Differentials

$$\frac{dz}{ds} = \frac{\partial z}{\partial x}\frac{dx}{ds} + \frac{\partial z}{\partial y}\frac{dy}{ds}$$

Chain rule

$$\nabla = \frac{\partial}{\partial x}\mathbf{i} + \frac{\partial}{\partial y}\mathbf{j} + \frac{\partial}{\partial z}\mathbf{k}$$

Differential operator

$$\nabla \phi = \frac{\partial \phi}{\partial x}\mathbf{i} + \frac{\partial \phi}{\partial y}\mathbf{j} + \frac{\partial \phi}{\partial z}\mathbf{k}$$

Gradient

$$\nabla \cdot \mathbf{F} = \frac{\partial M}{\partial x} + \frac{\partial N}{\partial y} + \frac{\partial O}{\partial z}$$

Divergence

$$\nabla \times \mathbf{v} = \begin{vmatrix} \mathbf{i} & \mathbf{j} & \mathbf{k} \\ \dfrac{\partial}{\partial x} & \dfrac{\partial}{\partial y} & \dfrac{\partial}{\partial z} \\ M & N & O \end{vmatrix}$$

Curl

2.4.1 Partial Differentiation

In many physical situations, a dependent variable is a function of two independent variables. This type of relationship is expressed as a surface in a three dimensional coordinate system.

To determine the rate of change of z as x and y vary, we hold one variable constant and differentiate with respect to the other. This is called **partial differentiation**.

$$z = f(x, y)$$

$$\frac{\partial z}{\partial x} = \lim_{x \to x_o} \frac{f(x, y_o) - f(x_o, y_o)}{x - x_o} \quad \text{holding } y \text{ constant at } y_o$$

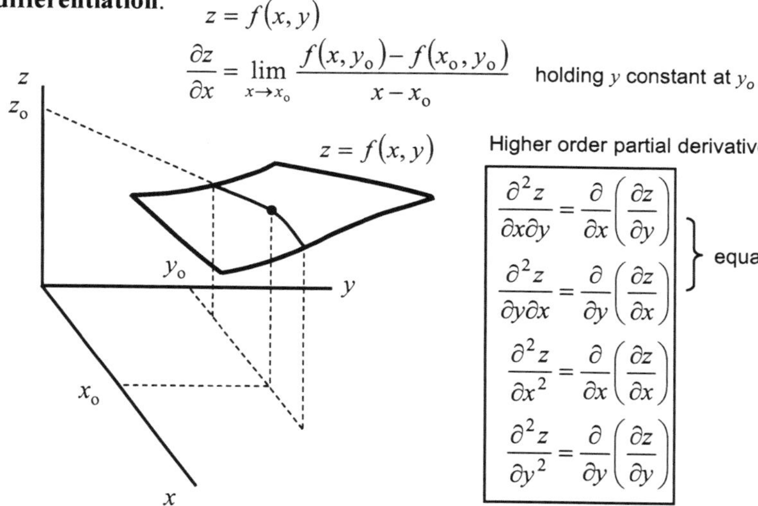

Higher order partial derivatives

$$\frac{\partial^2 z}{\partial x \partial y} = \frac{\partial}{\partial x}\left(\frac{\partial z}{\partial y}\right)$$

$$\frac{\partial^2 z}{\partial y \partial x} = \frac{\partial}{\partial y}\left(\frac{\partial z}{\partial x}\right)$$

(the two above are equal)

$$\frac{\partial^2 z}{\partial x^2} = \frac{\partial}{\partial x}\left(\frac{\partial z}{\partial x}\right)$$

$$\frac{\partial^2 z}{\partial y^2} = \frac{\partial}{\partial y}\left(\frac{\partial z}{\partial y}\right)$$

For example, if we wish to know the rate of change of z with respect to x at $x = x_o$, then we hold y constant at y_o.

$\frac{\partial z}{\partial x}$ gives the slope of the tangent to the surface in the x direction at x_o.

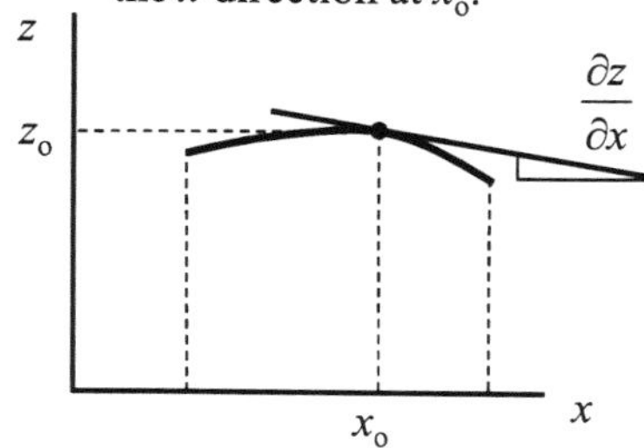

For example, if we wish to know the rate of change of z with respect to y at $y = y_o$, then we hold x constant at x_o.

$\frac{\partial z}{\partial y}$ gives the slope of the tangent to the surface in the y direction at y_o.

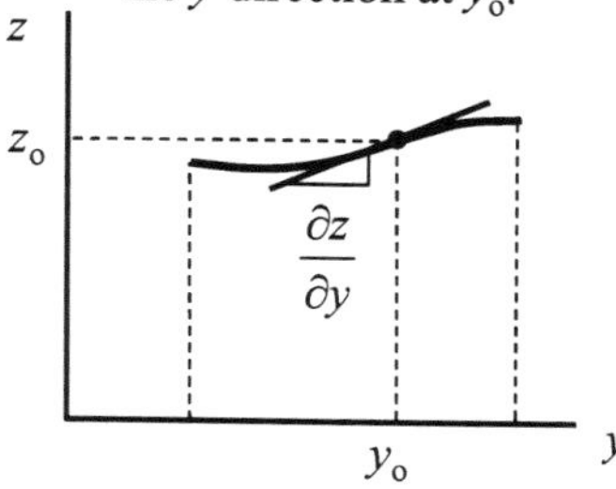

2.4.2 Chain Rule for Partial Derivatives

Let x be a function of two variables u and v, and y another function of the same variables u and v. If z is some function of x and y, then z is also a function of u and v.

$$x = x(u,v)$$
$$y = y(u,v)$$
$$z = z(x,y) = z(u,v)$$

The rate of change of z with respect to u, and the rate of change of z with respect to v, is given by the **chain rule**:

$$\frac{\partial z}{\partial u} = \frac{\partial z}{\partial x}\frac{\partial x}{\partial u} + \frac{\partial z}{\partial y}\frac{\partial y}{\partial u}$$

$$\frac{\partial z}{\partial v} = \frac{\partial z}{\partial x}\frac{\partial x}{\partial v} + \frac{\partial z}{\partial y}\frac{\partial y}{\partial v}$$

If $F(x,y) = 0$ where $y = f(x)$

then $\dfrac{dy}{dx} = -\dfrac{\partial F/\partial x}{\partial F/\partial y}$

If $F(x,y,z) = 0$ where $z = f(x,y)$

then $\dfrac{\partial z}{\partial x} = -\dfrac{\partial F/\partial x}{\partial F/\partial z}$

$$\frac{\partial z}{\partial y} = -\frac{\partial F/\partial y}{\partial F/\partial z}$$

Leibnitz's Rule (differentiation under the integral sign)

$$\frac{dF}{dt} = \frac{d}{dt}\int_a^b f(x,t)dx$$
$$= \int_a^b \frac{\partial}{\partial t} f(x,t)dx$$

2.4.3 Increments and Differentials

Let $z = f(x, y)$.

Consider points $f(x_1, y_1)$ and the corresponding increments Δx, Δy.

$$\Delta z = f(x_1 + \Delta x, y_1 + \Delta y) - f(x_1, y_1)$$

Add and subtract $f(x_1, y_1 + \Delta y)$

$$= \underbrace{\left(f(x_1 + \Delta x, y_1 + \Delta y) - f(x_1, y_1 + \Delta y)\right)}_{\substack{x_1 \text{ changes by } \Delta x \\ y \text{ remains constant at } y_1 + \Delta y}} - \underbrace{\left(f(x_1, y_1 + \Delta y) - f(x_1, y_1)\right)}_{\substack{y_1 \text{ changes by } \Delta y \\ x \text{ remains constant at } x_1}}$$

By the **mean value theorem**:

$$\frac{\left(f(x_1 + \Delta x, y_1 + \Delta y) - f(x_1, y_1 + \Delta y)\right)}{\Delta x} = \frac{\partial}{\partial x} f(c_x, y_1 + \Delta y)$$

and $$\frac{\left(f(x_1, y_1 + \Delta y) - f(x_1, y_1)\right)}{\Delta y} = \frac{\partial}{\partial y} f(x_1, c_y)$$

or

$$\left(f(x_1 + \Delta x, y_1 + \Delta y) - f(x_1, y_1 + \Delta y)\right) = \frac{\partial}{\partial x} f(c_x, y_1 + \Delta y)\Delta x$$

$$\left(f(x_1, y_1 + \Delta y) - f(x_1, y_1)\right) = \frac{\partial}{\partial y} f(x_1, c_y)\Delta y$$

Thus $$\Delta z = \frac{\partial}{\partial x} f(c_x, y_1 + \Delta y)\Delta x + \frac{\partial}{\partial y} f(x_1, c_y)\Delta y$$

As $\Delta x, \Delta y \to 0$ then $\frac{\partial}{\partial x} f(c_x, y_1 + \Delta y) \to \frac{\partial}{\partial x} f(x_1, y_1)$

$$\frac{\partial}{\partial y} f(x_1, c_y) \to \frac{\partial}{\partial y} f(x_1, y_1)$$

and $dx \approx \Delta x, dy \approx \Delta y, dz \approx \Delta z$

Thus $$\boxed{dz = \frac{\partial z}{\partial x} dx + \frac{\partial z}{\partial y} dy}$$

> **Mean value theorem**
>
>
>
> There is at least one value c between a and b such that
>
> $$\frac{f(b) - f(a)}{b - a} = f'(c)$$

The change in z is equal to the change in z due to Δx plus the change in z due to Δy.

2.4.4 Directional Derivatives

Let $z = f(x,y)$. The partial derivatives represent the rate of change of z with respect to x and y (i.e., z is a function of two independent variables). We may now wish to calculate the rate of change of z with respect to a distance ds in a direction at an angle θ with respect to the x axis. This **directional derivative** is given by:

$$\boxed{\frac{dz}{ds} = \frac{\partial z}{\partial x}\frac{dx}{ds} + \frac{\partial z}{\partial y}\frac{dy}{ds}}$$

If ds is chosen to be in the direction of the x axis, then $dy = 0$ and $dx = ds$ and we obtain the first partial derivative.

But, $\dfrac{dx}{ds} = \cos\theta$

and $\dfrac{dy}{ds} = \sin\theta$

Thus
$$\frac{dz}{ds} = \frac{\partial z}{\partial x}\cos\theta + \frac{\partial z}{\partial y}\sin\theta$$
$$= \left(\frac{\partial z}{\partial x}\mathbf{i} + \frac{\partial z}{\partial y}\mathbf{j}\right)\cdot(\cos\theta\mathbf{i} + \sin\theta\mathbf{j})$$
$$= \left(\frac{\partial z}{\partial x}\mathbf{i} + \frac{\partial z}{\partial y}\mathbf{j}\right)\cdot\mathbf{u}$$

where $\mathbf{u}$ is a unit vector in the direction ds.

We define the **vector differential operator** ∇ as $\nabla = \dfrac{\partial}{\partial x}\mathbf{i} + \dfrac{\partial}{\partial y}\mathbf{j}$.

Thus $\dfrac{dz}{ds} = \nabla z\cdot\mathbf{u}$. ∇z is called the **gradient** of z and is a vector.

The rate of change of z with respect to s (i.e., in the direction of $\mathbf{u}$) can be thus found from $\dfrac{dz}{ds} = |\nabla z||\mathbf{u}|\cos\phi$

which is a maximum at $\phi = 0$. That is, dz/ds is a maximum when the direction of $\mathbf{u}$ is the same as that of ∇z. If $\mathbf{u}$ is chosen perpendicular to ∇z, then $dz/ds = 0$.

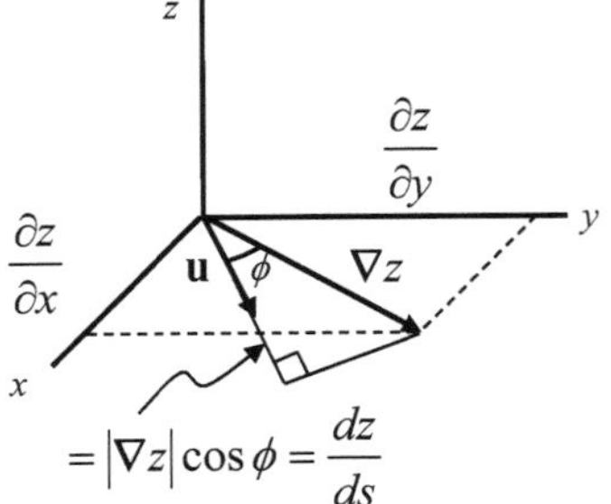

For functions of three variables $w(x,y,z)$, we have:
$$\nabla = \frac{\partial}{\partial x}\mathbf{i} + \frac{\partial}{\partial y}\mathbf{j} + \frac{\partial}{\partial z}\mathbf{k}$$
$$\frac{dw}{ds} = \nabla w\cdot\mathbf{u}$$

2.4.5 Tangent Planes and Normal Vector

The general form of a surface S in a three dimensional coordinate system is expressed $F(x,y,z) = k$ where k is some constant. The values of x, y and z can all be functions of some parameter t or s.

$$x = f(t); y = g(t); z = h(t)$$

Let $P_o(x_o,y_o,z_o)$ be a point on S. Points on the surface can be defined by their position vectors **r** which are referenced to the origin.

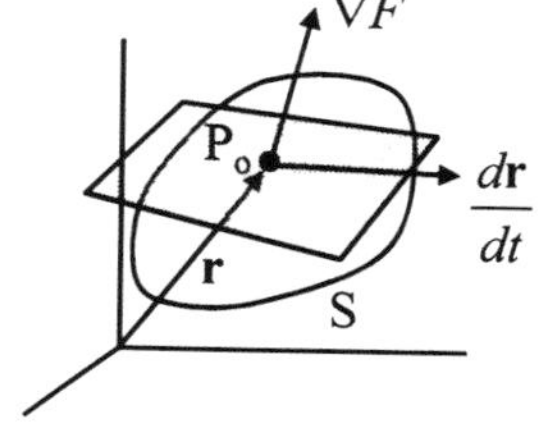

$$\mathbf{r} = x(t)\mathbf{i} + y(t)\mathbf{j} + z(t)\mathbf{k}$$

And the rate of change of **r** with respect to t is a vector tangent to the surface and is given by:

$$\frac{d\mathbf{r}}{dt} = \frac{dx}{dt}\mathbf{i} + \frac{dy}{dt}\mathbf{j} + \frac{dz}{dt}\mathbf{k}$$

Along points on the surface, the value of $F(x,y,z)$ remains at k, so $dF = 0$ for all values of t (i.e., there is no change of $F(x,y,z)$ as t varies). Thus:

$$0 = \frac{dF}{dt} = \frac{\partial F}{\partial x}\frac{dx}{dt} + \frac{\partial F}{\partial y}\frac{dy}{dt} + \frac{\partial F}{\partial z}\frac{dz}{dt}$$

$$= \left(\frac{\partial F}{\partial x}\mathbf{i} + \frac{\partial F}{\partial y}\mathbf{j} + \frac{\partial F}{\partial z}\mathbf{k}\right) \cdot \left(\frac{dx}{dt}\mathbf{i} + \frac{dy}{dt}\mathbf{j} + \frac{dz}{dt}\mathbf{k}\right)$$

$$= \nabla F \cdot \frac{d\mathbf{r}}{dt}$$

That is, ∇F is perpendicular to $d\mathbf{r}/dt$ and so ∇F must be a **normal vector** to the surface at $P_o(x_o,y_o,z_o)$.

The equation of a plane is written in terms of a normal vector **n**:

$$\mathbf{n} \cdot \langle x - x_o, y - y_o, z - z_o \rangle = 0$$

which, in terms of ∇F evaluated at $P_o(x_o,y_o,z_o)$ and position vectors **r** and $\mathbf{r}_o$ (for any two points P_o and P on the plane), becomes the **tangent plane**:

$$(\mathbf{r} - \mathbf{r}_o) \cdot \nabla F = 0$$

A flat plane surface is the special case of a general surface in which $F(x,y,z)$ takes the form $F = ax + by + cz + d = 0$.

$$\frac{\partial F}{\partial x} = a; \frac{\partial F}{\partial y} = b; \frac{\partial F}{\partial z} = c$$

$$\nabla F = a\mathbf{i} + b\mathbf{j} + c\mathbf{k}$$

2.4.6 Gradient, Divergence and Curl

The **vector differential operator "del"** is defined as:

$$\nabla = \frac{\partial}{\partial x}\mathbf{i} + \frac{\partial}{\partial y}\mathbf{j} + \frac{\partial}{\partial z}\mathbf{k}$$

The **gradient** is given by:

$$\nabla\phi = \frac{\partial\phi}{\partial x}\mathbf{i} + \frac{\partial\phi}{\partial y}\mathbf{j} + \frac{\partial\phi}{\partial z}\mathbf{k} \qquad \text{where } \phi = f(x, y, z)$$

The gradient of a function is a **vector** that gives the magnitude and direction of the maximum rate of change of that function.

The **divergence** is given by:

$$\nabla \cdot \mathbf{F} = \left(\frac{\partial}{\partial x}\mathbf{i} + \frac{\partial}{\partial y}j + \frac{\partial}{\partial z}\mathbf{k}\right)\cdot\left(M\mathbf{i} + N\mathbf{j} + O\mathbf{k}\right)$$

$$= \frac{\partial M}{\partial x} + \frac{\partial N}{\partial y} + \frac{\partial O}{\partial z}$$

where $F(x, y, z) = M(x, y, z)\mathbf{i} + N(x, y, z)\mathbf{j} + O(x, y, z)\mathbf{k}$

In electrostatics, if an electric field **E** has a non-zero **divergence** at a particular point, then this means that a charge exists there. The divergence is a **scalar** quantity.

Example:

The function V at a particular point in space is given by $V = 2x^2 + 4y^2 + 6z^2$.

(a) Find the rate of change of V at P(2,-2,-1) in a direction towards (0,0,0).
(b) Find the magnitude and direction of the maximum rate of change of V at P.

Solution:

Unit vector from P to (0,0,0) is: $\mathbf{u} = \dfrac{-2\mathbf{i} + 2\mathbf{j} + 1\mathbf{k}}{\sqrt{2^2 + 2^2 + 1^2}} = -\dfrac{2}{3}\mathbf{i} + \dfrac{2}{3}\mathbf{j} + \dfrac{1}{3}\mathbf{k}.$

(a) Rate of change is given by:

$$\nabla V = 4x\mathbf{i} + 8y\mathbf{j} + 12z\mathbf{k}$$
$$= 8\mathbf{i} - 16\mathbf{j} - 12\mathbf{k}$$
$$\nabla V \cdot \mathbf{u} = -\frac{16}{3} - \frac{32}{3} - \frac{12}{3}$$
$$\frac{dV}{ds} = -\frac{60}{3} = -20$$

(b) Maximum rate of change occurs in the direction of:

$$\nabla V = 8\mathbf{i} + -16\mathbf{j} - 12\mathbf{k}$$
$$|\nabla V| = \sqrt{8^2 + 16^2 + 12^2}$$
$$= 21.54$$

The **curl** is given by:

$$\nabla \times \mathbf{v} = \begin{vmatrix} \mathbf{i} & \mathbf{j} & \mathbf{k} \\ \frac{\partial}{\partial x} & \frac{\partial}{\partial y} & \frac{\partial}{\partial z} \\ M & N & O \end{vmatrix}$$

$$= \mathbf{i}\begin{vmatrix} \frac{\partial}{\partial y} & \frac{\partial}{\partial z} \\ N & O \end{vmatrix} - \mathbf{j}\begin{vmatrix} \frac{\partial}{\partial x} & \frac{\partial}{\partial z} \\ M & O \end{vmatrix} + \mathbf{k}\begin{vmatrix} \frac{\partial}{\partial x} & \frac{\partial}{\partial y} \\ M & N \end{vmatrix}$$

$$= \left(\frac{\partial O}{\partial y} - \frac{\partial N}{\partial z}\right)\mathbf{i} + \left(\frac{\partial M}{\partial z} - \frac{\partial O}{\partial x}\right)\mathbf{j} + \left(\frac{\partial N}{\partial x} - \frac{\partial M}{\partial y}\right)\mathbf{k}$$

where $F(x, y, z) = M(x, y, z)\mathbf{i} + N(x, y, z)\mathbf{j} + O(x, y, z)\mathbf{k}$

A physical interpretation of the curl can be obtained by consider a rotating body at point P. Angular velocity may be represented by a vector ω directed along the axis of rotation with magnitude equal to the angular speed.

The linear velocity of P is:

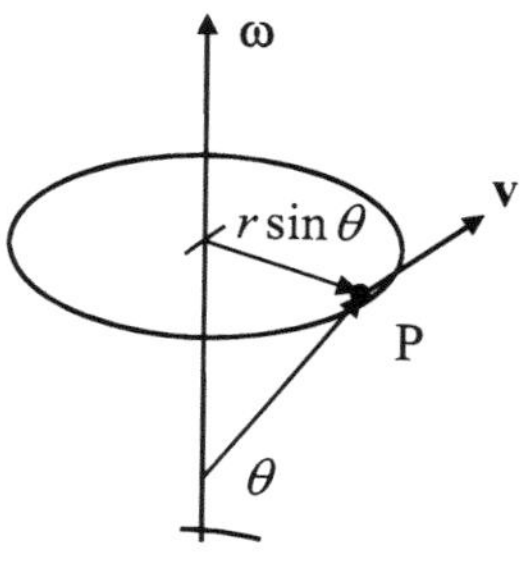

$$\mathbf{v} = \boldsymbol{\omega} \times \mathbf{r}$$

$$|\mathbf{v}| = \omega r \sin\theta$$

Now, $\nabla \times \mathbf{v} = \nabla \times (\boldsymbol{\omega} \times \mathbf{r})$

$$= (\nabla \cdot \mathbf{r})\boldsymbol{\omega} - (\boldsymbol{\omega} \cdot \nabla)\mathbf{r} \quad \text{if } \omega \text{ is a constant}$$

$$\nabla \cdot \mathbf{r} = \left(\frac{\partial}{\partial x}\mathbf{i} + \frac{\partial}{\partial y}\mathbf{j} + \frac{\partial}{\partial z}\mathbf{k}\right) \cdot (x\mathbf{i} + y\mathbf{k} + z\mathbf{j})$$

$$= 1 + 1 + 1 = 3$$

$$(\boldsymbol{\omega} \cdot \nabla)\mathbf{r} = \left(\omega_1 \frac{\partial}{\partial x}\mathbf{i} + \omega_2 \frac{\partial}{\partial y}\mathbf{j} + \omega_3 \frac{\partial}{\partial z}\mathbf{k}\right) \cdot (x\mathbf{i} + y\mathbf{k} + z\mathbf{j})$$

$$= \omega_1\mathbf{i} + \omega_2\mathbf{j} + \omega_3\mathbf{k}$$

$$= \boldsymbol{\omega}$$

Thus $\nabla \times \mathbf{v} = 3\boldsymbol{\omega} - \boldsymbol{\omega}$

$$= 2\boldsymbol{\omega}$$

And so $\boldsymbol{\omega} = \frac{1}{2}(\nabla \times \mathbf{v})$

For a point which is rotating about an axis of rotation, the physical significance of the **curl** is the angular velocity of that point (hence the name "curl").

2.4.7 Maxima and Minima

The function $z = f(x,y)$ has a relative maximum if, for sufficiently small increments h,k (both of which are not zero):

$$f(x_o, y_o) > f(x_o + h, y_o + k)$$

for all such neighbouring points. Similar expressions apply for relative minimum and points of inflection.
At a maximum or minimum, the slope of the tangents in the x and y direction is zero.

$$\frac{\partial z}{\partial x} = \frac{\partial z}{\partial y} = 0$$

To determine if a particular point is a maximum, minimum or a point of inflection, form the quantities:

$$A = \frac{\partial^2 z}{\partial x^2}; B = \frac{\partial^2 z}{\partial x \partial y}; C = \frac{\partial^2 z}{\partial y^2}$$

If: $B^2 - AC > 0$ Point of inflection

$B^2 - AC < 0$ $A < 0$ or $C < 0$ - Maximum

$B^2 - AC < 0$ $A > 0$ or $C > 0$ - Minimum

$B^2 - AC = 0$ Inconclusive result

Example:
Find the extrema of the surface $z = f(x,y) = y^2 + 4xy + 3x^2 + x^3$.

Solution:

$$\frac{\partial z}{\partial x} = 4y + 6x + 3x^2 = 0$$

$$\frac{\partial z}{\partial y} = 2y + 4x = 0$$

$$x = -\frac{1}{2}y$$

$$4y - \frac{6}{2}y + 3\frac{1}{4}y^2 = 0$$

$$y = -\frac{4}{3}; 0$$

$$x = \frac{2}{3}; 0$$

$$A = \frac{\partial^2 z}{\partial x^2} = 6 + 6x = 10; 6$$

$$B = \frac{\partial^2 z}{\partial x \partial y} = 4$$

$$C = \frac{\partial^2 z}{\partial y^2} = 2$$

$B^2 - AC = 16 - 2(10) = -4$
< 0, $A > 0$ } a minimum

$B^2 - AC = 16 - 2(6) = 4$
> 0, $A > 0$ } a saddle point

2.4.8 Lagrange Multipliers

It is often necessary to find the local maxima and minima of a function when the variables are restricted in some way. If the restriction can be expressed $g(x,y,z) = 0$, then we form the new function w such that:

$$w = f(x, y, z) + \lambda g(x, y, z)$$

The factor λ relates the constraint to the function whose minima and maxima are being determined and is called a **Lagrange multiplier**.

The condition for a maximum or minimum becomes:

$$\frac{\partial w}{\partial x} = \frac{\partial w}{\partial y} = \frac{\partial w}{\partial z} = \frac{\partial w}{\partial \lambda} = 0$$

Example:

Find positive values of (x,y,z) whose sum is 20 such that the function xyz^2 is a maximum.

Solution:

Express the function as $F(x, y, z) = xyz^2$ subject to $x + y + z - 20 = 0$.

We form the new function $w = xyz^2 + \lambda(x + y + z - 20)$

Thus:

$$\frac{\partial w}{\partial x} = yz^2 + \lambda = 0 \qquad (1)$$

$$\frac{\partial w}{\partial y} = xz^2 + \lambda = 0 \qquad (2)$$

$$\frac{\partial w}{\partial z} = 2xyz + \lambda = 0 \qquad (3)$$

$$x + y + z = 20 \qquad (4)$$

Four equations with four unknowns: x,y,z,λ

Add x(1)+y(2)+z(3): $4xyz^2 + (x + y + z)\lambda = 0$

By (4): $4xyz^2 + 20\lambda = 0$

$$xyz^2 = -5\lambda$$

By (1):

$$xyz^2 + x\lambda = 0$$

$$-5\lambda + x\lambda = 0$$

$$x = 5$$

By (2):

$$xyz^2 + y\lambda = 0$$

$$y = 5$$

By (3):

$$2xyz^2 + z\lambda = 0$$

$$z = 10$$

The maximum value of $F(x,y,z)$ is thus:

$$(5)(5)(10)^2 = 2500$$

subject to the constraint.

2.4.9 Multiple Least Squares Analysis

Let Z_i be a function that provides fitted values of a dependent variable y_i at each value of an independent variable x_i. Z_i can be a function of many parameters $Z_i = Z_i(x_i : a_1, a_2, \ldots a_j, \ldots a_r)$

We shall assume that initial values or estimates of these parameters are known and that we wish to calculate better values for them. The true value of the parameter a_j is found by adding an error term δa_j to the initial value a_j^{o}.

$$a_j = a_j^{\mathrm{o}} + \delta a_j$$

In terms of the initial guesses, we have $Z_i^{\mathrm{o}} = Z_i\left(x_i : a_1^{\mathrm{o}}, a_2^{\mathrm{o}}, a_3^{\mathrm{o}} \ldots a_r^{\mathrm{o}}\right)$

If the errors δa_j are small, then the function Z_i can be expressed as a Taylor series expansion:

$$Z_i = Z_i^{\mathrm{o}} + \sum_{j=1}^{r} \left(\frac{\partial Z_i^{\mathrm{o}}}{\partial a_j} \right) \delta a_j$$

This is a linear equation in δa_j and is thus amenable to multiple linear least squares analysis.

Now, by **least squares** theory, we wish to minimise the sum of the squares of the differences (or residuals) between the observed values y_i and the fitted values Z_i. The differences $y_i - Z_i$ at each data point i can be weighted by a factor w_i to reflect the error associated with the observed values y_i. Thus, the sum of the squares is expressed as:

$$\mathrm{X}^2 = \sum_{i=1}^{N} w_i [y_i - Z_i]^2$$

$$= \sum_{i=1}^{N} w_i \left[y_i - \left(Z_i^{\mathrm{o}} + \sum_{j=1}^{r} \left(\frac{\partial Z_i^{\mathrm{o}}}{\partial a_j} \right) \delta a_j \right) \right]^2$$

$$= \sum_{i=1}^{N} w_i \left[\left(y_i - Z_i^{\mathrm{o}} \right) - \sum_{j=1}^{r} \left(\frac{\partial Z_i^{\mathrm{o}}}{\partial a_j} \right) \delta a_j \right]^2$$

but $\Delta y_i = y_i - Z_i^{\mathrm{o}}$

thus $$\mathrm{X}^2 = \sum_{i=1}^{N} w_i \left[\Delta y_i - \sum_{j=1}^{r} \left(\frac{\partial Z_i^{\mathrm{o}}}{\partial a_j} \right) \delta a_j \right]^2$$

One possible **weighting factor** $w_i = 1/\sigma^2(y_i)$ – where $\sigma^2(y_i)$ is the standard error (or anything proportional to this) at each data point. This weighting would decrease the influence on the fit for data with the highest standard error.

N is the number of data points, and r is the number of fitting parameters.

The objective is to minimise this sum with respect to the values of the error terms δa_j, thus we set the derivative of X^2 with respect to δa_j to zero:

$$\frac{\partial \mathrm{X}^2}{\partial(\delta a_j)} = 0 = \sum_{i=1}^{N} w_i \left[\left(\Delta y_i - \sum_{j=1}^{r} \left(\frac{\partial Z_i^{\mathrm{o}}}{\partial a_j} \right) \delta a_j \right) \left(\frac{\partial Z_i^{\mathrm{o}}}{\partial a_j} \right) \right]$$

This expression can be expanded by considering a few examples of j. Letting $j = 1$, we obtain:

$$\sum_{i=1}^{N} w_i \Delta y_i \left(\frac{\partial Z_i^{o}}{\partial a_1} \right) = \delta a_1 \sum_{i=1}^{N} w_i \left(\frac{\partial Z_i^{o}}{\partial a_1} \right)^2 + \delta a_2 \sum_{i=1}^{N} w_i \left(\frac{\partial Z_i^{o}}{\partial a_1} \frac{\partial Z_i^{o}}{\partial a_2} \right) + \ldots. + \delta a_r \sum_{i=1}^{N} w_i \left(\frac{\partial Z_i^{o}}{\partial a_1} \frac{\partial Z_i^{o}}{\partial a_r} \right)$$

At j equal to some arbitrary value of k, we obtain:

$$\sum_{i=1}^{N} w_i \Delta y_i \left(\frac{\partial Z_i^{o}}{\partial a_k} \right) = \delta a_1 \sum_{i=1}^{N} w_i \left(\frac{\partial Z_i^{o}}{\partial a_k} \frac{\partial Z_i^{o}}{\partial a_1} \right) + \delta a_2 \sum_{i=1}^{N} w_i \left(\frac{\partial Z_i^{o}}{\partial a_k} \frac{\partial Z_i^{o}}{\partial a_2} \right) + \ldots \delta a_k \sum_{i=1}^{N} w_i \left(\frac{\partial Z_i^{o}}{\partial a_k} \right)^2 + \ldots + \delta a_r \sum_{i=1}^{N} w_i \left(\frac{\partial Z_i^{o}}{\partial a_k} \frac{\partial Z_i^{o}}{\partial a_r} \right)$$

In matrix notation, the sums for each error term δa from 1 to r is expressed:

$$\lfloor Y_j \rfloor = \lfloor A_{jk} \rfloor \lfloor X_j \rfloor$$

$$\begin{bmatrix} Y_1 \\ Y_2 \\ Y_k \\ \cdot \\ Y_r \end{bmatrix} = \begin{bmatrix} A_{11} & \cdot & \cdot & \cdot & A_{1r} \\ \cdot & \cdot & \cdot & \cdot & \cdot \\ \cdot & \cdot & A_{jk} & \cdot & \cdot \\ \cdot & \cdot & \cdot & \cdot & \cdot \\ A_{r1} & \cdot & \cdot & \cdot & A_{rr} \end{bmatrix} \begin{bmatrix} X_1 \\ X_2 \\ X_k \\ \cdot \\ X_r \end{bmatrix}$$

where $Y_j = \sum_{i=1}^{N} w_i \left[\Delta y_i \frac{\partial Z_i^{o}}{\partial a_j} \right]$

$$A_{jk} = A_{kj} = \sum_{i=1}^{N} w_i \left[\frac{\partial Z_i^{o}}{\partial a_j} \frac{\partial Z_i^{o}}{\partial a_k} \right]$$

$$X_j = \delta a_j$$

and $\Delta y_i = y_i - Z_i^{o}$

The solution is the matrix **X** that contains the error terms δa_j to be minimised. Thus:

$$\lfloor X_j \rfloor = \lfloor A_{jk} \rfloor^{-1} \lfloor Y_j \rfloor$$

When values of δa_j are calculated, they are added to the initial values a_j^{o} to give the fitted values a_j:

$$a_j^{1} = a_j^{o} + L \delta a_j$$

The process may then be repeated until the error terms δa_j become sufficiently small, indicating that the parameters a_j have converged to their optimum value. L in the above equation is a relaxation factor that is applied to error terms to prevent instability during the initial phases of the refinement process.

2.4.10 Constraints

It is sometimes necessary to constrain one or more of the fitting parameters in some manner. To see how this is done, consider the first two parameters to be fitted a_1 and a_2 out of a total of r parameters. We may decide that the change in a_2 is to be half of that of a_1. The constrained value of δa_2 is written δa_{r2} and is given by:

$$\delta a_{r1} = (1)\delta a_1$$
$$\delta a_{r2} = (0.5)\delta a_1$$

where the bracketed terms are **constraint factors** to be attached to each parameter. When the derivatives are found for parameter 1, we must calculate:

$$\frac{\partial Z_i{}^{o}}{\partial a_1} = \frac{\partial Z_i{}^{o}}{\partial a_{r1}}(1) + \frac{\partial Z_i{}^{o}}{\partial a_{r2}}(0.5)$$

The derivative associated with a_2 is added in with that for a_1. This has the effect of effectively reducing the number of independent variables. The above expression can more generally be written as:

$$\frac{\partial Z_i{}^{o}}{\partial a_j} = \frac{\partial Z_i{}^{o}}{\partial a_{r1}}\left(\frac{\partial a_{r1}}{\partial a_1}\right) + \frac{\partial Z_i{}^{o}}{\partial a_{r2}}\left(\frac{\partial a_{r2}}{\partial a_1}\right) + \ldots$$

Constraint factors: (1) (0.5)

where a parameter a_j now represents the combination of two or more other parameters, the value of the constrained parameters is found from:

$$\delta a_{rj} = \delta a_n \left(\frac{\partial a_{rj}}{\partial a_n}\right)$$

(where δa_n is as calculated from matrix inversion, $X_j = \delta a_j$)

In the present example,

$$\delta a_{r1} = \delta a_1 \left(\frac{\partial a_{r1}}{\partial a_1}\right) = (1)\delta a_1$$

$$\delta a_{r2} = \delta a_1 \left(\frac{\partial a_{r2}}{\partial a_1}\right) = (0.5)\delta a_1$$

$$a_j{}^{1} = a_j{}^{o} + L\delta a_{rj}$$

Example:

Data in an experiment is known to follow a power law of the form:

$$P = A(h - h_r)^m$$

Where P and h are experimentally observed quantities and A, h_r and m must be determined from the data. Using the notation given earlier, formulate the necessary equations to accomplish multiple least square analysis and write a computer program to accomplish the fitting procedure.

Solution:

The function is expressed in terms of Z_i, x_i, and the constants a_1, a_2 and a_3:

$$Z_i = a_1(x_i - a_2)^{a_3}$$

$$a_1 = A$$

$$a_2 = h_r$$

$$a_3 = m$$

The following equations are therefore required:

$$Z_i^{o} = a_1^{o}\left(x_i - a_2^{o}\right)^{a_3^{o}}$$

$$\frac{\partial Z_i^{o}}{\partial a_1} = \left(x_i - a_2^{o}\right)^{a_3^{o}}$$

$$\frac{\partial Z_i^{o}}{\partial a_2} = -a_3^{o} a_1^{o}\left(x_i - a_2^{o}\right)^{a_3^{o}-1}$$

$$\frac{\partial Z_i^{o}}{\partial a_3} = a_1^{o}\left(x_i - a_2^{o}\right)^{a_3^{o}} \ln\left(x_i - a_2^{o}\right)$$

```
For i = 1 To NumDataPoints
 Xi = XR(i)           ' raw data x
 yi = YR(i)           ' raw data y
 For j = 1 To 3
  dZda(j) = 0
 Next j
 a10 = Coefficients(1) ' initial values
 a20 = Coefficients(2)
 a30 = Coefficients(3)
 Zi0 = a10 * (Xi - a20) ^ a30
 dZda(1) = dZda(1) + ((Xi - a20) ^ a30)
 dZda(2) = dZda(2) + (-a30 * a10 * (Xi - a20) ^ (a30 - 1))
 dZda(3) = dZda(3) + (a10 * (Xi - a20) ^ a30 * Log( (Xi - a20)))
 Dyi = yi - Zi0
 For j = 1 To 3
  Y1(j) = Y1(j) + (Dyi * dZda(j))
   For k = 1 To R
    AR(j, k) = AR(j, k) +  dZda(j) * dZda(k)
   Next k
 Next j
Next i
' Invert and multiply
MatrixInvert(AR, AInv)
MatrixMultiply AInv, Y1, X1
' Update coefficients
For j = 1 To 3
 Coefficients(j) = Coefficients(j) + Relaxation(j) * (X1(j))
Next j
```

2.5 Multiple Integrals

Summary

$$\int_{P_1}^{P_2} f(x,y)\,ds = \int_C f(x,y)\,ds$$ Line integral

$$\iint_R f(x,y)\,dA = \int_{x_1}^{x_2}\int_{y_1(x)}^{y_2(x)} f(x,y)\,dy\,dx$$ Double integral

Triple integral
$$\iiint_V f(x,y,z)\,dV = \int_{x_1}^{x_2}\int_{y_1(x)}^{y_2(x)}\int_{z_1(x)}^{z_2(x)} f(x,y,z)\,dz\,dy\,dx$$

$$\iint_S g(x,y,z)\,dS = \iint_R g(x,y,z)\sec\gamma\,dx\,dy$$ Surface integral

$$\phi = \oiint_S \mathbf{E}\cdot d\mathbf{S} = \frac{q}{\varepsilon_0}$$ Gauss' law

$$\oiint_S \mathbf{F}\cdot\mathbf{n}\,dS = \iiint_V \nabla\cdot\mathbf{F}\,dV$$ Divergence theorem

$$\oint_C \mathbf{A}\cdot d\mathbf{r} = \iint_S (\nabla\times\mathbf{A})\cdot\mathbf{n}\,dS$$ Stokes' theorem

$$\oint_C M\,dx + N\,dy = \iint_R \left(\frac{\partial N}{\partial x} - \frac{\partial M}{\partial y}\right)dx\,dy$$ Green's theorem

2.5.1 Line Integrals

Consider a surface $z = f(x,y)$. Let a line increment on the curve C in the xy plane be given by Δs.

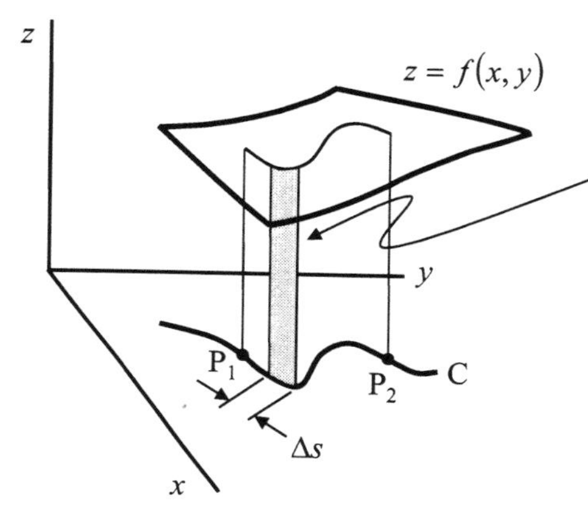

The area of the band formed from the line increment to the surface is given by the product:

$$f(x, y)\Delta s$$

The total area under the surface along the curve from P_1 to P_2 is thus the limit of the sum as Δs goes to zero.

$$\lim_{\substack{n \to \infty \\ \Delta s \to 0}} \sum f(x, y)\Delta s$$

This limit is called the **line integral** of $f(x,y)$ along the curve C from P_1 to P_2 and is written:

$$\int_{P_1}^{P_2} f(x, y)ds = \int_C f(x, y)ds$$

→ Short-hand notation indicating an interval (or group of intervals) along the curve C

If the curve C can be represented as a **vector function** in terms of the path or arc length parameter s $\quad \mathbf{r}(s) = x(s)\mathbf{i} + y(s)\mathbf{j}$

Then the line integral is written:

$$\int_{P_1}^{P_2} f(x(s), y(s))ds$$

If the curve C is given in terms of the parameter t such that $\mathbf{r}(t) = x(t)\mathbf{i} + y(t)\mathbf{j}$ then:

$$\int_C f(x, y)ds = \int_{P_1}^{P_2} f(x(t), y(t))\frac{ds}{dt}dt$$

$$\text{where } \frac{ds}{dt} = \sqrt{\frac{d\mathbf{r}}{dt} \cdot \frac{d\mathbf{r}}{dt}} = \sqrt{\left(\frac{dx}{dt}\right)^2 + \left(\frac{dy}{dt}\right)^2}$$

The line integral can be extended to three dimensions where $F = f(x,y,z)$:

$$\int_C f(x, y, z)ds$$

It is possible to express the line integral of $F(x,y)$ along C with respect to x (or y) instead of the arc length s.

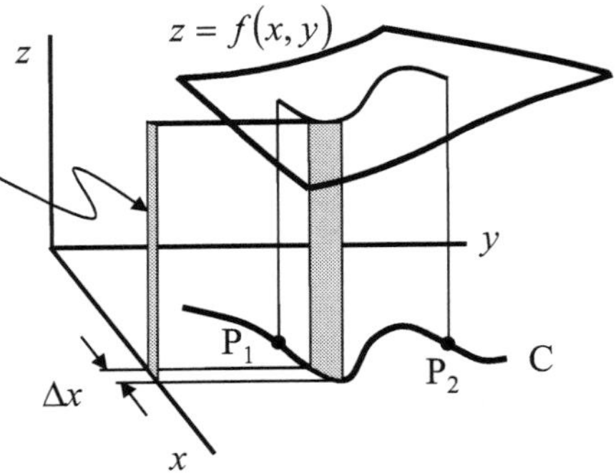

This gives the projected area on to the xz plane (for Δx) or the yz plane (for Δy).

$z = f(x, y)$, z, y, x, Δx, P_1, P_2, C

In this case, the surface z may be represented by the function $f(x,y)$ in the x direction and $g(x,y)$ in the y direction. The line integral becomes:

$$\int_C F(x,y)ds = \int_C f(x,y)dx + \int_C g(x,y)dy$$

$$= \int_C \left(f(x,y,z)dx + g(x,y,z)dy\right)$$

In the general case for three dimensions, the line integral is given by the sum of the projected line integrals and the functions f, g and h:

$$\int_L F(x,y,z)ds = \int_C f(x,y,z)dx + \int_C g(x,y,z)dy + \int_C h(x,y,z)dz$$

$$= \int_C \left(f(x,y,z)dx + g(x,y,z)dy + h(x,y,z)dz\right)$$

The functions f, g and h may be components of a vector function $\mathbf{F}$:

$$\mathbf{F} = f(x,y,z)\mathbf{i} + g(x,y,z)\mathbf{j} + h(x,y,z)\mathbf{k}$$
$$= F_1\mathbf{i} + F_2\mathbf{j} + F_3\mathbf{k}$$

Now, $F_1 dx + F_2 dy + F_3 dz = \left(F_1 \frac{dx}{ds} + F_2 \frac{dy}{ds} + F_3 \frac{dz}{ds}\right)ds$

$$= \left(\mathbf{F} \cdot \frac{d\mathbf{r}}{ds}\right)ds$$

$$d\mathbf{r} = dx\mathbf{i} + dy\mathbf{j} + dz\mathbf{k}$$

$$\frac{d\mathbf{r}}{ds} = \frac{dx}{ds}\mathbf{i} + \frac{dy}{ds}\mathbf{j} + \frac{dz}{ds}\mathbf{k}$$

Thus, the line integral $\int_C \mathbf{F} \cdot \frac{d\mathbf{r}}{ds}ds$ can be written: $\int_C \mathbf{F} \cdot d\mathbf{r}$

If the integral is taken around a closed curve, then we write $\oint_C \mathbf{F} \cdot d\mathbf{r}$

This integral is sometimes referred to as the **circulation** of **F** around C.

In general, the value of the line integral depends upon the path of integration. If, however, the **vector field F** can be represented by:

$$\mathbf{F} = \nabla \phi$$

then the line integral is independent of the path. In this case, ϕ is called the **scalar potential**. The field **F** is said to be a **conservative field** and $d\phi$ is called an **exact differential**.

Let $\mathbf{F} = \nabla \phi$

$$\text{Then } \int \mathbf{F} \cdot d\mathbf{r} = \int_{P_1}^{P_2} \nabla \phi \cdot d\mathbf{r} \qquad d\mathbf{r} = dx\mathbf{i} + dy\mathbf{j} + dz\mathbf{k}$$

$$= \int_{P_1}^{P_2} \left(\frac{\partial \phi}{\partial x}\mathbf{i} + \frac{\partial \phi}{\partial y}\mathbf{j} + \frac{\partial \phi}{\partial z}\mathbf{k} \right) \cdot \left(dx\mathbf{i} + dy\mathbf{j} + dz\mathbf{k} \right)$$

$$= \int_{P_1}^{P_2} \frac{\partial \phi}{\partial x} dx + \frac{\partial \phi}{\partial y} dy + \frac{\partial \phi}{\partial z} dz$$

$$= \int_{P_1}^{P_2} d\phi$$

$$= \phi(x_2, y_2, z_2) - \phi(x_1, y_1, z_1)$$

The integral depends only on the end points P_1 and P_2 and not on the path joining them (since the form or equation of the curve C is not given in the above).

If the path of integration C is a **closed curve**, *and* if **F** is conservative, then:

$$\oint_C \mathbf{F} \cdot d\mathbf{r} = 0$$ where the circle in the integral sign denotes integration around a closed curve.

If **F** is conservative, then: $\nabla \times \mathbf{F} = 0$.

$$\begin{vmatrix} \mathbf{i} & \mathbf{j} & \mathbf{k} \\ \dfrac{\partial}{\partial x} & \dfrac{\partial}{\partial y} & \dfrac{\partial}{\partial z} \\ F_1 & F_2 & F_3 \end{vmatrix} = 0$$ i.e., a conservative field has no **curl**.

2.5.2 Electrical Potential

Consider a positive charge q_2 which is moved from A to B in an electric field created by q_1 located at the origin. We wish to compute the work done on (positive) or by (negative) the charge during this movement.

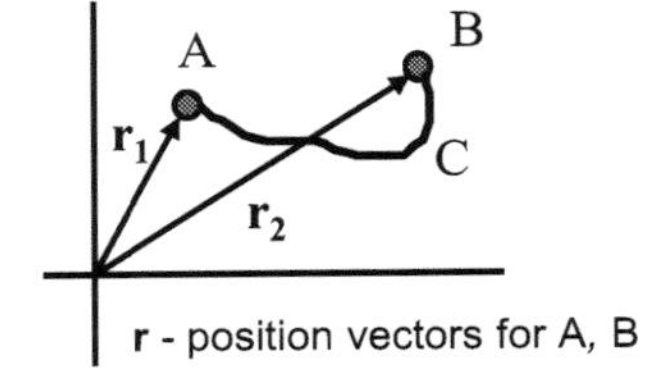

$$\mathbf{F} = \frac{q_1 q_2 \mathbf{u}}{4\pi\varepsilon_o r^2} = q_2\mathbf{E}$$

← unit vector in direction of *r* **u**=**u**(*r*).

$$\mathbf{E} = \frac{q_1}{4\pi\varepsilon_o r^2}\mathbf{u}$$ Newtons/Coulomb

$\Delta V = -\mathbf{E}.d\mathbf{l}$ By definition, ΔV is the work per unit charge in moving the charge through a distance *dl*.

$V_2 - V_1 = -\int_A^B \mathbf{E}.d\mathbf{l}$ $V_2 - V_1$ is the change in electrical potential is the line integral from A to B along C.

$$= -\int_A^B \frac{q_1}{4\pi\varepsilon_o r^2}\mathbf{u}.d\mathbf{l}$$

$$= -\frac{q_1}{4\pi\varepsilon_o}\int_A^B \frac{1}{r^2}\mathbf{u}.d\mathbf{l}$$

$$\mathbf{u}.d\mathbf{l} = dl\cos\theta$$

$$= -\frac{q_1}{4\pi\varepsilon_o}\int_A^B \frac{1}{r^2}\cos\theta dl$$

$$d\mathbf{r} = dl\cos\theta$$

$$= -\frac{q_1}{4\pi\varepsilon_o}\int_{r_1}^{r_2} \frac{1}{r^2}dr$$

$$= \frac{q_1}{4\pi\varepsilon_o}\left[\frac{1}{r_2} - \frac{1}{r_1}\right]$$ which shows that ΔV is independent of path, but only depends on r_1 and r_2 (i.e., the end points A and B)

If the end points are the same (i.e., the line integral is taken around a closed curve then $\oint \mathbf{E}.d\mathbf{l} = 0$

The field E is a **conservative field** and V is the **gradient** of the scalar potential. $\mathbf{E} = -\nabla V$

E is conservative.
V is a "scalar potential."

If *dl* is in the same direction as ∇V, then *dV*/*dl* is a maximum

2.5.3 Work Done by a Force

Consider a force $\mathbf{F}(x,y,z)$ acting on a particle which moves along a curve C. We may wish to calculate the work done by the force on the particle.

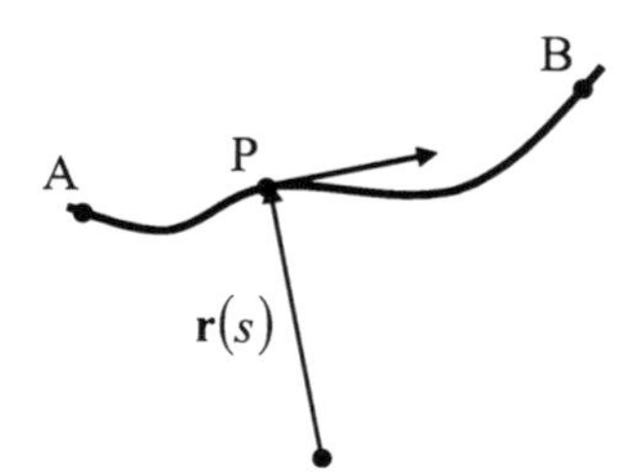

We divide the curve up into n sub-intervals of length Δs. The position vector $\mathbf{r}(s)$ gives location of the points on the curve in terms of the arc length s.

$$\mathbf{r}(s_o + \Delta s) = \mathbf{r}(s_o) + \Delta \mathbf{r}$$

For some small increment of arc length Δs, $\mathbf{F}$ can be assumed to be approximately constant and so the work done on the particle over that increment is:

$$W = \left(\mathbf{F} \cdot \frac{d\mathbf{r}}{ds} \right) \Delta s$$

the component of $\mathbf{F}$ along the curve times the distance Δs along the curve

The total work is the summation of the work done in each increment for the path from a point at A to another point at B:

$$W = \lim_{\Delta s \to 0} \sum_{j=1}^{n} \left(\mathbf{F} \cdot \frac{d\mathbf{r}}{ds} \right) \Delta s$$

$$= \int_{\mathrm{A}}^{\mathrm{B}} \left(\mathbf{F} \cdot \frac{d\mathbf{r}}{ds} \right) ds$$

$$= \int_{\mathrm{C}} \mathbf{F} \cdot d\mathbf{r}$$

If some other parameter (other than s) is used, then (e.g., t for time) we obtain:

$$W = \int_{\mathrm{C}} \mathbf{F} \cdot d\mathbf{r} = \int_{t_o}^{t_1} \left(\mathbf{F} \cdot \frac{d\mathbf{r}}{dt} \right) dt$$

$$= \int_{t_o}^{t_1} \mathbf{F} \cdot \mathbf{v}\, dt$$

2.5.4 Double Integral

Consider a surface $z = f(x,y)$. Let an incremental area of a region R in the *xy* plane be given by ΔA.

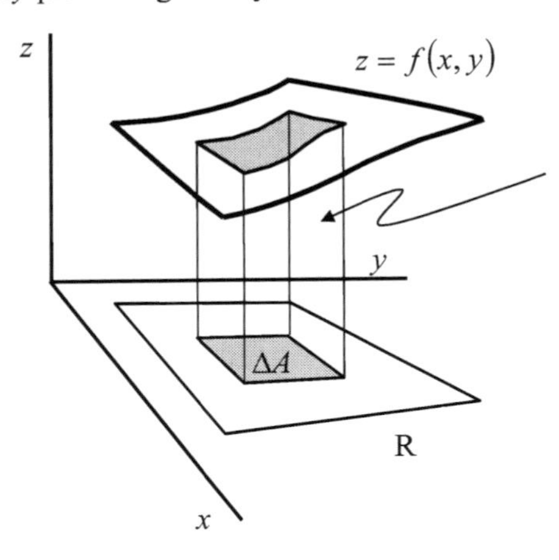

The volume of the column from the surface *z* over the area ΔA is given by the product: $f(x,y)\Delta A$

The total volume under the surface is thus the limit of the sum as ΔA goes to zero.

$$\lim_{\substack{n\to\infty \\ \Delta A\to 0}} \sum f(x,y)\Delta A$$

This limit is called the **double integral** of $f(x,y)$ over the region R.

$$\iint_R f(x,y)dA$$

Expressed in terms of increments in the *x* and *y* axes directions, we obtain:

$$\iint_R f(x,y)dA = \int_{x_1}^{x_2}\int_{y_1(x)}^{y_2(x)} f(x,y)dydx$$

This integral is evaluated by holding *x* fixed and performing the inside sum with respect to *dy*, and then integrating that expression with respect to *x*. In double integrals, the integration is performed over a region R.

A physical interpretation of the **double integral** is thus the **volume** under the surface over the region of integration on the *xy* plane.

2.5.5 Triple Integral

For functions of three variables, the region of integration is a volume V so that the **triple integral** is given by:

$$\iiint_V f(x,y,z)dV = \int_{x_1}^{x_2}\int_{y_1(x)}^{y_2(x)}\int_{z_1(x)}^{z_2(x)} f(x,y,z)dz\,dy\,dx$$

If $\rho(x,y,z)$ is a **density function** giving the density of a solid at any point, then the **mass** of a volume element is given by: $dM = \rho(x,y,z)dV$.

The total mass is: $M = \iiint_V dM = \iiint_V \rho(x,y,z)dV$

Example:

Find the volume and centre of mass of the solid of density ρ described by $z = 4 - x^2$ and the planes x = 0, y = 6 and z = 0

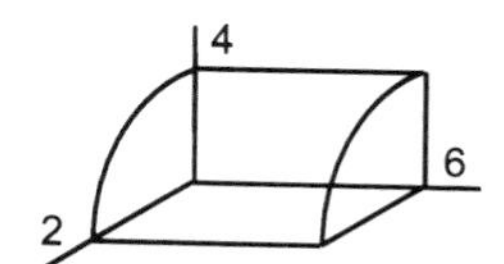

Solution:

$$dM = \rho(x,y,z)dV$$

$$M = \iiint_V dM = \iiint_V \rho dV = \rho\int_0^2\int_0^6\int_0^{4-x^2} dxdydz = \rho\int_0^2\int_0^6 (4-x^2)dxdy = \rho\int_0^2 24 - 6x^2 dx$$

$$= 32\rho$$

$$M\bar{x} = \iiint_S xdM$$

$$= \rho\int_0^2\int_0^6\int_0^{4-x^2} x\,dz\,dy\,dx$$

$$= \rho\int_0^2\int_0^6 4x - x^3 dzdydx$$

$$= \rho\int_0^2 24x - 6x^3 dx$$

$$= 24\rho$$

$$\bar{x} = \frac{24}{32} = \frac{3}{4}$$

$$M\bar{z} = \rho\iiint zdM$$

$$\rho\int_0^2\int_0^6\int_0^{4-x^2} z\,dz\,dy\,dx$$

$$= \rho\int_0^2\int_0^6 \frac{1}{2}(4-x^2)^2 dydx$$

$$= \rho\int_0^2 3(4-x^2)^2 dx$$

$$= 51.2\rho$$

$$\bar{z} = \frac{51.2}{32}$$

$$M\bar{y} = \iiint ydM$$

$$= \rho\int_0^2\int_0^6 4y - x^2y\,dydx$$

$$= \rho\int_0^2\left[\frac{4}{2}y^2 - \frac{x^2}{2}y^2\right]_0^6 dx$$

$$= \rho\int_0^2 72 - 18x^2 dx$$

$$= 96\rho$$

$$\bar{y} = 3$$

2.5.6 Surface Integrals

A **surface integral** is a generalisation of the **double integral**. Double integrals are integrals over a flat surface (a region R of integration in the *xy* plane). Surface integrals are double integrals over a curved surface S.

In this case, S is a curved surface which has a projection R onto the *xy* plane. Let the equation of the surface S be: $z = f(x, y)$

Let $\phi(x,y,z)$ be the function to be integrated (not shown in the figure) over the surface. We divide *S* into increments ΔS and let ΔT be a tangent plane to ΔS at $P(x,y,z)$.

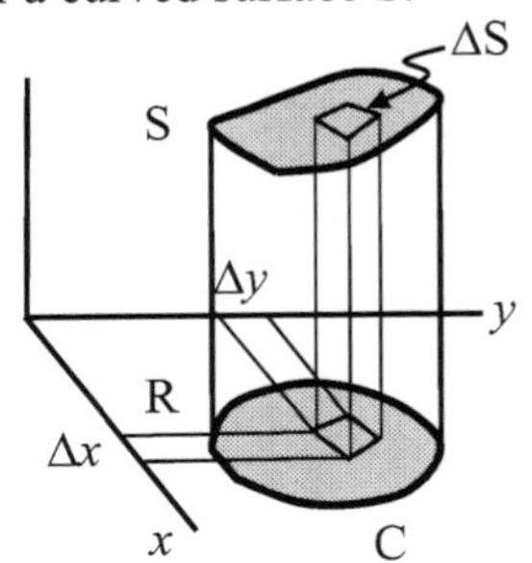

Consider a point P and let **a** be a vector tangent to the surface on the *xz* plane. Let **b** be a vector tangent to the surface on the *yz* plane.

$$\mathbf{a} = \Delta x\mathbf{i} + \Delta z\mathbf{k}$$

But: $$\frac{\Delta z}{\Delta x} = \frac{\partial z}{\partial x}$$

$$\Delta z = \frac{\partial z}{\partial x}\Delta x$$

Thus: $$\mathbf{a} = \Delta x\mathbf{i} + \frac{\partial z}{\partial x}\Delta x\mathbf{k} \quad \text{and} \quad \mathbf{b} = \Delta y\mathbf{j} + \frac{\partial z}{\partial y}\Delta y\mathbf{k}$$

The area of $\Delta T = |\mathbf{a} \times \mathbf{b}|$

$$\mathbf{a} \times \mathbf{b} = \begin{vmatrix} \mathbf{i} & \mathbf{j} & \mathbf{k} \\ \Delta x & 0 & \frac{\partial z}{\partial x}\Delta x \\ 0 & \Delta y & \frac{\partial z}{\partial y}\Delta y \end{vmatrix}$$

$$\mathbf{a} \times \mathbf{b} = \left(-\frac{\partial z}{\partial x}\Delta x\Delta y\right)\mathbf{i} + \left(\frac{\partial z}{\partial y}\Delta x\Delta y\right)\mathbf{j} + (\Delta x\Delta y)\mathbf{k}$$

$$= \sqrt{\left(\frac{\partial z}{\partial x}\right)^2 \Delta x^2\Delta y^2 + \left(\frac{\partial z}{\partial y}\right)^2 \Delta x^2\Delta y^2 + \Delta x^2\Delta y^2}$$

$$= \left(\sqrt{\left(\frac{\partial z}{\partial x}\right)^2 + \left(\frac{\partial z}{\partial y}\right)^2 + 1}\right)\Delta x\Delta y$$

$= \Delta T \approx \Delta S$ if Δx and Δy are small

$$dS = \left(\sqrt{\left(\frac{\partial z}{\partial x}\right)^2 + \left(\frac{\partial z}{\partial y}\right)^2 + 1}\right) dxdy \text{ as } \Delta x \text{ and } \Delta y \to 0$$

Thus, *dS* is expressed in terms of *dx* and *dy*.

The **surface integral** of $\phi(x,y,z)$ over S can thus be written:

$$\oiint_S \phi(x,y,z)dS = \iint_R \phi(x,y,z)\sqrt{\left(\frac{\partial z}{\partial x}\right)^2+\left(\frac{\partial z}{\partial y}\right)^2+1}\ dxdy$$

where R is the projection of S on the xy plane.

Now, the vector result of $\mathbf{a}\times\mathbf{b}$ is a vector $\perp$ to ΔT, the magnitude of which is the area of ΔT. Let $\mathbf{n}$ be an outward normal unit vector in this same direction.

$$\mathbf{n}=\frac{\mathbf{a}\times\mathbf{b}}{|\mathbf{a}\times\mathbf{b}|}=\frac{-\frac{\partial z}{\partial x}\Delta x\Delta y\mathbf{i}-\frac{\partial z}{\partial y}\Delta x\Delta y\mathbf{j}+\Delta x\Delta y\mathbf{k}}{\left(\sqrt{\left(\frac{\partial z}{\partial x}\right)^2+\left(\frac{\partial z}{\partial y}\right)^2+1}\right)(\Delta x\Delta y)}$$

n

k

γ

ΔS

S

But $dS=\left(\sqrt{\left(\frac{\partial z}{\partial x}\right)^2+\left(\frac{\partial z}{\partial y}\right)^2+1}\right)dxdy$ and also $\mathbf{n}\cdot\mathbf{k}=\dfrac{1}{\sqrt{\left(\frac{\partial z}{\partial x}\right)^2+\left(\frac{\partial z}{\partial y}\right)^2+1}}$

Thus: $dS=\dfrac{1}{\mathbf{n}\cdot\mathbf{k}}dxdy$

$$\iint_S \phi(x,y,z)dS=\iint_R \phi(x,y,z)\frac{1}{\mathbf{n}\cdot\mathbf{k}}dxdy$$

The normal unit vector $\mathbf{n}$ may also be represented by $\mathbf{n}=l\mathbf{i}+m\mathbf{j}+n\mathbf{k}$

$$|\mathbf{n}\cdot\mathbf{k}|=|\mathbf{n}||\mathbf{k}|\cos\gamma$$

$$=\cos\gamma \quad \text{since } |\mathbf{n}|,\ |\mathbf{k}|=1$$

$$\frac{1}{|\mathbf{n}\cdot\mathbf{k}|}=\sec\gamma$$

where γ is the angle between the outward normal vector $\mathbf{n}$ and $\mathbf{k}$ (which is in the direction of the z axis).

So:

$$\cos\gamma=\frac{1}{\sqrt{\left(\frac{\partial z}{\partial x}\right)^2+\left(\frac{\partial z}{\partial y}\right)^2+1}}$$

Thus: $$\iint_S \phi(x,y,z)dS=\iint_R \phi(x,y,z)\sec\gamma\, dxdy$$

Similar treatments hold for a projection of S onto the yz or xz planes if that is desired.

In the case of **vector functions** that require integrating over a surface,

$$\mathbf{F} = M(x,y,z)\mathbf{i} + N(x,y,z)\mathbf{j} + O(x,y,z)\mathbf{k}$$

If **n** is an outward normal unit vector at $P(x,y,z)$, then for the special case of integrating the normal component of **F** over the surface S, we have:

$\mathbf{F} \cdot \mathbf{n}$ is the normal component of **F** at P(x,y,z)

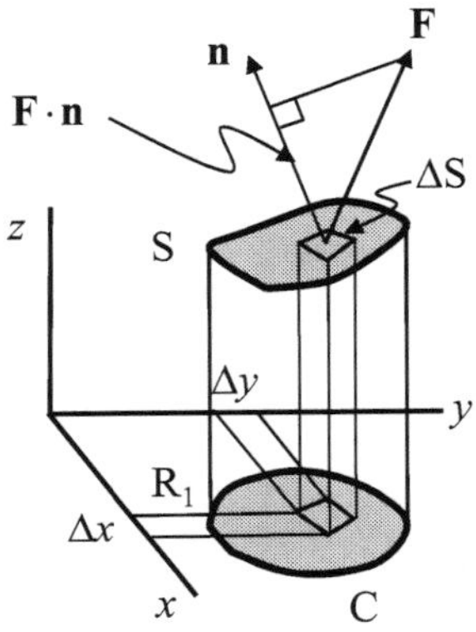

The **surface integral** of the normal component of **F** over S is:

$$\iint_S \mathbf{F} \cdot \mathbf{n} dS = \iint_S (Ml + Nm + On) dS$$

$$= \iint_S (M\cos\alpha + N\cos\beta + O\cos\gamma) dS$$

$$= \iint_{R_3} M dy dx + \iint_{R_2} N dx dz + \iint_{R_1} O dx dy$$

$\longrightarrow \cos\gamma \sec\gamma = 1$

where: R_3 is the projection on the yz plane
R_2 is the projection on the xz plane
R_1 is the projection on the xy plane

This integral is sometimes referred to as the **flux** of **F** over S.

The physical interpretation of surface integrals depends upon the nature of the function being integrated. For example:

Area of the surface $A = \iint_S dS$

Mass of the surface $M = \iint_S \rho(x,y,z) dS$

Moment of inertia $I = \iint_S (x^2 + y^2) \rho dS$

Surface integrals are generally found by solving double integrals of parametric equations of the surface.

2.5.7 Gauss' Law

1. Calculation of the field around a point charge:

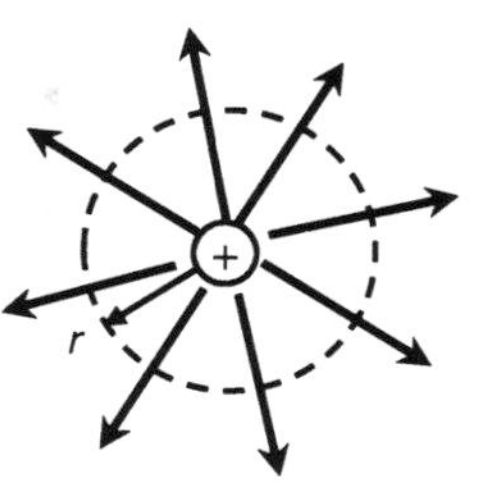

$A = 4\pi r^2$ Surface area of a sphere radius r

$E \propto \frac{N}{A}$ by definition

$EA \propto N$ Electric flux

$$E = \frac{1}{4\pi\varepsilon_o}\frac{q}{r^2}$$

$$EA = \frac{1}{4\pi\varepsilon_o}\frac{q}{r^2}\left(4\pi r^2\right)$$

$= \frac{q}{\varepsilon_o}$ ← independent of r but proportional to N

$= \phi$ Electric flux

2. Consider now the portion ΔS of non-spherical surface of area S:

The elemental area on a spherical surface surrounding the charge is given by:

$$\Delta S\cos\theta$$

Total flux (i.e., number of flux lines) over the whole surface S is thus the same as that over the spherical surface A:

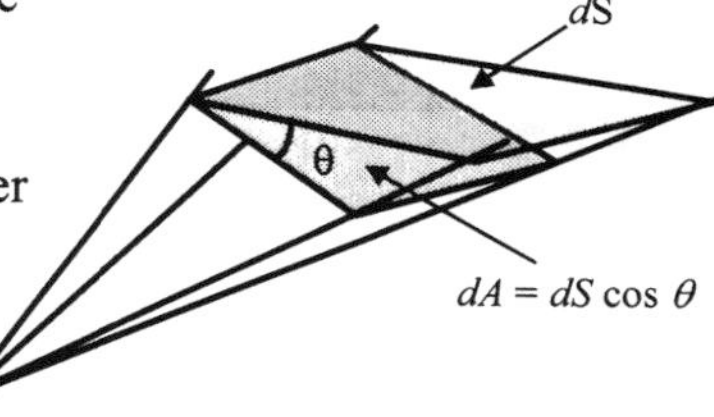

$$\phi = \oint\!\!\oint_S \frac{1}{4\pi\varepsilon_o}\frac{q}{r^2}\cos\theta dS$$

$$= \frac{1}{4\pi r^2}\frac{q}{\varepsilon_o}\oint\!\!\oint_S \cos\theta dS$$

$$= \frac{1}{4\pi r^2}\frac{q}{\varepsilon_o}\oint\!\!\oint_A dA$$

$$= \frac{1}{4\pi r^2}\frac{q}{\varepsilon_o}4\pi r^2$$ since the surface area of a sphere $\oint\!\!\oint_A dA = 4\pi r^2$

$$= \frac{q}{\varepsilon_o}$$

Integral form of **Gauss' law**

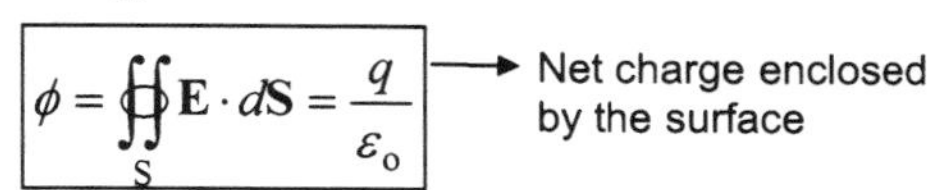

$$\boxed{\phi = \oint\!\!\oint_S \mathbf{E}\cdot d\mathbf{S} = \frac{q}{\varepsilon_o}}$$ → Net charge enclosed by the surface

2.5.8 Divergence Theorem

Let a small elemental volume ΔV have a charge distributed uniformly within it. The total charge within the volume is given by Gauss' law:

$$q = \iint_S \varepsilon_0 \mathbf{E} \cdot d\mathbf{S} = \rho \Delta V$$

where ρ is the charge density within the volume ΔV. The charge density at a single point in space is therefore expressed as the limit:

$$\rho = \lim_{\Delta V \to 0} \frac{\iint_S \varepsilon_0 \mathbf{E} \cdot d\mathbf{S}}{\Delta V}$$

This limit is called the divergence of the field **E**. If a field has a **divergence** at a particular point, then this means that a charge exists there. Thus:

$$\rho = \nabla \cdot \varepsilon \mathbf{E}$$

which is called the differential form of **Gauss' law**.

The total charge q within a total volume V can also be expressed in terms of the summation of the charge densities over all the volume elements within it.

$$q = \iiint_V \rho dV$$

and so:

$$q = \iint_S \varepsilon_0 \mathbf{E} \cdot d\mathbf{S} = \iiint_V \nabla \cdot \varepsilon_0 \mathbf{E}\, dV$$

More generally, we can say: If $\mathbf{F} = M\mathbf{i} + N\mathbf{j} + O\mathbf{k}$, then

$$\boxed{\oiint_S \mathbf{F} \cdot \mathbf{n} dS = \iiint_V \nabla \cdot \mathbf{F}\, dV}$$

n is a positive unit outer normal vector.
S is a closed surface.

or

$$\iint_S (M\cos\alpha + N\cos\beta + O\cos\gamma) dS = \iiint_V \left(\frac{\partial M}{\partial x} + \frac{\partial N}{\partial y} + \frac{\partial O}{\partial z} \right) dV$$

The **flux** of **F** over S is equal to the triple integral of the **divergence** of **F** over V. This is referred to as the **divergence theorem** or **Gauss' theorem**.

2.5.9 Stokes' Theorem

Let S be an open, two-sided surface bounded by a closed curve C with a normal unit vector **n**. Stokes' theorem says that:

$$\oint_C \mathbf{A}\cdot d\mathbf{r} = \iint_S (\nabla\times\mathbf{A})\cdot\mathbf{n}\,dS$$ where C is traversed in an anti-clockwise direction.

The line integral of the tangential component of a vector **A** taken around a simple closed curve C is equal to the surface integral of the normal component of curl **A** taken over any surface having C as its boundary.

$$\mathbf{A} = A_1 i + A_2 j + A_3 k$$

$$\mathbf{n} = \cos\alpha\,\mathbf{i} + \cos\beta\,\mathbf{j} + \cos\gamma\,\mathbf{k}$$

$$\nabla\times\mathbf{A} = \left(\frac{\partial A_3}{\partial y} - \frac{\partial A_2}{\partial z}\right)\mathbf{i} + \left(\frac{\partial A_1}{\partial z} - \frac{\partial A_3}{\partial x}\right)\mathbf{j} + \left(\frac{\partial A_2}{\partial x} - \frac{\partial A_1}{\partial y}\right)\mathbf{k}$$

$$(\nabla\times\mathbf{A})\cdot\mathbf{n} = \left(\frac{\partial A_3}{\partial y} - \frac{\partial A_2}{\partial z}\right)\cos\alpha + \left(\frac{\partial A_1}{\partial z} - \frac{\partial A_3}{\partial x}\right)\cos\beta + \left(\frac{\partial A_2}{\partial x} - \frac{\partial A_1}{\partial y}\right)\cos\gamma$$

$$\mathbf{A}\cdot d\mathbf{r} = A_1 dx + A_2 dy + A_3 dz$$

Thus: $\oint_C A_1 dx + A_2 dy + A_3 dz =$

$$\iint_S \left(\left(\frac{\partial A_3}{\partial y} - \frac{\partial A_2}{\partial z}\right)\cos\alpha + \left(\frac{\partial A_1}{\partial z} - \frac{\partial A_3}{\partial x}\right)\cos\beta + \left(\frac{\partial A_2}{\partial x} - \frac{\partial A_1}{\partial y}\right)\cos\gamma\right) dS$$

Example:

Evaluate $\iint_S (\nabla\times\mathbf{A})\cdot\mathbf{n}\,dS$ where S is the surface $x^2 + y^2 + z^2 = 16$ of the hemisphere above the xy plane and $\mathbf{A} = (x^2 + y - 4)\mathbf{i} + 3xy\mathbf{j} + (2xz + z^2)\mathbf{k}$

Solution: by Stokes' theorem

$$\oint_C \mathbf{A}\cdot d\mathbf{r} = \oint_C (x^2 + y - 4)dx + \oint_C 3xy\,dy$$

Let $x = 4\cos\theta$ $\quad y = 4\sin\theta$

$dx = -4\sin\theta d\theta$ $\quad dx = 4\cos\theta d\theta$

$$= \int_0^{2\pi} \left[(16\cos^2\theta + 4\sin\theta - 4)(-4\sin\theta) + (12\cos\theta)(4\sin\theta)(4\cos\theta)\right]d\theta$$

$$= \int_0^{2\pi} \left[8(16)\cos^2\theta\sin\theta - 8 + 8\cos 2\theta + 16\sin\theta\right]d\theta$$

$$= \left[\frac{-8(16)}{3}\cos^3\theta - 8\theta + 4\sin 2\theta - 16\cos\theta\right]_0^{2\pi}$$

$$= -16\pi$$

$C: x^2 + y^2 = 16$

$\theta : 0 \to 2\pi$

2.5.10 Green's Theorem

Let $M(x,y)$ and $N(x,y)$ be continuous functions over a region R. If R is a closed region of the xy plane bounded by the closed curve C, then **Green's theorem** is:

$$\oint_C Mdx + Ndy = \iint_R \left(\frac{\partial N}{\partial x} - \frac{\partial M}{\partial y} \right) dxdy$$

where C is traversed in an anti-clockwise direction.

If $\dfrac{\partial N}{\partial x} = \dfrac{\partial M}{\partial y}$ then $d\phi = Mdx + Ndy$ and the integral does not depend upon the path.

Green's theorem allows us to calculate integrals either as a line or as a surface integral, depending on which is the easier.

Example:

Evaluate $\oint_C \left(3x^2 - 8y^2\right)dx + \left(4y - 6xy\right)dy$

where C is the boundary of the region enclosed by the curves: $y = x^2$, $y = \sqrt{x}$

(i) By line integral

(a) $x = t^2 \quad 0 \le t \le 1$

$y = t \quad dx = 2tdt; dy = dt$

$$\int_1^0 \left(3t^4 - 8t^2\right)(2t)dt + \int_1^0 \left(4t - 6t^3\right)dt$$

$$= \left[t^6 - 4t^4 + 2t^2 - \frac{6}{4}t^4 \right]_1^0$$

$$= 2.5$$

(b) $x = t \quad 0 \le t \le 1$

$y = t^2 \quad dx = dt; dy = 2tdt$

$$\int_0^1 \left(3t^2 - 8t^4\right)dt + \int_0^1 \left(4t^2 - 6t^3\right)(2t)dt$$

$$= \left[t^3 - \frac{8}{5}t^5 + 2t^4 - \frac{12}{5}t^5 \right]_0^1$$

$$= -1$$

Total: $2.5 - 1 = 1.5$

(ii) By Green's theorem

$$\oiint_R \left(-6y + 16y\right)dxdy$$

$$= \int_0^1 \int_{x^2}^{\sqrt{x}} \left(-6y + 16y\right)dydx$$

$$= \int_0^1 \left[5y^2\right]_{x^2}^{\sqrt{x}} dx$$

$$= 5\int_0^1 x - x^4 dx$$

$$= 5\left[\frac{x^2}{2} - \frac{x^5}{5} \right]_0^1 = 1.5$$

Note: Path of integration must go anti-clockwise around the loop.

2.5.11 Vector Representations of Green's Theorem

(i) $$\oint_C Mdx + Ndy = \iint_R \left(\frac{\partial N}{\partial x} - \frac{\partial M}{\partial y}\right) dxdy$$

Let $\mathbf{A} = M\mathbf{i} + N\mathbf{j}$ where M and N are functions of x and y

Then $\mathbf{A} \cdot d\mathbf{r} = Mdx + Ndy$ since $\mathbf{r} = x\mathbf{i} + y\mathbf{j}$

$$d\mathbf{r} = dx\mathbf{i} + dy\mathbf{j}$$

Now,

$$\nabla \times \mathbf{A} = \begin{vmatrix} \mathbf{i} & \mathbf{j} & \mathbf{k} \\ \frac{\partial}{\partial x} & \frac{\partial}{\partial y} & \frac{\partial}{\partial z} \\ M & N & O \end{vmatrix} = \mathbf{i}\left(-\frac{\partial N}{\partial z}\right) - \mathbf{j}\left(\frac{\partial M}{\partial z}\right) + \mathbf{k}\left(\frac{\partial N}{\partial x} - \frac{\partial M}{\partial y}\right)$$

$$= \mathbf{k}\left(\frac{\partial N}{\partial x} - \frac{\partial M}{\partial y}\right)$$

$$\left(\frac{\partial N}{\partial x} - \frac{\partial M}{\partial y}\right) = (\nabla \times \mathbf{A}) \cdot \mathbf{k}$$ the z component of $\nabla \times \mathbf{A}$

(the $\perp$ component to the xy plane containing R)

Thus $$\oint_C Mdx + Ndy = \iint_R (\nabla \times \mathbf{A}) \cdot \mathbf{k}\, dxdy$$

$$\boxed{\oint_C \mathbf{A} \cdot d\mathbf{r} = \iint_R (\nabla \times \mathbf{A}) \cdot \mathbf{k}\, dA}$$

(ii) Let $\mathbf{B} = N\mathbf{i} + -M\mathbf{j} + O\mathbf{k}$

Now $$(\nabla \cdot \mathbf{B}) = \frac{\partial N}{\partial x} + -\frac{\partial M}{\partial y}$$

Thus $$\iint \frac{\partial N}{\partial x} - \frac{\partial M}{\partial y} dxdy = \iint \nabla \cdot \mathbf{B}\, dxdy$$

If C is a curve in the xy plane, the unit tangent to C is: $\mathbf{T} = \frac{dx}{ds}\mathbf{i} + \frac{dy}{ds}\mathbf{j}$

$$\mathbf{n} \cdot \mathbf{T} = 0$$

$$\mathbf{n} = \frac{dy}{ds}\mathbf{i} - \frac{dx}{ds}\mathbf{j}$$

$$Mdx + Ndy = \mathbf{B} \cdot \mathbf{n}\, ds$$

$$\boxed{\oint_C \mathbf{B} \cdot \mathbf{n}\, ds == \iint_R \nabla \cdot \mathbf{B}\, dxdy}$$

2.5.12 Application of Green's Theorem

An application of Green's theorem is the procedure for finding the area of a plane region by performing a line integral over the boundary of the region.

In Green's theorem, let $M = 0$ and $N = x$

$$\oint_C x\,dy = \iint_R \frac{\partial}{\partial x}(x)\,dxdy$$

$$= \iint_R dxdy \qquad \text{(i)}$$

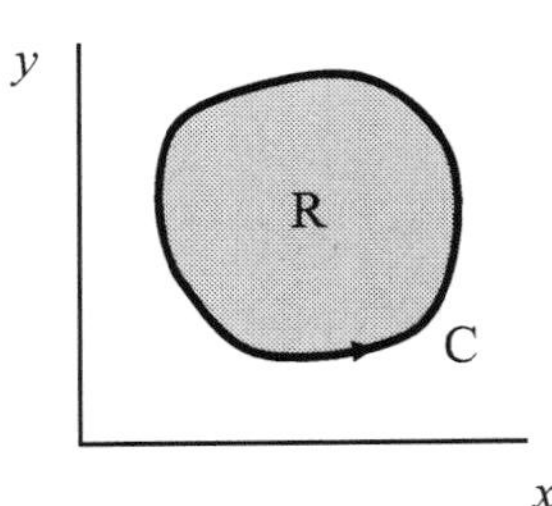

In Green's theorem, let $M = -y$ and $N = 0$

$$-\oint y\,dx = \iint_R dxdy \qquad \text{(ii)}$$

Adding (i) and (ii)

$$2A = \oint_C x\,dy - y\,dx$$

$$A = \frac{1}{2}\oint_C x\,dy - y\,dx$$

where the integration is taken in the anti-clockwise direction.

Green's theorem can be expressed:

$$\oint_C \mathbf{B}\cdot\mathbf{n}\,ds == \iint_R \nabla\cdot\mathbf{B}\,dxdy$$

and is concerned with the line integral around the boundary of a flat plane. If the region of integration is now a volume (rather than an area), and the boundary is the closed surface S, we obtain the **divergence theorem**:

$$\oiint_S \mathbf{F}\cdot\mathbf{n}dS = \iiint_V \nabla\cdot\mathbf{F}\,dV$$

Green's theorem in the plane can also be expressed:

$$\oint_C \mathbf{A}\cdot d\mathbf{r} = \iint_R (\nabla\times\mathbf{A})\cdot\mathbf{k}\,dA$$

which is a special case of Stokes' theorem with the normal vector to the surface in the direction of **k** or z axis.

2.5.13 Maxwell's Equations (integral form)

The preceding theorems and equations form the basis of our understanding of electromagnetism.

Gauss' law: (Electric charge)

$$\oiint_S \mathbf{E} \cdot d\mathbf{S} = \frac{q}{\varepsilon_o} = \phi$$

Gauss' law: (Magnetism)

$$\oiint_S \mathbf{B} . d\mathbf{S} = 0$$

Faraday's law:

$$\oint \mathbf{E} . d\mathbf{l} = -\frac{d\Phi}{dt}$$

but $\Phi = \iint \mathbf{B} \cdot d\mathbf{S}$

thus

$$\oint \mathbf{E} . d\mathbf{l} = -\frac{d}{dt} \iint \mathbf{B} \cdot d\mathbf{S} = Emf$$

Ampere's law:

$$\oint \mathbf{B} . d\mathbf{l} = \mu_o I + \mu_o \varepsilon_o \frac{\partial \phi}{\partial t}$$

but $\phi = \oiint \mathbf{E} \cdot d\mathbf{S}$

thus

$$\oint \mathbf{B} \cdot d\mathbf{l} = \mu_o I + \mu_o \varepsilon_o \oiint \frac{d\mathbf{E}}{dt} \cdot d\mathbf{S}$$

These four equations are known as **Maxwell's equations** and can be used to quantify all aspects of electricity and magnetism, including the existence and properties of electromagnetic waves. The equations shown here apply where there is no polarisable or magnetic material (μ_r, $\varepsilon_r = 1$).

2.5.14 Maxwell's Equations (differential form)

Maxwell's equations can also be expressed in differential form:

Gauss' law: (Electric charge) $q = \varepsilon_0 \oiint \mathbf{E} \cdot d\mathbf{S}$

$$= \varepsilon_0 \iiint \nabla \cdot \mathbf{E} dV \quad \text{by Divergence theorem}$$

but $q = \iiint \rho dV$ ρ is charge density

thus $\boxed{\nabla \cdot \mathbf{E} = \frac{\rho}{\varepsilon_0}}$

Gauss' law: (Magnetism) $\oiint_S \mathbf{B}.d\mathbf{S} = 0$

therefore $\iiint \nabla \cdot \mathbf{B} dV = 0$ by divergence theorem

$$\boxed{\nabla \cdot \mathbf{B} = 0}$$

Faraday's law: $\oint \mathbf{E}.d\mathbf{l} = -\frac{\partial \Phi}{\partial t}$

$$\iint (\nabla \times \mathbf{E}) \cdot d\mathbf{S} = -\frac{\partial}{\partial t} \iint \mathbf{B} \cdot d\mathbf{S}$$

by Stokes' theorem and $\Phi = \iint \mathbf{B} \cdot d\mathbf{S}$

$$\boxed{\nabla \times \mathbf{E} = -\frac{\partial}{\partial t} \mathbf{B}}$$

Ampere's law: $\oint \mathbf{B}.d\mathbf{l} = \mu_0 I + \mu_0 \varepsilon_0 \frac{\partial \phi}{\partial t}$

$$\frac{1}{\mu_0} \oint \mathbf{B} \cdot d\mathbf{l} = I + \varepsilon_0 \frac{\partial}{\partial t} \oiint \mathbf{E} \cdot d\mathbf{S}$$

$$= \frac{1}{\mu_0} \iint (\nabla \times \mathbf{B}) \cdot d\mathbf{S} \quad \text{by Stokes' theorem}$$

$$\frac{1}{\mu_0} \iint (\nabla \times \mathbf{B}) \cdot d\mathbf{S} = I + \varepsilon_0 \frac{\partial}{\partial t} \oiint \mathbf{E} \cdot d\mathbf{S}$$

$$= \oiint \mathbf{J} \cdot d\mathbf{S} + \varepsilon_0 \frac{\partial}{\partial t} \oiint \mathbf{E} \cdot d\mathbf{S}$$

J is current density

$$\boxed{\nabla \times \mathbf{B} = \mu_0 \mathbf{J} + \mu_0 \varepsilon_0 \frac{\partial}{\partial t} \mathbf{E}}$$

2.6 Fourier Series

Summary

Fourier series

$$f(t) = A_o + \sum_{n=1}^{\infty}\left[A_n \cos n\omega_o t + B_n \sin n\omega_o t\right]$$

$$A_o = \frac{1}{T_o}\int_0^{T_o} f(t)\,dt;\quad A_n = \frac{2}{T_o}\int_0^{T_o} f(t)\cos n\omega_o t\,dt$$

$$B_n = \frac{2}{T_o}\int_0^{T_o} f(t)\sin n\omega_o t\,dt$$

$$f(t) = \int_0^{\infty}\left(A(\omega)\cos \omega t + B(\omega)\sin \omega t\right)d\omega$$

Fourier integral

$$A(\omega) = \frac{1}{\pi}\int_{-\infty}^{\infty} f(t)\cos \omega t dt$$

$$B(\omega) = \frac{1}{\pi}\int_{-\infty}^{\infty} f(t)\sin \omega t dt$$

$$\frac{F(\omega_k)}{\Delta t} = \sum_{i=0}^{N-1} y_i(i\Delta t)e^{-j2\pi ik/N}$$

$$= C(k)$$

Discrete Fourier transform

$$C(k) = \sum_{i=0}^{N-1} y_i(i\Delta t)\left[\cos\frac{2\pi ik}{N} - j\sin\frac{2\pi ik}{N}\right]$$

2.6.1 Fourier Series

For a periodic function $f(t)$ of period T_o with frequency ω_o, the function may be represented by a **Fourier series** which is written:

$$f(t) = A_o + \sum_{n=1}^{\infty}\left[A_n \cos n\omega_o t + B_n \sin n\omega_o t\right]$$

$$T_o = \frac{2\pi}{\omega_o}$$

with

$$A_o = \frac{1}{T_o}\int_0^{T_o} f(t)\,dt;\quad A_n = \frac{2}{T_o}\int_0^{T_o} f(t)\cos n\omega_o t\,dt;\quad B_n = \frac{2}{T_o}\int_0^{T_o} f(t)\sin n\omega_o t\,dt$$

"DC" term (A_o)

amplitude terms for component frequency $n\omega_o$ (A_n, B_n)

Euler's formula shows that a sinusoidal function can be represented by a pair of exponential functions:

$$\cos\omega t = \frac{1}{2}\left[e^{j\omega t} + e^{-j\omega t}\right];\ \sin\omega t = -j\frac{1}{2}\left[e^{j\omega t} - e^{-j\omega t}\right]$$

Note, we use $j^2 = -1$ in this chapter to avoid confusion with the counter variable i.

Substituting into the Fourier series we obtain:

$$f(t) = \sum_{n=-\infty}^{\infty} C_n e^{nj\omega_o t}$$

$$C_n = \frac{1}{T_o}\int_0^{T_o} f(t)e^{-nj\omega_o t}\,dt$$

where C_n is a complex number; the real part contains the amplitude of the cos terms, and the imaginary part the amplitude of the sin terms.

$$\begin{aligned} C_n &= \frac{1}{2}(A_n - jB_n) & n > 0 \\ &= A_o & n = 0 \\ &= \frac{1}{2}(A_n + jB_n) & n < 0 \end{aligned}$$

These relationships connect the real coefficients with the complex coefficients.

A plot of C_n vs frequency is an **amplitude spectrum** of the signal. For example, if $f(t) = A \cos \omega_o t$, then the amplitude spectrum consists of two lines of height $A/2$ located at $\pm\omega_o$.

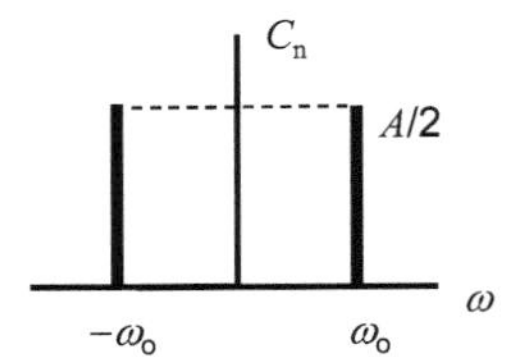

Negative frequencies are a result of the representation of sinusoidal signals by two exponential functions. There is no "DC" term since the average value of this function (cos) is zero.

The amplitude spectrum represents the magnitude of the exponential components of the signal. An amplitude spectrum using trigonometric coefficients would be a single line of height A at ω_o.

2.6.2 Fourier Transform

Non-periodic functions can be considered as functions with an infinite period and zero frequency.

As $T_o \to \infty$, $C_n \to 0$, which means that the amplitude of the spectral lines becomes vanishingly small as the spectral lines merge into a continuum. But, the integral is finite, hence, the product C_nT_o can be written:

$$C_nT_o = \int_{-\infty}^{\infty} f(t)e^{-nj\omega_o t}dt$$

The product $n\omega_o$ becomes the continuous variable ω, hence:

$$C_nT_o = \int_{-\infty}^{\infty} f(t)e^{-j\omega t}dt$$

$$= F(\omega)$$

$$\text{or: } C_n = \frac{F(\omega)}{T_o} = \frac{\omega_o}{2\pi}F(\omega)$$

Replacing the product $n\omega_o$ in the periodic case with the continuous variable ω, we obtain:

$$f(t) = \sum_{n=-\infty}^{\infty} \frac{\omega_o}{2\pi}F(\omega)e^{j\omega t} = \frac{1}{2\pi}\int_{-\infty}^{\infty} F(\omega)e^{j\omega t}d\omega \qquad \omega_o \to 0$$

Fourier integral.

which can be written:

$$f(t) = \int_0^{\infty} \underbrace{(A(\omega)\cos\omega t + B(\omega)\sin\omega t)d\omega}_{\text{Even function, hence we can halve the interval to 0 to } \infty.}$$

where

$$A(\omega) = \frac{1}{\pi}\int_{-\infty}^{\infty} f(t)\cos\omega t dt$$

$$B(\omega) = \frac{1}{\pi}\int_{-\infty}^{\infty} f(t)\sin\omega t dt$$

The function $F(\omega)$ is called the **Fourier transform** of $f(t)$ and is written $F[f(t)]$.
The Fourier transform is a continuous function of ω. Because of the continuous nature of ω, the amplitude of the frequency component of the signal at any one particular frequency approaches zero. $F(\omega)$ is thus more correctly thought of as a frequency density function:

$$F(\omega) = C_nT_o = C_n\frac{2\pi}{\omega_o}$$

The function $f(t)$ is called the **inverse Fourier transform** of $F(\omega)$

2.6.3 Sampling

If a continuous signal $y(t)$ is sampled N times at equal time intervals Δt, then the **sampling frequency** is: $\omega_s = 2\pi/\Delta t$.

The sampled data sequence has a total duration $T_o = N\Delta t$.

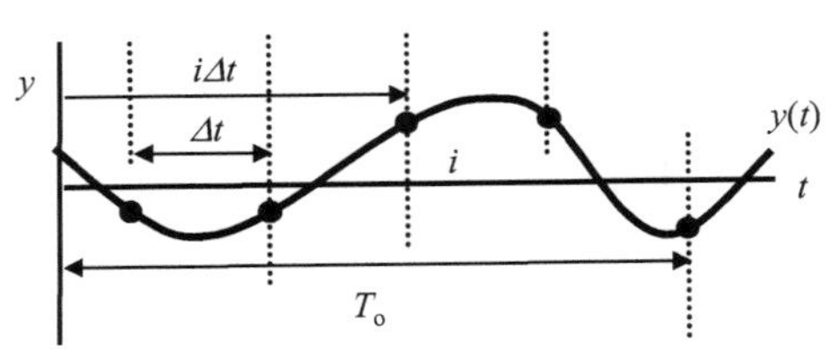

We can call this duration the "period" of the non-periodic sequence (as if it were repeated over and over again). For a regular periodic signal, the frequency spectrum of the signal consists of lines spaced $\omega_o = 2\pi/T_o$ apart where ω_o is the fundamental frequency of the signal. For a non-periodic signal, we say that a **non-periodic signal of finite length** is periodic with a period equal to the length of the signal. The component frequencies consist of equally spaced intervals:

$$\Delta\omega = \frac{2\pi}{T_o} = \frac{2\pi}{N\Delta t} = \frac{\omega_s}{N}$$

The frequency spectrum of a **non-periodic** signal can be completely specified by a set of regularly spaced frequencies a minimum of $\Delta\omega = 2\pi/T_o$ radians/sec apart.

We might ask how many component frequencies are required to represent the data? The spectrum of a digital signal is always periodic with the same set of frequencies repeating with a frequency period of $2\pi/\Delta t$. If we sample at intervals of $\Delta\omega = 2\pi/N\Delta t$, then the total number of frequencies per period is:

$$\frac{2\pi}{\Delta t}\frac{N\Delta t}{2\pi} = N$$

The frequency components contained within the N data points are:

$$\omega_k = \frac{2\pi k}{N\Delta t} = \frac{k}{N}\omega_s \quad \text{where } k \text{ goes from 0 to } N-1$$

The **frequency resolution** $\Delta\omega$ is ω_s/N. The greater the value for N, the finer the resolution of the frequency bins or **channels** used to represent the original signal.

2.6.4 Discrete Fourier Transform

Now, in general, $F(\omega)$ is given by the integral: $F(\omega)=\int_{-\infty}^{\infty} f(t)e^{-j\omega t}dt$

Expressed in terms of a finite sum, this becomes:

$$F(\omega)=\sum_{i=0}^{N-1} y_i(i\Delta t)e^{-j\omega(i\Delta t)}\Delta t$$

N is the total number of equally spaced data points and $y_i(i\Delta t)$ is the actual data at i recorded at time $i\Delta t$.

But, the circular frequency ω in this formula is a continuous variable; however, for a discrete number of samples we can say: $\omega_k = \dfrac{2\pi k}{N\Delta t}$

Further, since $F(\omega)$ is actually a frequency density function, we can thus write the actual **amplitude spectrum** of the signal as:

$$\frac{F(\omega_k)}{\Delta t}=\sum_{i=0}^{N-1} y_i(i\Delta t)e^{-j2\pi ik/N}$$

$$= C(k) \quad \text{where } k \text{ goes from 0 to } N-1$$

Making use of Euler's formula, we can express this in terms of sines and cosines:

$$C(k)=\sum_{i=0}^{N-1} y_i(i\Delta t)\left[\cos\frac{2\pi ik}{N} - j\sin\frac{2\pi ik}{N}\right]$$

Each value of $C(k)$ is a complex number of the form $A - jB$. The complete array of N complex numbers comprising this series is called a **discrete Fourier transform** or DFT. The values of C_n are found by dividing C_k by N. The magnitude of the complex numbers C_n is the amplitude spectrum of the signal.

Recalling that: $C_n = \frac{1}{2}(A_n - jB_n)$

we obtain values for A_n and B_n by multiplying the real part of C_n by 2 and the imaginary part by –2 respectively.

The **inverse transform** is found from:

$$y_i(i\Delta t)=\frac{1}{N}\sum_{k=0}^{N-1} C(k)\left[\cos\frac{2\pi ik}{N} + j\sin\frac{2\pi ik}{N}\right]$$

2.6.5 Odd and Even Functions

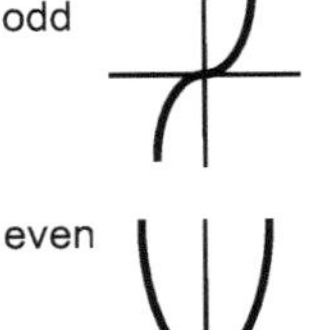

A function is said to be **odd** if: $f(-x) = -f(x)$.

A function is said to be **even** if $f(-x) = f(x)$.

Properties of odd and even functions:

- The product of two even functions is an even function.
- The product of two odd functions is an even function.
- The product of an even function and an odd function is an odd function.

In terms of a Fourier series, an odd function has only sine terms present. For an even function, only the cosine terms (and A_0) are present. Each of these is referred to as a **half-range Fourier series.**

Odd

$$f(t) = \sum_{n=1}^{\infty}\left[B_n \sin n\omega_o t\right]$$

$$B_n = \frac{2}{T_o}\int_0^{T_o} f(t)\sin n\omega_o t\, dt$$

Even

$$f(t) = A_o + \sum_{n=1}^{\infty}\left[A_n \cos n\omega_o t\right]$$

$$A_o = \frac{1}{T_o}\int_0^{T_o} f(t)\,dt; \quad A_n = \frac{2}{T_o}\int_0^{T_o} f(t)\cos n\omega_o t\, dt$$

Example:

Find the Fourier series for the function $f(x) = 0 \quad -c < x < 0$

$= 1 \quad 0 < x < c$

$T = 2c$

−c c

$$\omega_o = \frac{2\pi}{2c} = \frac{\pi}{2}$$

Solution:

$$A_o = \frac{1}{2c}\int_{-c}^{c} f(x)dx = \frac{1}{2c}\left[\int_{-c}^{0} 0dx + \int_0^c 1dx\right]$$

$$= \frac{1}{2c}[x]_0^c = \frac{1}{2}$$

$$A_n = \frac{2}{2c}\int_{-c}^{c} f(x)\cos\frac{n\pi x}{c}dx$$

$$= \frac{2}{2c}\left[\int_{-c}^{0} 0dx + \int_0^c \cos\frac{n\pi x}{c}dx\right]$$

$$= \frac{2}{2c}\left[\frac{c}{n\pi}\sin\frac{n\pi x}{c}\right]_0^c$$

$$= 0$$

$$B_n = \frac{2}{2c}\int_{-c}^{c} f(x)\sin\frac{n\pi x}{c}dx$$

$$= \frac{1}{c}\left[\int_0^c \sin\frac{n\pi x}{c}dx\right]$$

$$= \frac{1}{c}\left[\frac{-c}{n\pi}\cos\frac{n\pi x}{c}\right]_0^c$$

$$= \frac{-1}{n\pi}\left((-1)^n - 1\right)$$

$$f(x) = 1 + \sum_{n=1}^{\infty}\frac{1}{n\pi}\left((-1)^n - 1\right)\sin\frac{n\pi x}{c}$$

2.6.6 Convolution

Fourier transforms of data can be used to apply digital filtering to reduce high or low frequency noise from the signal. The amplitude spectrum of a filter function $H(\omega)$ could take the form:

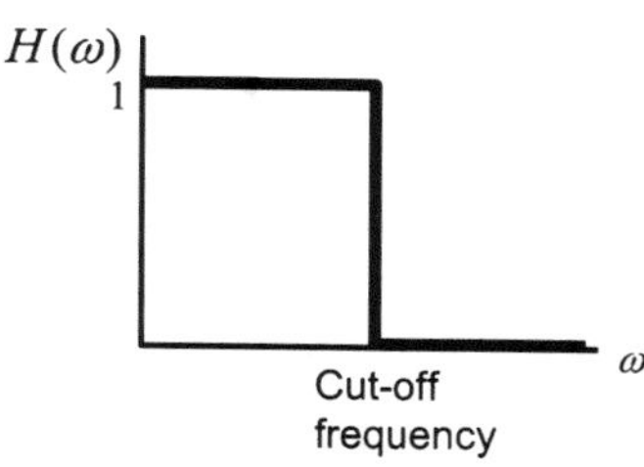

When this filter amplitude spectrum is multiplied into the signal amplitude spectrum, we obtain a filtered spectrum. The inverse transform then is a filtered representation $y_{out}(t)$ of the original data.

Let
$$y_{out}(t) = \int_{-\infty}^{\infty} H(\omega)F(\omega)e^{j\omega t}\,d\omega$$

$y_{out}(t)$ is the inverse Fourier transform of the product $H(\omega)F(\omega)$.

where
$$F(\omega) = \int_{-\infty}^{\infty} y_{in}(t)e^{-j\omega t}\,dt$$

$$H(\omega) = 1 \qquad \omega \le \omega_c$$
$$= 0 \qquad \omega > \omega_c$$

If the functions $h(t)$ and $y_{in}(t)$ are the original functions of the filter and the signal in the time domain, we say that the output signal $y_{out}(t)$ is the **convolution** (*) of these two functions.

$$y_{out}(t) = \int_{-\infty}^{\infty} F[y_{in}(t) * h(t)]e^{j\omega t}\,d\omega$$

$$F[y_{in}(t) * h(t)] = F(\omega)H(\omega)$$
$$= F[y_{in}(t)]F[h(t)]$$

A convolution in the time domain is a multiplication in the frequency domain.

A convolution is a type of **superposition** in the time domain. It is equivalent to a multiplication in the frequency domain. Mathematically, it is often easier to perform multiplications in the frequency domain than convolutions in the time domain, hence the popularity of Fourier analysis in **digital signal processing**.

Modulation is another type of superposition used in digital signal processing. Consider the formula:

$$y(t) = Ae^{j\omega t}$$

The frequency term is modulated by the amplitude term (**amplitude modulation**). Modulation is a multiplication in the time domain which is equivalent to a convolution in the frequency domain.

Example:

Write a computer program to provide a Fourier transform of an array of numbers accumulated at regular time intervals.

Solution:

In this program, raw data is supplied in the first dimension of a three dimensional array whose 3 rows carry the series and whose *n* columns carry the data points. The second dimension will receive the magnitude of the C_n terms. The third dimension will receive the real and imaginary components of C_k.

```
Function Fourier(ByRef Array)
 n = UBound(Array, 2) - 1
 ws = 2 * gpi / (Array(1, n + 1, 1) - Array(1, (n + 1) - 1, 1))
 For k = 0 To n - 1
  Array(1, k + 2, 3) = 0
  Array(2, k + 2, 3) = 0
  Array(3, k + 2, 3) = 0
  For i = 0 To n - 1
   Ai = Array(2, i + 2, 1) * Cos(2 * gpi * k * (i) / n)
   Array(2, k + 2, 3) = Array(2, k + 2, 3) + Ai   'Ck Re
   Bi = -1 * Array(2, i + 2, 1) * Sin(2 * gpi * k * (i) / n)
   Array(3, k + 2, 3) = Array(3, k + 2, 3) + Bi   ' Ck im
  Next i
  Array(1, k + 2, 3) = (k * (ws / n)) / (2 * gpi)      ' Hz
 Next k
 For i = 1 To n
  Array(1, i + 1, 2) = Array(1, i + 1, 3) ' Hz
  Array(2, i + 1, 2) = Magnitude(Array(2, i + 1, 3) / n, _
   Array(3, i + 1, 3) / n) ' amplitude |Cn| goes into series 2. Cn = Ck/N
 Next i
End Function
```

The amplitude spectrum is given by the values $|C_n|$. The coefficients A_n and B_n are given by 2 times C_n real and −2 times C_n imag, respectively.

```
Function InverseFourier(ByRef Array)
n = UBound(Array, 2) - 1
For i = 1 To n
 RealSum = 0
 ImagSum = 0
 For k = 0 To n - 1
  CkReal = Array(2, k + 2, 3)
  CkImag = Array(3, k + 2, 3)
  Ai = Cos(2 * gpi * k * (i - 1) / n)
  Bi = Sin(2 * gpi * k * (i - 1) / n)
  Call ComplexMultiply(CkReal, CkImag, Ai, Bi, Real, Imag)
  RealSum = RealSum + Real
  ImagSum = ImagSum + Imag
 Next k
 Array(2, i + 1, 1) = RealSum   ' Magnitude
 Array(3, i + 1, 1) = ImagSum  ' Phase
Next i
```

2.7 Partial Differential Equations

Summary

$$\frac{\partial^2 y}{\partial t^2} = a^2 \frac{\partial^2 y}{\partial x^2}$$ Wave equation

$$\frac{\partial u}{\partial t} = \frac{K}{c\rho}\frac{\partial^2 u}{\partial x^2}$$ 1D Heat conduction equation

$$y = \psi(x + at) + \phi(x - at)$$ d'Alembert's solution to the wave equation

2.7.1 Partial Differential Equations

A linear **partial differential equation** of order 2 in two independent variables x and y has the form:

$$A\frac{\partial^2 u}{\partial x^2}+B\frac{\partial^2 u}{\partial xy}+C\frac{\partial^2 u}{\partial y^2}+D\frac{\partial u}{\partial x}+Fu=G$$

where the coefficients A, B, C, etc may be functions of x and y, but not u. If the coefficients are constants, then the general solution of the equation may be found by assuming that:

$$u=e^{ax+by}$$

where a and b are constants to be determined. As an example, consider the partial differential equation:

$$\frac{\partial^2 u}{\partial x^2}+3\frac{\partial^2 u}{\partial xy}+2\frac{\partial^2 u}{\partial y^2}=0$$

Let the solutions have the form $u=e^{ax+by}$.

Now, $\quad u=e^{ax+by}$

Thus: $\quad \frac{\partial u}{\partial x}=ae^{ax+by} \qquad \frac{\partial u}{\partial y}=be^{ax+by} \qquad \frac{\partial^2 u}{\partial x\partial y}=abe^{ax+by}$

$$\frac{\partial^2 u}{\partial x^2}=a^2e^{ax+by} \qquad \frac{\partial^2 u}{\partial y^2}=b^2e^{ax+by}$$

And the equation becomes $\quad a^2e^{ax+by}+3abe^{ax+by}+2b^2e^{ax+by}=0$

$$\left(a^2+3ab+2b^2\right)e^{ax+by}=0$$

$$a^2+3ab+2b^2=0$$

$$(a+b)(a+2b)=0$$

$$a=-b$$

$$a=-2b$$

If $a=-b$, then we have $\quad u=e^{-bx+by}=e^{b(y-x)}$.

If $a=-2b$, then we have $u=e^{-2bx+by}=e^{b(y-2x)}$.

By the superposition of the two solutions, the general solution is expressed: $\quad u=F(y-x)+G(y-2x)$

2.7.2 General Wave Equation

Partial differential equations have wide application to physical phenomena. For example, consider the **vibration of a stretched string** of length L under tension T. At any time, the displacement at any point on the string is a function of the position from the end of the string and the time after the string was plucked: $y = y(x,t)$.

Consider a portion or element of the string:

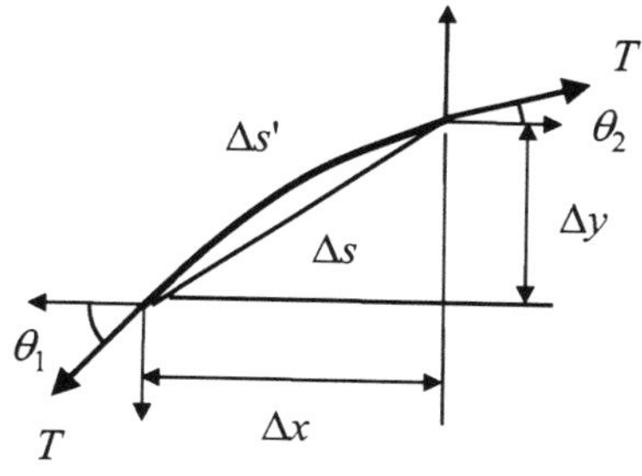

There is no movement of elements in the string in the horizontal direction, so:

$$T\cos\theta_1 = T\cos\theta_2$$

There is movement of elements in the string in the vertical direction, so the net unbalanced force is given by:

$$F_u = T\sin\theta_2 - T\sin\theta_1$$

The actual value of θ depends upon where we are on the string, that is: $\theta = \theta(x)$.

$$F_u = T\sin\theta(x+\Delta x) - T\sin\theta(x)$$

For an arc element: $\Delta s' \approx \Delta s$

The mass of the element is given by $\Delta m = \rho\Delta s$

And so any unbalanced forced on the element is found from

$$F_u = \rho\Delta s\frac{\partial^2 y}{\partial t^2} \quad \text{(Newton's 2}^{\text{nd}}\text{ law } F = ma\text{)}$$

Thus:

$$\rho\Delta s\frac{\partial^2 y}{\partial t^2} = T\sin\theta(x+\Delta x) - T\sin\theta(x)$$

In the limit of very small Δs, $\sin\theta = \tan\theta$ and so we can write:

$$\rho\Delta s\frac{\partial^2 y}{\partial t^2} = T\tan\theta(x+\Delta x) - T\tan\theta(x)$$

$$\rho\frac{\Delta s}{\Delta x}\frac{\partial^2 y}{\partial t^2} = \frac{T\tan\theta(x+\Delta x) - T\tan\theta(x)}{\Delta x}$$

$$\lim_{\Delta x\to 0}\rho\frac{\Delta s}{\Delta x}\frac{\partial^2 y}{\partial t^2} = \lim_{\Delta x\to 0}\frac{T\tan\theta(x+\Delta x) - T\tan\theta(x)}{\Delta x}$$

$$\rho\frac{\partial^2 y}{\partial t^2} = \frac{\partial}{\partial x}T\frac{\partial y}{\partial x} \quad \text{as } \Delta x \to 0 \quad \text{and} \quad \tan\theta(x) = \frac{\Delta y}{\Delta x}$$

$$\Delta x \to \Delta s \qquad = \frac{\partial y}{\partial x}$$

$$\boxed{\frac{\partial^2 y}{\partial t^2} = \frac{T}{\rho}\frac{\partial^2 y}{\partial x^2}}$$ 1D general **wave equation**

2.7.3 Solution to the General Wave Equation

To solve the general wave equation, we need to consider the initial conditions and then use the method of **separation of variables.**

The initial conditions can be described as an initial displacement in the centre of the string:

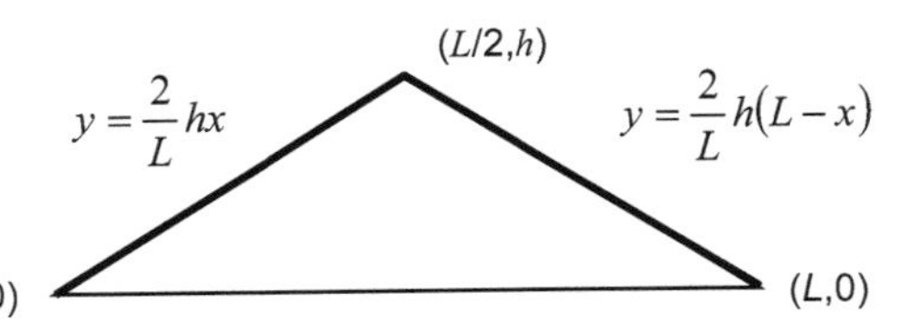

The initial conditions are thus:

$y(0,t)=0$
$y(L,t)=0$ — Fixed points at the end of the string at all t

$$y\left(\frac{L}{2},0\right)=h$$

$$\left.\begin{aligned} y(x,0)&=\frac{2}{L}hx & 0<x<\frac{L}{2} \\ &=\frac{2}{L}h(L-x) & \frac{L}{2}<x<L \end{aligned}\right\} f(x,t)$$

Initial shape of the string (string is plucked rather than struck)

$\frac{\partial}{\partial t}y(x,0)=0$ — Released from rest at $t=0$

For convenience, we let $\alpha^2=\frac{T}{\rho}$

We will assume that the solution to the wave equation can be expressed in terms of two functions X and T such that: $y(x,t)=X(x)T(t)$.

$$\frac{\partial^2 y}{\partial t^2}=\alpha^2\frac{\partial^2 y}{\partial x^2}$$

$$XT''=\alpha^2 X''T \quad \text{where} \quad T''=\frac{\partial^2}{\partial t^2};X''=\frac{\partial^2}{\partial x^2}$$

We thus separate the variables:

$$\frac{T''}{T}=-\lambda^2 \qquad\qquad \frac{X''}{X}=-\lambda^2$$

(a convenient arbitrary constant)

$$T''(t)+\alpha^2\lambda^2T(t)=0 \qquad\qquad X''(x)+\lambda^2X=0$$

$$\frac{d^2T}{dt^2}+\alpha^2\lambda^2T=0$$ Function of t alone

$$\frac{d^2X}{dx^2}+\lambda^2X=0$$ Function of x alone

The problem is now one of two ordinary differential equations:

$$0 = m^2 + \alpha^2\lambda^2 \qquad\qquad 0 = m^2 + \lambda^2 \quad \text{Auxiliary equations}$$

$$m = \pm\alpha\lambda i \qquad\qquad m = \pm\lambda i$$

See 1.1.2

$$T = C_1 e^{\alpha\lambda it} + C_2 e^{-\alpha\lambda it} \qquad\qquad X = C_1 e^{\lambda ix} + C_2 e^{-\lambda ix}$$

$$= (C_1 + C_2)\cos\alpha\lambda t + i(C_1 - C_2)\sin\alpha\lambda t \quad = (C_1 + C_2)\cos\lambda x + i(C_1 - C_2)\sin\lambda x$$

Now, the above equations, when combined $y(x,t) = XT$ represents the motion of the string but there is a difficulty. Intuitively we know that the motion involves real displacements, but both X and T above contain imaginary parts. This difficulty is overcome if we let $C_1 = C_2$, but then we have only one arbitrary constant for each of the 2nd order equations, so this is not a complete general solution. A better approach is to assume that C_1 and C_2 are both complex. For the case of T, we have:

$$C_1 = a_1 + ib_1$$
$$C_2 = a_2 + ib_2$$

$$T = (C_1 + C_2)\cos\alpha\lambda t + i(C_1 - C_2)\sin\alpha\lambda t$$
$$= (a_1 + ib_1 + a_2 + ib_2)\cos\alpha\lambda t + i(a_1 + ib_1 - a_2 - ib_2)\sin\alpha\lambda t$$
$$= (a_1 + a_2)\cos\alpha\lambda t - (b_1 - b_2)\sin\alpha\lambda t + i[(b_1 + b_2)\cos\alpha\lambda t + (a_1 - a_2)\sin\alpha\lambda t]$$

For T to be real, then: $a_1 = a_2$

$$b_1 = -b_2$$

and so: and similarly:

$$T = 2a_1\cos\alpha\lambda t - 2b_1\sin\alpha\lambda t \qquad\qquad X = 2a_1\cos\lambda x - 2b_1\sin\lambda x$$
$$= A_1\cos\alpha\lambda t - B_1\sin\alpha\lambda t \qquad\qquad = A_2\cos\lambda x - B_2\sin\lambda x$$

Thus, the general solution is written $y(x,t) = XT$:

$$y(x,t) = (A_1\cos\alpha\lambda t - B_1\sin\alpha\lambda t)(A_2\cos\lambda x - B_2\sin\lambda x)$$

We now make use of the initial conditions:

$$y(0,t) = 0$$
$$0 = (A_1\cos\alpha\lambda t - B_1\sin\alpha\lambda t)(B_2)$$
$$\therefore B_2 = 0$$

$$y(L,t) = 0$$
$$0 = (A_1\cos\alpha\lambda t - B_1\sin\alpha\lambda t)(B_2\sin\lambda L)$$
$$B_2\sin\lambda L = 0$$
$$\therefore \lambda = \frac{n\pi}{L}$$

Combining the constants A_1, B_1 and B_2, the solution can now be expressed:

$$y(x,t)=\left(A\cos\alpha\frac{n\pi}{L}t-B\sin\alpha\frac{n\pi}{L}t\right)\left(\sin\frac{n\pi}{L}x\right)$$

Now we consider the other initial conditions: $y(x,0)=f(x)$; $\frac{\partial}{\partial t}y(x,0)=0$

$$y(x,t)=\left(A\cos\left(\alpha\frac{n\pi}{L}t\right)+B\sin\left(\alpha\frac{n\pi}{L}t\right)\right)\left(\sin\left(\frac{n\pi}{L}x\right)\right)$$

$$\frac{\partial}{\partial t}y(x,t)=\left(-A\alpha\frac{n\pi}{L}\sin\left(\alpha\frac{n\pi}{L}t\right)+B\frac{n\pi}{L}\cos\left(\alpha\frac{n\pi}{L}t\right)\right)\left(\sin\left(\frac{n\pi}{L}x\right)\right)=0$$

$$\frac{\partial}{\partial t}y(x,0)=B\frac{n\pi}{L}\left(\sin\left(\frac{n\pi}{L}x\right)\right)=0$$

$$\therefore B=0$$

The general solution to the wave equation is now written:

$$y(x,t)=A\sin\left(\frac{n\pi}{L}x\right)\cos\left(\alpha\frac{n\pi}{L}t\right)$$

Or, by the principle of superposition, we have as a general solution:

$$y(x,t)=\sum_{n=1}^{\infty}A_n\sin\left(\frac{n\pi}{L}x\right)\cos\left(\alpha\frac{n\pi}{L}t\right)$$

But, $y(x,0)=f(x)$

Thus: $f(x)=\sum_{n=1}^{\infty}A_n\sin\frac{n\pi x}{L}$

which is a Fourier sine series.

In a Fourier series, A_n must equal: $A_n=\frac{2}{L}\int_0^L f(x)\sin\frac{n\pi x}{L}dx$

Note: In this treatment we have effectively arranged for the coefficients of the general solution to be real in line with our expectations for the motion of a stretched string. In general, the coefficients are complex. The imaginary parts of the complex coefficients may have physical significance. For example, when describing the deformation of materials, the real parts correspond to elastic behaviour while the imaginary parts correspond to viscous, or fluid-like behaviour of the material

and so the final solution can be written as a series:

$$y(x,t)=\sum_{n=1}^{\infty}\left(\frac{2}{L}\int_0^L f(x)\sin\left(\frac{n\pi x}{L}\right)dx\right)\left(\sin\frac{n\pi x}{L}\cos\alpha\frac{n\pi t}{L}\right)$$

The terms in this series correspond to the natural frequency of the string. The natural or **harmonic frequencies** are found from: $\omega=\frac{n\pi\alpha}{L}$

2.7.4 d'Alembert's Solution to the Wave Equation

The general wave equation is written: $\dfrac{\partial^2 y}{\partial t^2} = \alpha^2 \dfrac{\partial^2 y}{\partial x^2}$

Let: $y = y(u,v)$ where $u = x + \alpha t;\; v = x - \alpha t$

Therefore: $\dfrac{\partial y}{\partial x} = \dfrac{\partial y}{\partial u}\dfrac{\partial u}{\partial x} + \dfrac{\partial y}{\partial v}\dfrac{\partial v}{\partial x}$

$= \dfrac{\partial y}{\partial u} + \dfrac{\partial y}{\partial v}$

Thus: $\dfrac{\partial}{\partial x} = \dfrac{\partial}{\partial u} + \dfrac{\partial}{\partial v}$

Now: $\dfrac{\partial^2 y}{\partial x^2} = \dfrac{\partial}{\partial x}\left(\dfrac{\partial y}{\partial x}\right)$ where: $\dfrac{\partial}{\partial x} = \dfrac{\partial}{\partial u} + \dfrac{\partial}{\partial v}$

Therefore: $\dfrac{\partial^2 y}{\partial x^2} = \dfrac{\partial}{\partial u}\left(\dfrac{\partial y}{\partial x}\right) + \dfrac{\partial}{\partial v}\left(\dfrac{\partial y}{\partial x}\right)$

$= \dfrac{\partial}{\partial u}\left(\dfrac{\partial y}{\partial u} + \dfrac{\partial y}{\partial v}\right) + \dfrac{\partial}{\partial v}\left(\dfrac{\partial y}{\partial u} + \dfrac{\partial y}{\partial v}\right)$

$= \dfrac{\partial^2 y}{\partial u^2} + 2\dfrac{\partial^2 y}{\partial u \partial v} + \dfrac{\partial^2 y}{\partial v^2}$

$= y_{uu} + 2y_{uv} + y_{vv}$

$\dfrac{\partial y}{\partial t} = \alpha\left(\dfrac{\partial y}{\partial u} - \dfrac{\partial y}{\partial v}\right)$

$\dfrac{\partial^2 y}{\partial t^2} = \alpha\dfrac{\partial}{\partial u}\left(\dfrac{\partial y}{\partial t}\right) - \alpha\dfrac{\partial}{\partial v}\left(\dfrac{\partial y}{\partial t}\right)$

$\dfrac{\partial^2 y}{\partial t^2} = \alpha^2\left(y_{uu} - 2y_{uv} + y_{vv}\right)$

Substituting into the original equation:

$\alpha^2\left(y_{uu} - 2y_{uv} + y_{vv}\right) = \alpha^2\left(y_{uu} + 2y_{uv} + y_{vv}\right)$

Thus $y_{uv} = 0$

or $\dfrac{\partial^2 y}{\partial u \partial v} = 0$

$\dfrac{\partial}{\partial v}\dfrac{\partial y}{\partial u} = 0$

Therefore $\dfrac{\partial y}{\partial u} = f(u)$ only

$y = \int f(u)du + \phi(v)$ since $y = y(u,v)$

$= \psi(u) + \phi(v)$

$= \psi(x + \alpha t) + \phi(x - \alpha t)$ is the general solution

2.7.5 Heat Conduction Equation

Another important partial differential equation in physics and engineering is the **heat conduction equation**. Consider a length $L = \Delta x$ of thin rod of unit cross-section. There is an initial temperature distribution along the rod – it might be a contant, or some varying function of x. Then, at $t = 0$, the ends of the rod are set to temperatures u_1 and u_2.

u_1 ———————— u_2

x $x+\Delta x$

Heat will flow within the rod and after a long time, a steady state temperature distribution will be established. Our concern here is with the diffusion of heat energy during that time from the initial state to the final steady state.

In general, the rate of heat flow is given by:

$$\frac{dQ}{dt} = K\frac{\Delta u}{L}$$ where K is the coefficient of thermal conductivity ($\mathrm{W\,m^{-1}\,K^{-1}}$) and L is the length of the rod.

The energy into, or out from the rod at x during period Δt is $\Delta Q_x = K\frac{\partial u}{\partial x}\Delta t.$

The energy into, or out from the rod at $x+\Delta x$ during period Δt is

$$\Delta Q_{x+\Delta x} = K\frac{\partial u}{\partial x}\Delta t.$$

The difference is the heat energy within the rod and is thus given by:

$$\begin{aligned}\Delta Q &= K\left(\frac{\partial u}{\partial x}_x - \frac{\partial u}{\partial x}_{x+\Delta x}\right)\Delta t \\ &= K\left(f'_x(x,t) - f'_x(x+\Delta x,t)\right)\Delta t \quad \text{since } u = f(x,t)\end{aligned}$$

Now, this heat raises temperature of the mass of the rod by Δu,: $\Delta Q = mc\Delta u$ where c is the specific heat and the mass m can be expressed in terms of the density ρ per unit length: $m = \rho\Delta x$

$$\begin{aligned}\text{Thus, } \Delta Q &= \rho\Delta x c\Delta u \\ &= K\left(f'_x(x,t) - f'_x(x+\Delta x,t)\right)\Delta t\end{aligned}$$

$$\frac{\Delta u}{\Delta t} = \frac{K}{c\rho}\frac{\left(f'_x(x,t) - f'_x(x+\Delta x,t)\right)}{\Delta x}$$

$$\boxed{\frac{\partial u}{\partial t} = \kappa\frac{\partial^2 u}{\partial x^2}}$$

$$\kappa = \frac{K}{c\rho}$$

Heat conduction equation in 1D

The solution to this equation gives the temperature at any point x along the rod at time t. The constant κ is called the **diffusivity**.

2.7.6 Solution to the Heat Conduction Equation

To solve the heat conduction equation, we need to consider the initial conditions. Let them be that both ends of the rod are kept at 0°C and there is some initial temperature distribution $f(x)$ along the rod. We wish to know what the temperature at any point along the rod might be as a function of x and t.

$$\frac{\partial u}{\partial t} = K\frac{\partial^2 u}{\partial x^2} \quad \text{where } u(0,t)=0; u(L,t)=0; u(x,0)=f(x)$$

Let: $u = XT$ where X is a function of x only and T is a function of t only.

$$XT' = KX''T$$

$$\frac{1}{K}\frac{T'}{T} = \frac{X''}{X} = -\lambda^2$$

Auxiliary equations:

$$m + \lambda^2 K = 0 \qquad m + \lambda^2 = 0$$

$$m = -\lambda^2 K \qquad m = \pm\lambda i$$

$$T = e^{-\lambda^2 KT} \qquad X = A\cos\lambda x + B\sin\lambda x$$

The solution equation is thus: $u(x,t) = (A\cos\lambda x + B\sin\lambda x)e^{-\lambda^2 Kt}$

Consider the initial conditions: $u(0,t) = 0 = Ae^{-\lambda^2 Kt}$

$$A = 0$$

$$u(L,t) = 0 = (B\sin\lambda L)e^{-\lambda^2 Kt}$$

$$\lambda = \frac{n\pi}{L}$$

Thus: $u(x,t) = B\sin\frac{n\pi x}{L}e^{-\lambda^2 KT}$

$$= \sum_{n=1}^{\infty} B\sin\frac{n\pi x}{L}e^{-\lambda^2 KT}$$

$u(x,0) = f(x)$ Temperature distribution at $t = 0$

$f(x) = \sum_{n=1}^{\infty} B_n \sin\frac{n\pi x}{L}$ Fourier sine series

thus $B_n = \frac{2}{L}\int_0^L f(x)\sin\frac{n\pi x}{L}dx$

Final solution: $u(x,t) = \sum_{n=1}^{\infty}\left(\frac{2}{L}\int_0^L f(x)\sin\frac{n\pi x}{L}dx\right)\sin\frac{n\pi x}{L}e^{-\lambda^2 Kt}$

2.7.7 Heat Equation for a Thin Rod of Infinite Length

The general heat flow equation can be extended to the case of a rod of infinite length. The equation, and associated boundary conditions, are:

$$\frac{\partial^2 u}{\partial x^2} = \frac{1}{K}\frac{\partial u}{\partial t} \qquad -\infty < x < \infty$$

$$t > 0$$

$$u(x,0) = f(x)$$

We let $u(x,t) = XT$ as before

$$X = A\cos\lambda x + B\sin\lambda x$$

$$T = e^{-\lambda^2 Kt}$$

$$u(x,t) = e^{-\lambda^2 Kt}\left(A\cos\lambda x + B\sin\lambda x\right)$$

Now, λ may take on any value, so we replace A and B by $A(\lambda)$, $B(\lambda)$ and integrate the series from 0 to ∞

$$u(x,t) = \int_0^\infty e^{-\lambda^2 Kt}\left(A(\lambda)\cos\lambda x + B(\lambda)\sin\lambda x\right)d\lambda$$

Making use of the boundary conditions: $u(x,0) = f(x)$

$$f(x) = \int_0^\infty \left(A(\lambda)\cos\lambda x + B(\lambda)\sin\lambda x\right)d\lambda \qquad \textbf{Fourier integral}$$

where $A(\lambda) = \frac{1}{\pi}\int_{-\infty}^{\infty} f(x)\cos\lambda x dx; B(\lambda) = \frac{1}{\pi}\int_{-\infty}^{\infty} f(x)\sin\lambda x dx$

Therefore, the final answer is:

$$u(x,t) = \frac{1}{\pi}\int_0^\infty e^{-\lambda^2 Kt}\left(\left(\int_{-\infty}^{\infty} f(\alpha)\cos\lambda\alpha d\alpha\right)\cos\lambda x + \left(\int_{-\infty}^{\infty} f(\alpha)\sin\lambda\alpha d\alpha\right)\sin\lambda x\right)d\lambda$$

$$= \frac{1}{\pi}\int_0^\infty \int_{-\infty}^{\infty} f(\alpha)\left[\cos\lambda\alpha\cos\lambda x + \sin\lambda\alpha\sin\lambda x\right]e^{-K\lambda^2 t}d\alpha d\lambda$$

$$= \frac{1}{\pi}\int_{-\infty}^{\infty} f(\alpha)\int_0^\infty \left[\cos\lambda(\alpha - x)\right]e^{-K\lambda^2 t}d\lambda d\alpha$$

α is a dummy variable whereby $f(\alpha)$ is the temperature distribution at $t = 0$.

$$= \frac{1}{\pi}\int_{-\infty}^{\infty} f(\alpha)\left(\sqrt{\frac{\pi}{4Kt}}e^{-\frac{(\alpha - x)^2}{4Kt}}\right)d\alpha$$

2.8 Numerical Methods

Summary

$$x_{n+1} = x_n - \frac{f(x_n)}{f'(x_n)}$$ Newton's method

Newton–Gregory forward polynomial

$$P_n(x) = fo + s\Delta^1 f_o + \frac{s(s-1)}{2!}\Delta^2 f_o + \frac{s(s-1)(s-2)}{3!}\Delta^3 f_o + \ldots$$

Linear least squares

$$a = \frac{N\sum xy - \sum x \sum y}{N\sum x^2 - \left(\sum x\right)^2} \quad b = \frac{\sum y \sum x^2 - \sum x \sum xy}{N\sum x^2 - \left(\sum x\right)^2}$$

$$\int_a^b f(x) \approx \sum_{i=1}^{N} f(x_i)\Delta x$$ Mid-point rule

Trapezoidal rule

$$\int_a^b f(x)dx \approx \frac{b-a}{2N}\left(f(x_o) + 2f(x_1) + 2f(x_2)\ldots + 2f(x_{N-1}) + f(x_N)\right)$$

Simpson's rule

$$\int_a^b f(x)dx \approx \frac{b-a}{3N}\begin{bmatrix} \left(f(x_o) + f(x_N)\right) + \\ 4\left(f(x_1) + f(x_3) + \ldots + f(x_{N-1})\right) + \\ 2\left(f(x_2) + f(x_4) + f(x_6) + \ldots + f(x_{N-2})\right) \end{bmatrix}$$

Runge–Kutta method (4th order)

$$\left.\begin{aligned} k_1 &= hf(x_o, y_o) \\ k_2 &= hf\left(x_o + \frac{h}{2}, y_o + \frac{k_1}{2}\right) \\ k_3 &= hf\left(x_o + \frac{h}{2}, y_o + \frac{k_2}{2}\right) \\ k_4 &= hf(x_o + h, y_o + k_3) \end{aligned}\right\} \quad f(x) = y_o + \frac{1}{6}(k_1 + 2k_2 + 2k_3 + k_4)$$

2.8.1 Newton's Method

It is often required to find the root of a polynomial function. For a first degree polynomial (i.e., a straight line), this involves finding x when $y = 0$.

$$y = ax + b$$

$$x = -\frac{b}{a} \quad \text{at } y = 0$$

For a second degree polynomial, the quadratic formula can be used:

$$y = ax^2 + bx + c$$

$$x = \frac{-b \pm \sqrt{b^2 - 4ac}}{2a}$$

For higher degree polynomials, numerical methods which provide an approximate answer are usually required. These methods usually involve an estimation or guess of an initial starting value from which to begin.

In **Newton's method**, an initial estimate of the root x_1 is made. A tangent is drawn and the value of x_2 where this meets the x axis is determined. From geometry, we then have:

$$\tan\theta = \frac{f(x_1)}{x_1 - x_2}$$

$$= f'(x)$$

Thus:

$$x_2 = x_1 - \frac{f(x)}{f'(x_1)}$$

first approximation to the root

The process is repeated for as many times as needed until the change in the estimated value of the root is as small as desired. In general, the approximation to the root is given by:

$$\boxed{x_{n+1} = x_n - \frac{f(x_n)}{f'(x_n)}} \qquad n = 1,2,3\ldots$$

2.8.2 Interpolating Polynomial

It is often desirable to represent a set of data with a polynomial function since then values of the function at intermediate values of the independent variable can be predicted. A matrix method could be used to determine the coefficients of an (n-1)th degree polynomial from n data points, but this can become unwieldy when there are a large number of data points to fit. A numerical method based upon a **difference table** often provides a more tractable solution. For the case of equally spaced data we have:

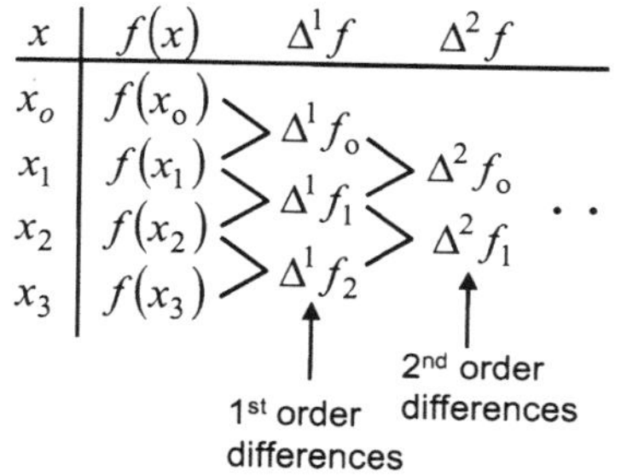

x	$f(x)$	$\Delta^1 f$	$\Delta^2 f$	
x_o	$f(x_o)$			
		$\Delta^1 f_o$		
x_1	$f(x_1)$		$\Delta^2 f_o$	
		$\Delta^1 f_1$		. . .
x_2	$f(x_2)$		$\Delta^2 f_1$	
		$\Delta^1 f_2$		
x_3	$f(x_3)$			
		1st order differences	2nd order differences	

We then form an **interpolating polynomial** or **Newton–Gregory forward polynomial** of degree n:

$$P_n(x) = f_o + s\Delta^1 f_o + \frac{s(s-1)}{2!}\Delta^2 f_o + \frac{s(s-1)(s-2)}{3!}\Delta^3 f_o + \ldots$$

$$\text{where } s = \frac{x - x_o}{\Delta x}$$

If the data is an exact fit to a polynomial of degree n, then it is found that the nth order differences $\Delta^n f(x)$ are all equal and the next column of differences $\Delta^{n+1} f(x)$ is, as a consequence, zero.

When the data is not equally spaced, the **divided differences method** gives:

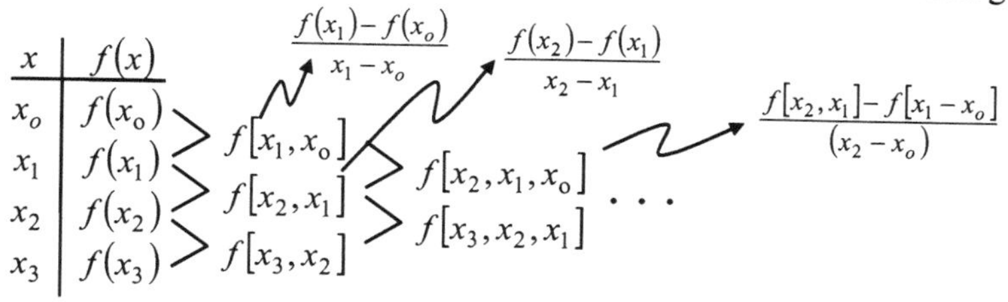

x	$f(x)$			
x_o	$f(x_o)$			
		$f[x_1, x_o]$		
x_1	$f(x_1)$		$f[x_2, x_1, x_o]$	
		$f[x_2, x_1]$		. . .
x_2	$f(x_2)$		$f[x_3, x_2, x_1]$	
		$f[x_3, x_2]$		
x_3	$f(x_3)$			

where $f[x_1, x_o] = \dfrac{f(x_1) - f(x_o)}{x_1 - x_o}$, $f[x_2, x_1] = \dfrac{f(x_2) - f(x_1)}{x_2 - x_1}$, $f[x_2, x_1, x_o] = \dfrac{f[x_2, x_1] - f[x_1 - x_o]}{(x_2 - x_o)}$

and the interpolating polynomial is expressed:

$$P_n(x) = f[x_o] + (x - x_o)f[x_1, x_o] + (x - x_o)(x - x_1)f[x_2, x_1, x_o] + (x - x_o)(x - x_1)(x - x_2)f[x_3, x_2, x_1, x_o] + \ldots$$

2.8.3 Linear Least Squares

(i) Linear least squares

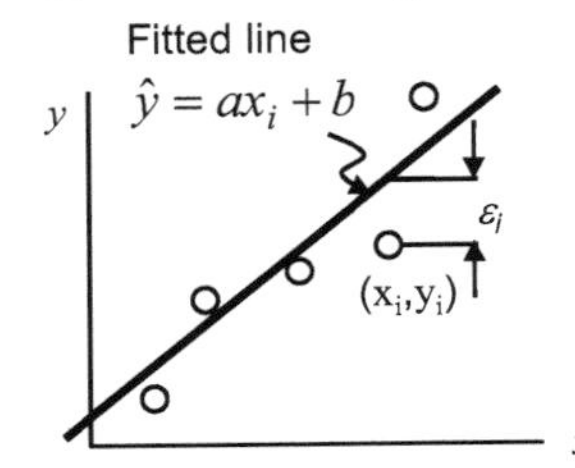

$\hat{y} = ax_i + b$ Expected value of y at x_i

y_i = Observed value of y at x_i

$\varepsilon_i = y - y_i$ Difference between the fitted and observed values at x_i

The method of least squares requires that the fitted line be such that the sum of the squares of the differences between the fitted and observed values is a minimum. $S = \sum \varepsilon_i^2 = \sum (y_i - \hat{y})^2 = \sum (y_i - ax_i - b)^2$ is a minimum

Minimum occurs at:

$$\frac{\partial S}{\partial a} = 0 \qquad\qquad \frac{\partial S}{\partial b} = 0$$

$$0 = \frac{\partial}{\partial a}\sum (y_i - ax_i - b)^2 \qquad 0 = \frac{\partial}{\partial b}\sum (y_i - ax_i - b)^2$$

$$= \sum 2(y_i - ax_i - b)(-x_i) \qquad = \sum -2(y_i - ax_i - b)$$

$$= \sum -x_i y_i + ax_i^2 + bx_i \qquad = \sum y_i - ax_i - b$$

$$\sum x_i y_i = a\sum x_i^2 + b\sum x_i \qquad \sum y_i = a\sum x_i + Nb$$ **Normal equations**

Thus: $a = \dfrac{N\sum xy - \sum x \sum y}{N\sum x^2 - \left(\sum x\right)^2}$ $\quad b = \dfrac{\sum y \sum x^2 - \sum x \sum xy}{N\sum x^2 - \left(\sum x\right)^2}$

(ii) Exponential: $\hat{y}_i = ab^{x_i}$

Normal equations: $\sum \log y = N\log a + (\log b)\sum x$

$$\sum (x\log y) = (\log a)\sum x + (\log b)\sum x^2$$

(iii) Power function: $\hat{y}_i = ax_i^{\,b}$

Normal equations: $\sum \log y = N\log a + b\sum \log x$

$$\sum (\log x \log y) = (\log a)\sum \log x + b\sum \log^2 x$$

2.8.4 Non-Linear Least Squares

Similar procedures can be followed for data that is non-linear. In many cases, a polynomial of sufficiently high degree may be found to fit the data reasonably well. Consider a polynomial of degree n:

$$\hat{y}_i = a_o + a_1 x_i + a_2 x_i^{\,2} + a_3 x_i^{\,3} + \ldots + a_n x_i^{\,n}$$

$$\varepsilon_i = y_i - \hat{y}_i$$

$$S = \sum_{i=1}^{N} \left(y_i - a_o - a_1 x_i - a_2 x_i^{\,2} - a_3 x_i^{\,3} - \ldots a_n x_i^{\,n}\right)^2$$

$$0 = \frac{\partial S}{\partial a_o} = \sum_{i=1}^{N} 2\left(y_i - a_o - a_1 x_i - a_2 x_i^{\,2} - a_3 x_i^{\,3} - \ldots a_n x_i^{\,n}\right)\left(-x_i^{\,0}\right)$$

$$0 = \frac{\partial S}{\partial a_1} = \sum_{i=1}^{N} 2\left(y_i - a_o - a_1 x_i - a_2 x_i^{\,2} - a_3 x_i^{\,3} - \ldots a_n x_i^{\,n}\right)\left(-x_i^{\,1}\right)$$

.

.

number of data pairs > n+1

$$0 = \frac{\partial S}{\partial a_n} = \sum_{i=1}^{N} 2\left(y_i - a_o - a_1 x_i - a_2 x_i^{\,2} - a_3 x_i^{\,3} - \ldots a_n x_i^{\,n}\right)\left(-x_i^{\,n}\right)$$

With some rearrangement, this becomes a matrix equation:

$$\begin{bmatrix} N & \sum x_i & \sum x_i^{\,2} & . & \sum x_i^{\,n} \\ \sum x_i & \sum x_i^{\,2} & \sum x_i^{\,3} & . & \sum x_i^{\,n+1} \\ \sum x_i^{\,2} & \sum x_i^{\,3} & \sum x_i^{\,4} & . & \sum x_i^{\,n+2} \\ . & . & . & . & . \\ \sum x_i^{\,n} & \sum x_i^{\,n+1} & \sum x_i^{\,n+2} & . & \sum x_i^{\,2n} \end{bmatrix} \begin{bmatrix} a_o \\ a_1 \\ a_2 \\ . \\ a_n \end{bmatrix} = \begin{bmatrix} \sum y_i \\ \sum x_i y_i \\ \sum x_i^{\,2} y_i \\ . \\ \sum x_i^{\,n} y_i \end{bmatrix}$$

Note that the quantity x_i^{2n} needs to be calculated, which can potentially lead to significant round-off or truncation errors when this computation is done by computer. This matrix equation can then be solved using the usual methods.

2.8.5 Error Propagation through Equations

Consider $y = f(x)$. At any single measurement at x_i, we have:

$$\begin{aligned} y_i &= f(x_i) \\ &= f(x_i - \bar{x} + \bar{x}) \\ &= f(\bar{x}) + (x_i - \bar{x})\frac{dy}{dx} + \frac{(x_i - x)^2}{2}\frac{d^2y}{dx^2} \quad \text{if } (x_i - \bar{x}) \text{ is small} \end{aligned}$$

Taylor series

Now, considering all the y_i's:

$$\begin{aligned} \bar{y} &= \frac{\sum \bar{y}_i}{N} \\ &= \frac{1}{N}\left(\sum f(\bar{x}) + (x_i - \bar{x})\frac{dy}{dx} + \frac{(x_i - x)^2}{2}\frac{d^2y}{dx^2}\right) \\ &= f(\bar{x}) + \frac{dy}{dx}\underbrace{\sum \frac{x_i - \bar{x}}{N}}_{=0} + \frac{1}{2}\frac{d^2y}{dx^2}\underbrace{\sum \frac{(x_i - \bar{x})^2}{N}}_{=\sigma^2(x)} \end{aligned}$$

$$\bar{y} = f(\bar{x}) + \frac{1}{2}\sum \frac{d^2y}{dx^2}\sigma^2(x) \qquad \text{Note: } \int \frac{1}{2}\frac{d^2y}{dx^2} = \left(\frac{dy}{dx}\right)^2$$

$$\sigma^2(y) = \left(\frac{dy}{dx}\right)^2 \sigma^2(x)$$

For $z = f(x,y)$ then:

$$\sigma^2(z) = \left(\frac{\partial z}{\partial x}\right)^2_{\bar{x},\bar{y}} \sigma^2(x) + \left(\frac{\partial z}{\partial y}\right)^2_{\bar{x},\bar{y}} \sigma^2(y)$$

In general:

$$\sigma^2(z) = \sum_1^r \left(\frac{\partial z}{\partial x_i}\right)^2 \sigma^2(x_i)$$

Example:
For linear least squares, using the methods above gives the standard errors for a and b as:

$$\sigma^2(a) = \frac{\sigma^2}{N\left(\overline{x^2} - (\bar{x})^2\right)} \qquad \sigma^2(b) = \frac{\sigma^2}{N}\left(1 + \frac{\bar{x}^2}{N\left(\overline{x^2} - (\bar{x})^2\right)}\right)$$

2.8.6 Cubic Spline

A **cubic spline** fit attempts to fit the data exactly with a cubic equation in each interval such that at each point, neighbouring cubics have the same values of y_i, and the same values of the first and second derivatives with respect to x_i. Such a fitting procedure may avoid large excursions in the fitted line when higher order polynomial fitting does not adequately represent the data.

The end points of a cubic spline fit require special consideration. There are a number of choices:

(i) Natural spline: $\dfrac{d^2y}{dx^2} = 0$. End cubics are linear.

(ii) Parabolic spline: $\dfrac{d^2y}{dx^2} = C$. End cubics are parabolas.

(ii) Specify: $\dfrac{dy}{dx}$ at the end points explicitly.

For the interior points:

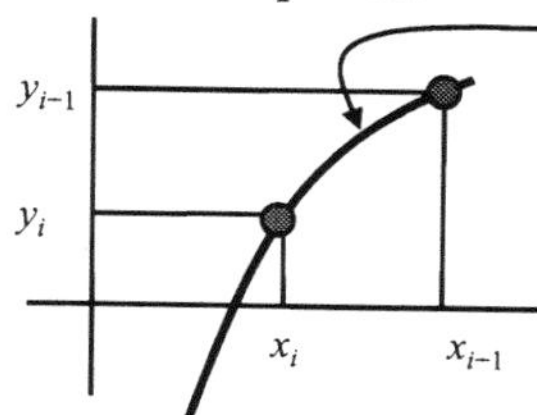

Each segment is a cubic

$$y_i = a_i(x - x_i)^3 + b_i(x - x_i)^2 + c_i(x - x_i) + d_i$$

for: $x_i \le x \le x_{i+1}$

At $x = x_i$, $y_i = d_i$

At $x = x_{i+1}$, $y_{i+1} = a_i(x_{i+1} - x_i)^3 + b_i(x_{i+1} - x_i)^2 + c_i(x_{i+1} - x_i) + y_i$

The procedure is then to build up equations so that compatibility between the first and second derivatives is maintained at adjoining segments. The objective is to find the coefficients a_i, b_i, c_i and d_i for each segment. The equation to be solved for each point becomes:

$$\frac{dy}{dx} = 3a_i(x - x_i)^2 + 2b_i(x - x_i) + c_i$$

$$\frac{d^2x}{dy^2} = 6a_i(x - x_i) + 2b_i$$

$$\Delta x_{i-1}\frac{d^2y}{dx^2}_{i-1} + (2\Delta x_{i-1} + 2\Delta x_i)\frac{d^2y}{dx^2}_{i} + \Delta x_i\frac{d^2y}{dx^2}_{i+1} = 6\left(\frac{y_{i+1} - y_i}{\Delta x_i} - \frac{y_i - y_{i-1}}{\Delta x_{i-1}}\right)$$

When conditions for the end points are included, we obtain n equations in n unknowns (n is the total number of points) which can be solved by matrix methods. The complexity of the calculation is simplified when the points are equally spaced.

2.8.7 Differentiation

Differentiation of numerical or tabular data can take many forms and the selection of the method depends very much on the character of the data.

For a series of y and corresponding x values (not necessarily equally spaced), the simplest calculation of the derivative at any interval is:

$$\frac{dy}{dx} \approx \frac{y_{i+1} - y_i}{x_{i+1} - x_i}$$

Naturally, the error in the estimation of dy/dx using this method will be greatly increased if the data contains a significant amount of uncertainty or noise.

If it is known that the data is not exact, then the derivative of a fitted polynomial or cubic spline to the data may offer more accurate results. Fitting a smooth curve is essentially filtering out the high frequency "noise" from the data. This is important for differentiation because numerical differentiation amplifies the noise in the y values.

If the data contains high frequency noise, say $y = \sin\omega t$, then any differentiation of this is such that:

$$\frac{d}{dt}\sin\omega t = \omega\cos\omega t$$

If ω is large, then the contribution to the derivative of the data from the noise will be large – the noise is "amplified" in the derivative since ω is a product in the derivative.

In general, the following choices are generally appropriate:

- Use difference equations (interpolating polynomials) and then differentiate.
- Fit the data with a cubic spline (if data is to be exact) and differentiate.
- Fit data using least squares method and differentiate (inexact data) the fitted equation analytically.
- Use a Fourier transform and discard high frequency components, then differentiate analytically.

2.8.8 Integration

Midpoint rule

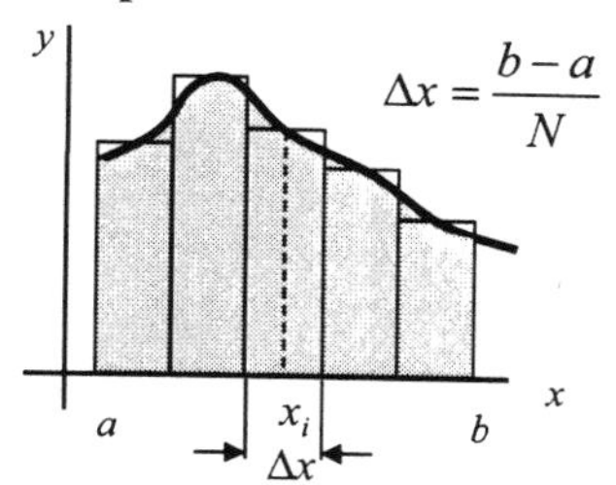

$$\int_a^b f(x) \approx \sum_{i=1}^{N} f(x_i)\Delta x$$

Trapezoidal rule

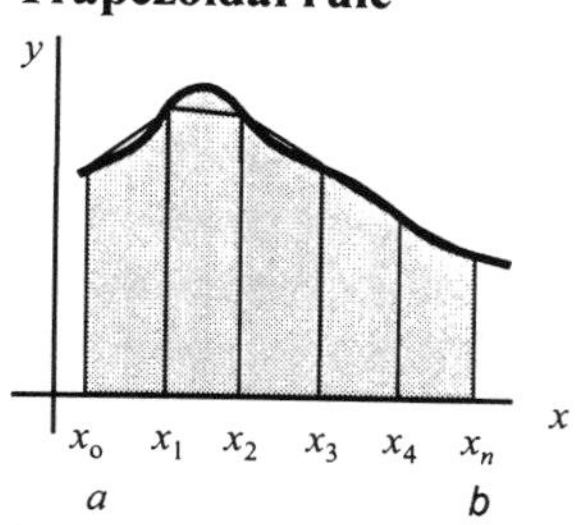

$$\int_a^b f(x)dx \approx \frac{b-a}{2N}\left(f(x_o)+2f(x_1)+2f(x_2)\ldots+2f(x_{N-1})+f(x_N)\right)$$

Simpson's rule

$y = ax^2 + bx + c$

$y = f(x)$

x_o x_1 x_2 x_3 x_4 x_n

a b

$$\int_a^b f(x)dx \approx \frac{b-a}{3N}\left[\begin{array}{l}(f(x_o)+f(x_N))+4(f(x_1)+f(x_3)+f(x_5)+\ldots+f(x_{N-1}))+\\ 2(f(x_2)+f(x_4)+f(x_6)+\ldots+f(x_{N-2}))\end{array}\right]$$

Integration of tabulated data

- If the x's are equally spaced, then use Simpson's rule.
- If the x's are not equally spaced, then use the trapezoidal rule.
- Spline fit and then integrate (inexact).
- Least squares fit and then integrate.

If the data contains high frequency noise, say $y = \sin \omega t$, then any integration of this serves to attenuate this noise:

$$\int \sin \omega t dt = \frac{1}{\omega}\cos \omega t$$

If ω is large, then the contribution to the integral of the data from the noise will be small due to the factor $1/\omega$ – the noise is "attenuated" in the integral.

2.8.9 First Order Ordinary Differential Equations

Consider a first order **differential equation** (higher order equations are treated as systems of first order equations): $\dfrac{dy}{dx} = f(x, y)$ with $f(x_o) = y_o$

The solution to this equation is the function $f(x)$ which connects y and x (i.e., $y = f(x)$). A value of y for any value of x can be obtained by travelling along from the known value of y at x_0 on the assumption that the slope does not change very much at each increment.

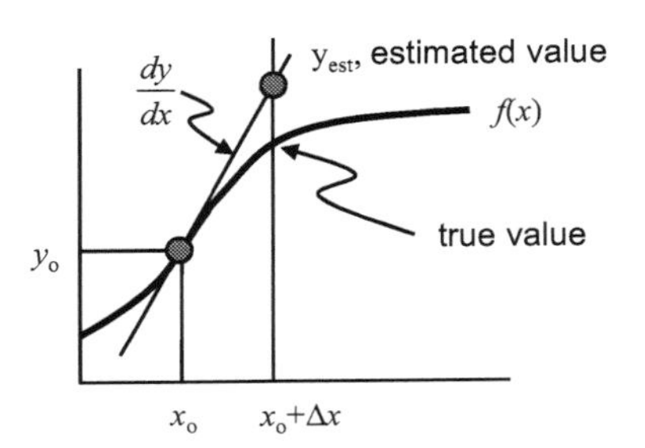

An estimated value of y at $x_0+\Delta x$ is calculated from the slope at x_0, y_0, which is known from the boundary conditions. A new slope is then calculated from the initial equation at $x_0+\Delta x$, y_{est} and then another increment Δx is taken and the procedure repeated until we arrive at the desired value of x.

This procedure is known as the **Euler method**. A more precise estimation of the true value of the function would be obtained if the average slope (taken at the beginning and end points of the interval) is used. Since the slope at the end point is initially unknown, a predicted value of y at the end point is first used and the procedure repeated until the desired accuracy is obtained.

Alternately, a **Taylor series** (with coefficients as yet undetermined) can be used to represent the value of y for a particular value x given an initial condition of y_0 at a value x_0:

$$f(x) = \underset{\displaystyle \downarrow \atop \displaystyle y_0}{f(x_0)} + \frac{f'(x_0)}{1!}(x - x_0) + \frac{f''(x_0)}{2!}(x - x_0)^2 + \ldots + \frac{f^n(x_0)}{n!}(x - x_0)^n + \ldots$$

Now, $f'(x)$ is given by the initial equation being solved. This then is the coefficient of the second term in the expansion. Higher order derivatives of the initial equation, evaluated at $x = x_0$, provide coefficients for the remaining terms in the series. The number of terms depends on the desired level of error in the series expansion, which is usually when the contribution to the series sum becomes very small.

2.8.10 Runge–Kutta Method

A 4th order **Runge–Kutta method** simulates a Taylor series but does not require the calculation of higher order derivatives (which is a convenient advantage). Letting $h = x - x_o$, we form the values:

$$
\left.
\begin{aligned}
k_1 &= hf\left(x_o, y_o\right) \\
k_2 &= hf\left(x_o + \frac{h}{2}, y_o + \frac{k_1}{2}\right) \\
k_3 &= hf\left(x_o + \frac{h}{2}, y_o + \frac{k_2}{2}\right) \\
k_4 &= hf\left(x_o + h, y_o + k_3\right)
\end{aligned}
\right\}
$$

In each step, the derivative is calculated 4 times and a weighted average used to compute $f(x)$ for an increment $x - x_o$.

$$f(x) \approx y_o + \frac{1}{6}\left(k_1 + 2k_2 + 2k_3 + k_4\right)$$

Estimated value of y at $x = x_o + h$

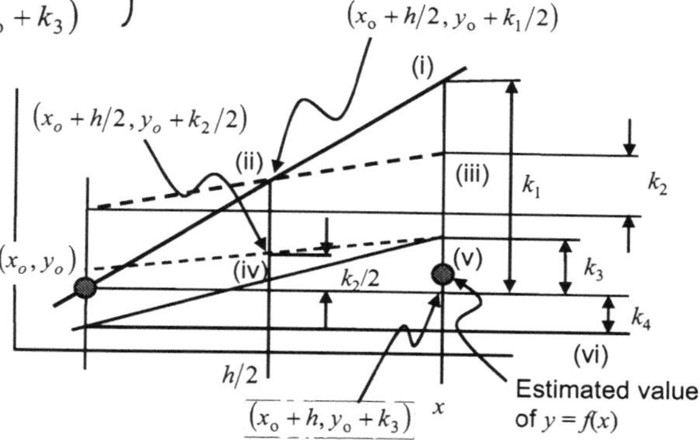

How it works:

(i) Starting from the initial point (x_o, y_o), go to $x = x_o + h$ and go up an amount given by $y_o + k_1$ where k_1 is found from the slope of the function $y = f(x)$. i.e., from the original differential equation: $\left.\frac{dy}{dx}\right|_{x_o, y_o} = f\left(x_o, y_o\right) = \frac{k_1}{h}$

(ii) Go to $x = h/2$ and draw a line up to the line drawn in (i). The slope of $y = f(x)$ (the unknown function) at this point is given by $f(x_o + h/2, y_o + k_1/2)$ – i.e., from the original differential equation. Thus, k_2 is given by:

$$f\left(x_o + h, y_o + k_1/2\right) = \frac{k_2}{h}$$

(iii) Draw the line of slope given by (ii) and measure off a value for k_2 at x.

(iv) Go to a point $(x_o + h/2, y_o + k_2/2)$ and calculate the slope of the function at that point from:

$$f\left(x_o + h, y_o + k_2/2\right) = \frac{k_3}{h}$$

(v) Draw the line of slope determined in (iv) and read of a value of k_3 at x.

(vi) At x, calculate the slope of the line for $y_o + k_3$ and draw this line thus obtaining a value for k_4. Then calculate $f(x + h)$ from the formula.

2.8.11 Finite Element Method

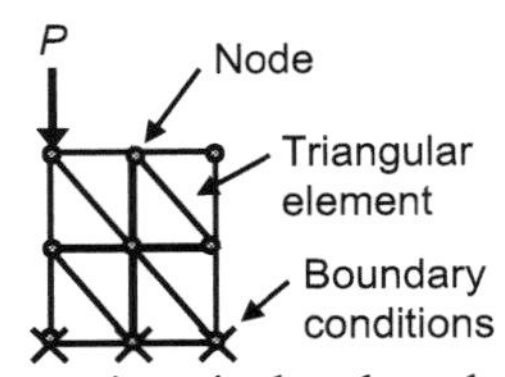

Consider the deformation of an elastic body under load P. The geometry of the body is divided into a series of small elements connected by nodes.

To solve the deformation of the body, a set of matrix equations is developed for one element and then all the matrix equations assembled into a system of equations which represent the geometry of the specimen as a whole.

Let ε be the strain associated with node displacements u. In matrix notation, we have:

$$\boldsymbol{\varepsilon} = \begin{bmatrix} \partial/\partial x & 0 \\ 0 & \partial/\partial y \\ \partial/\partial y & \partial/\partial x \end{bmatrix} \mathbf{u}$$

ε = strains in x and y directions

Stresses $\boldsymbol{\sigma}$ and strains $\boldsymbol{\varepsilon}$ are connected by Hooke's law:

$$\boldsymbol{\sigma} = \frac{E}{(1+\nu)(1-2\nu)} \begin{bmatrix} 1-\nu & & \\ \nu & 1-\nu & \\ & & (1-2\nu)/2 \end{bmatrix} \boldsymbol{\varepsilon}$$

$$= \mathbf{E}\boldsymbol{\varepsilon}$$

If the entire structure is in equilibrium, then so is each element. Thus, for mechanical equilibrium $\partial W = \partial U$, where ∂W is the virtual work done by external loads and ∂U is the work done by internal forces. If ∂u_i is the displacement of a node i acted on by a force F_i, then at equilibrium:

> The f.e. method can be used to solve a wide variety of physical problems involving ordinary and partial differential equations with multiple variables and complicated geometry.

$$\delta \mathbf{u}_i^T \mathbf{F}_i = \int_V \delta \boldsymbol{\varepsilon}^T \boldsymbol{\sigma}\, dV = \int_V \delta \mathbf{u}_i^T \mathbf{B}^T \mathbf{E} \mathbf{B}\, \mathbf{u}_i\, dV$$

$$\mathbf{F}_i = \int_V \mathbf{B}^T \mathbf{E} \mathbf{B}\, \mathbf{u}_i\, dV$$

Element stiffness matrix

$$\mathbf{F}_i = \mathbf{K} \mathbf{u}_i$$ Element matrix equation

$$\mathbf{P}_i = \mathbf{K} \mathbf{u}_i$$ Structure matrix equation

The structure **stiffness matrix** is formed by successively adding stiffness terms from each element into appropriate locations in the stiffness matrix. The matrices P are the applied loads. The procedure for the whole structure is thus:

i. Assemble stiffness matrix **K**
ii. Apply boundary constraints
iii. Solve matrix equation for the unknown displacements $\boxed{\mathbf{u}_i = \mathbf{K}^{-1} \mathbf{P}_i}$

It can be appreciated that the solution for a **finite element analysis** involves the inversion of a matrix, which can be quite a large undertaking. It is no surprise that the popularity of the finite element method has coincided with the development of readily available computers for this purpose.

Part 3

Applications

3.1 Capacitance
3.2 Solid Mechanics
3.3 Signal Processing
3.4 Fourier Optics
3.5 Quantum Mechanics

3.1 Capacitance

Summary

$\varepsilon_r = \dfrac{\varepsilon}{\varepsilon_o}$	Relative permittivity
$\varepsilon^* = \varepsilon' - j\varepsilon''$	Complex relative permittivity
$Z = R + j(-X_C) = R + \left(\dfrac{1}{j\omega C}\right)$	Series impedance
$\dfrac{1}{Z} = \dfrac{1}{R} + j\left(\dfrac{1}{X_C}\right) = \dfrac{1}{R} + j\omega C$	Parallel impedance
$Z_1 Z_3 = Z_2 Z_4$	General bridge equation
$r = \dfrac{C_4}{C_3} R_2$	Dynamic resistance from Schering bridge
$C = \dfrac{R_4}{R_2} C_3$	Total capacitance from Schering bridge

3.1.1 Permittivity

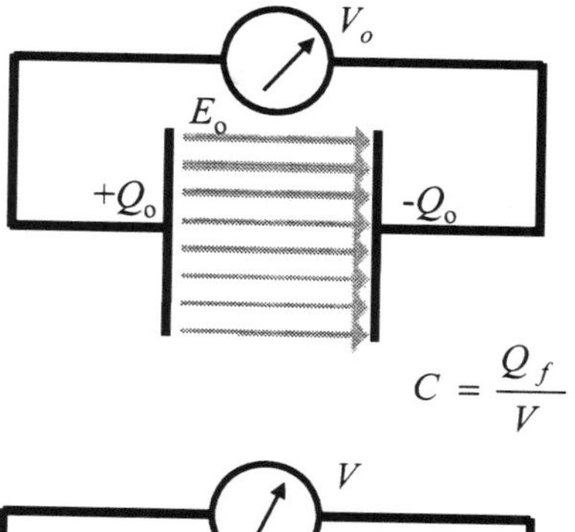

Consider a parallel plate capacitor charged to a voltage V_o and then the voltage source is removed. A voltmeter placed across the capacitor terminals will register the voltage V_o used to charge the capacitor.

$$C = \frac{Q_f}{V}$$

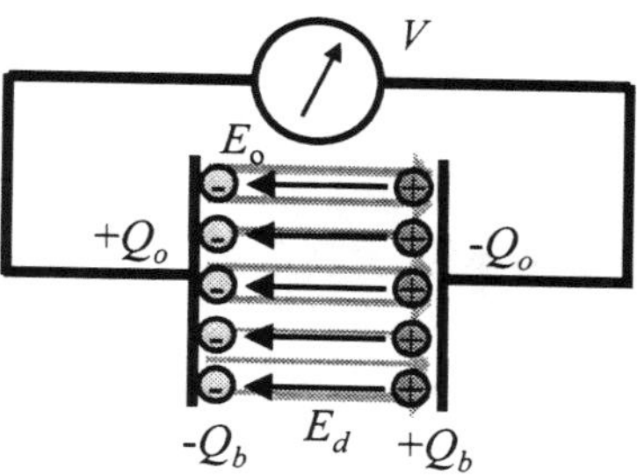

When a dielectric is inserted, dipoles are created within the **dielectric** material because of the field E_o, leading to a net additional charge Q_b (i.e., in addition to Q_o) being deposited on each plate.

Since the total net charge on each plate is now reduced in magnitude, the voltage recorded on a voltmeter across the plates is reduced to V and the total average field between the plates is reduced from E_o to E. But, the total *free* charge on each plate (i.e., the charge initially drawn from the voltage source) remains the same: $Q_f = Q_o$.

Since the distance d is a constant, the ratio of voltage V_o/V gives the relative change in capacitance of the plates. By definition, the **dielectric constant**, or **relative permittivity**, ε_r, of an insulating material is the capacitance with the dielectric material inserted divided by the capacitance when the plates are in a vacuum.

$$\varepsilon_r = \frac{C}{C_o} = \frac{V_o}{V}$$

Expressed in terms of the fields E_o and E, we have:

$$\varepsilon_r = \frac{E_o}{E}$$

The **relative permittivity** is a macroscopic measure of the amount by which a dielectric material is polarised by an electric field. The permittivity of an actual dielectric is found from:

$$\varepsilon_r = \frac{\varepsilon}{\varepsilon_o}$$

ε_o → Permittivity of free space (vacuum) 8.85×10^{-12} F m^{-1}

3.1.2 Complex Permittivity

An ideal dielectric is non-conducting, but there are always losses in a real dielectric (e.g., **leakage**). For a capacitor, this is represented by a **loss resistor** R_L:

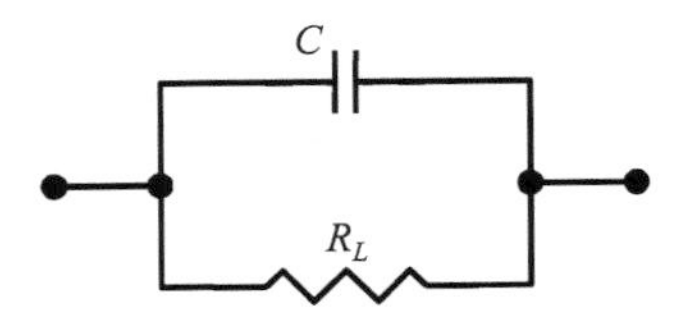

Without the loss resistor, the capacitance of the capacitor (with a dielectric) is:

$$C = \varepsilon_r \left[\varepsilon_o \frac{A}{d} \right]$$

where $C_o = \varepsilon_o \dfrac{A}{d}$

With the loss resistor, the impedance of the capacitor is found from:

$$V = IZ$$

where $\dfrac{1}{Z} = \dfrac{1}{R_L} + j\omega C$

Now, the resistance R_L can be expressed in terms of the **conductivity** σ of the dielectric:

resistivity

$$R_L = \rho \frac{d}{A}$$

$$\frac{1}{R_L} = \sigma \frac{A}{d}$$

conductivity

Note: R is small when A/d is large. That is, for a given d, large capacitors have greater losses.

The impedance of the capacitor becomes:

$$\begin{aligned}
\frac{1}{Z} &= \sigma \frac{A}{d} + j\omega C \\
&= \sigma \frac{A}{d} + j\omega \varepsilon_r C_o \\
&= j\omega C_o \left(\varepsilon_r - j\sigma \frac{A}{d} \frac{1}{\omega C_o} \right) \\
&= j\omega C_o \left(\varepsilon_r - j\sigma \frac{A}{d} \frac{1}{\omega \varepsilon_o} \frac{d}{A} \right) \\
&= j\omega C_o \left(\varepsilon_r - j \frac{\sigma}{\omega \varepsilon_o} \right) \\
&= j\omega C_o (\varepsilon' - j\varepsilon'')
\end{aligned}$$

where $\varepsilon' = \varepsilon_r$

$$\varepsilon'' = \frac{\sigma}{\omega \varepsilon_o}$$

Complex relative permittivity

$$\varepsilon^* = \varepsilon' - j\varepsilon''$$

This term reflects the resistive or conduction losses in a capacitor. It may also contain terms associated with frictional losses in dipolar materials.

The **phase angle** δ is a measure of the quality of a dielectric. The **loss factor** D is given by:

$$D = \tan \delta = \frac{1}{R_L \omega C} = \frac{\varepsilon''}{\varepsilon'}$$

- Ideal capacitor: $D = 0$
- High quality dielectric: $D \approx 10$
- Lossy capacitor: $D \approx 0.05$

When R_L and C are constant, the D is inversely proportional to ω.

Time constant τ of a dielectric is:

$$\tau = R_L C = \rho \varepsilon_o \varepsilon_r$$

τ is independent of the dimensions of the capacitor and only depends on the nature of the dielectric.

3.1.3 Series Impedance

For a series electrical circuit, the **impedance** is the vector sum of the resistances and the reactances within the circuit.

$$Z = \frac{V}{I} \quad \text{by definition}$$

(*V*, *I* can be either peak or rms. Note, all quantities are vectors.)

$$V = V_R + j(V_L - V_C)$$

$$\frac{V}{I} = \frac{V_R}{I} + j\frac{(V_L - V_C)}{I}$$

but

$R = \frac{V_R}{I}$ Resistance

$X_C = \frac{V_C}{I}$ Capacitive reactance

$X_L = \frac{V_L}{I}$ Inductive reactance

Thus

$$Z = R + j(X_L - X_C) = R + j\left(\omega L - \frac{1}{\omega C}\right)$$

Note: $|Z| = \frac{|V|}{|I|}$ Magnitudes of the peaks or rms voltages. For series or parallel circuits, we cannot simply add peak or rms values.

If $X_C > X_L$, circuit is **capacitively reactive**.
If $X_C < X_L$, circuit is **inductively reactive**.
If $X_C = X_L$ - then **resonant**.

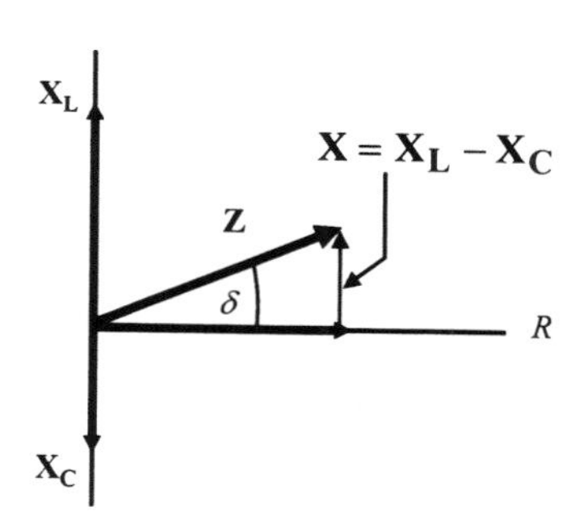

Modulus of *Z*

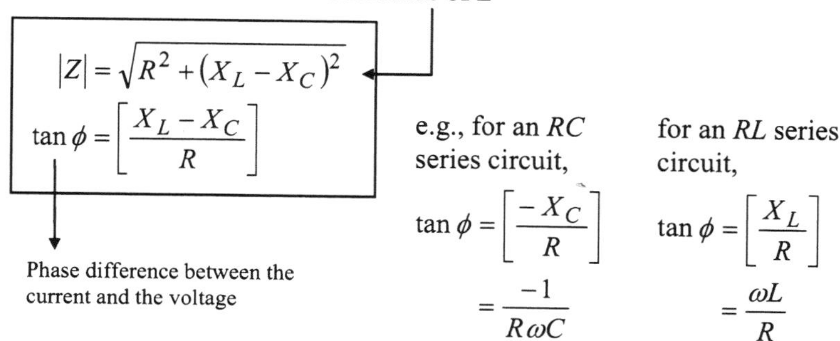

$$|Z| = \sqrt{R^2 + (X_L - X_C)^2}$$

$$\tan\phi = \left[\frac{X_L - X_C}{R}\right]$$

Phase difference between the current and the voltage

e.g., for an *RC* series circuit,

$$\tan\phi = \left[\frac{-X_C}{R}\right] = \frac{-1}{R\omega C}$$

for an *RL* series circuit,

$$\tan\phi = \left[\frac{X_L}{R}\right] = \frac{\omega L}{R}$$

3.1.4 Parallel Impedance

A **parallel circuit** is one in which the same voltage appears across all components. For parallel circuits, the voltage is the common point of reference (rather than the current, as was the case in series circuits).

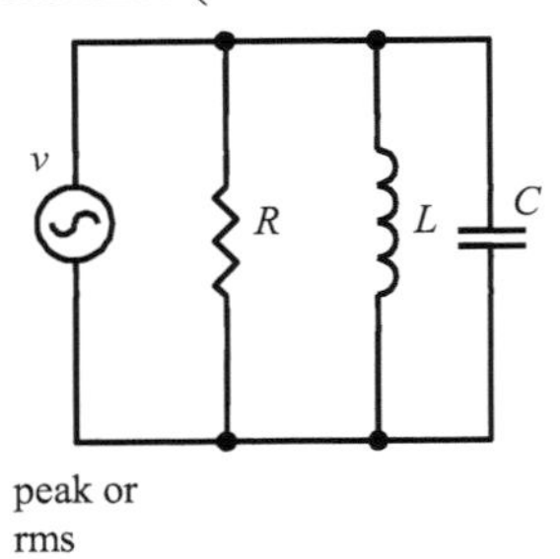

In a parallel circuit, the instantaneous current across the resistor is always in phase with the instantaneous voltage; thus, for the capacitor and the inductor:

- I_C always precedes I_R by $\pi/2$; thus I_C is upwards on the vertical axis;
- I_L always follows I_R by $\pi/2$; thus I_L is downwards on the vertical axis.

peak or rms

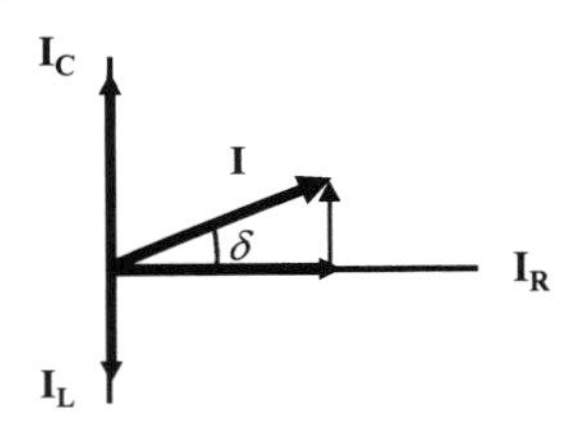

Voltage is the common point of reference for a parallel circuit:

- same voltage across all components;
- vector sum of rms or peak currents must equal the rms or peak current;
- algebraic sum of instantaneous currents equals the total instantaneous current.

In complex number form:

$$\frac{1}{Z} = \frac{1}{R} + j\left(\frac{1}{X_C} - \frac{1}{X_L}\right)$$

Note: this formula is consistent with addition of parallel impedances Z. The "j" has been moved to the numerator by multiplying through by j/j and remembering that $j^2 = -1$ (hence positions of X_L and X_C reversed).

At **resonance**, $X_C = X_L$ and total current i is a minimum since Z is a maximum.

$$\omega_R{}^2 = \frac{1}{LC}$$

Parallel circuits exhibit high impedance at resonance.

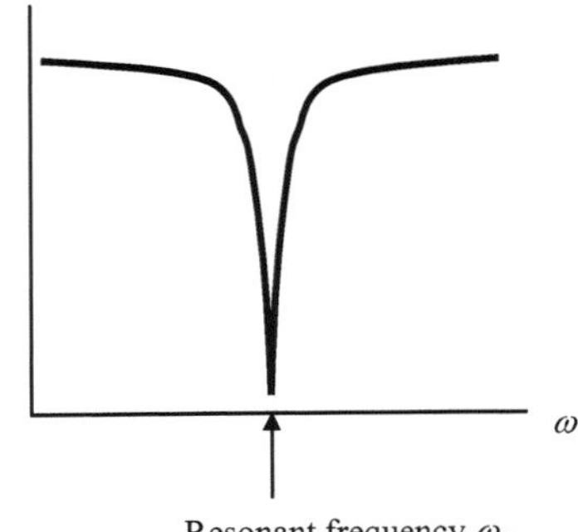

3.1.5 AC Bridge

For the galvanometer G to read zero, the voltage across its terminals must be zero. Thus, the voltage across Z_1 and Z_3 must be equal in magnitude and phase.

$$V_1 = V_3$$

$$I_1 Z_1 = I_3 Z_3$$

At balance condition, no current flows through the galvanometer; thus:

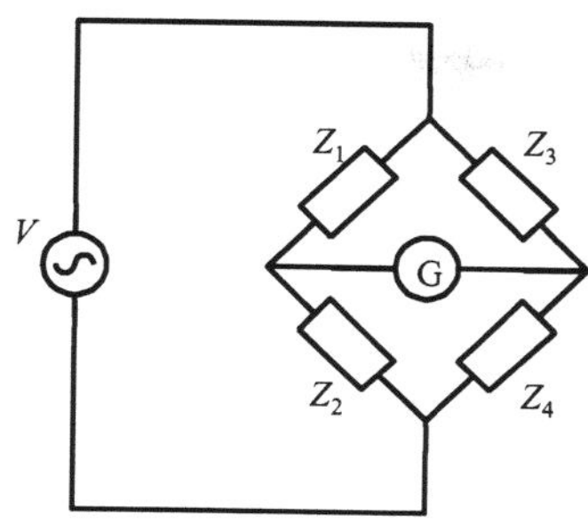

$$V = I_1(Z_1 + Z_2)$$

$$= I_3(Z_3 + Z_4)$$

$$I_1 = \frac{V}{Z_1 + Z_2}; \quad I_3 = \frac{V}{Z_3 + Z_4}$$

$$I_1 Z_1 = \frac{Z_1 V}{Z_1 + Z_2}; \quad I_3 Z_3 = \frac{Z_3 V}{Z_3 + Z_4}$$

$$\frac{Z_3 V}{Z_3 + Z_4} = \frac{Z_1 V}{Z_1 + Z_2}$$

⇩

$$\boxed{Z_1 Z_4 = Z_2 Z_3}$$

General bridge equation at balance condition

Example

$$Z_1 Z_4 = R_1\left(R_4 - \frac{1}{\omega C_4} j\right)$$

$$= R_1 R_4 - \frac{R_1}{\omega C_4} j$$

$$Z_2 Z_3 = \left(0 - \frac{1}{\omega C_3} j\right)(R_u + \omega L_u j)$$

$$= \frac{L_u}{C_3} + \frac{-R_u}{\omega C_3} j$$

$$Z_1 = R_1$$

$$Z_2 = 0 - \frac{1}{\omega C_3 j}$$

$$Z_3 = R_u + \omega L_u$$

$$Z_4 = R_4 - \frac{1}{\omega C_4}$$

$$R_1 R_4 - \frac{R_1}{\omega C_4} j = \frac{L_u}{C_3} + \frac{-R_u}{\omega C_3} j$$

$$\boxed{R_1 R_4 = \frac{L_u}{C_3} \qquad \frac{R_1}{C_4} = \frac{R_u}{C_3}}$$

3.1.6 Capacitor Equivalent Circuit

A parallel plate capacitor consists of an equivalent circuit comprising a leakage resistance R_B, the capacitance of the dielectric test material C_B, and the capacitance of any air gap C_A between the material and the parallel plates or that might exist within the dielectric.

$$Z_T = Z_A + Z_B$$

$$Z_A = \frac{1}{j\omega C_A}$$

$$\frac{1}{Z_B} = \frac{1}{R_B} + j\omega C_B$$

$$Z_B = \frac{1}{\frac{1}{R_B} + j\omega C_B}$$

$$= \frac{R_B}{1 + j\omega R_B C_B}$$

$$= \frac{R_B\left(1 - j\omega R_B C_B\right)}{1 + \omega^2 R_B{}^2 C_B{}^2}$$

$$= \frac{R_B}{1 + \omega^2 R_B{}^2 C_B{}^2} + \frac{1}{j\omega C_B}\frac{\omega^2 R_B{}^2 C_B{}^2}{1 + \omega^2 R_B{}^2 C_B{}^2}$$

$$Z_T = \underbrace{\frac{R_B}{1 + \omega^2 R_B{}^2 C_B{}^2}}_{r} + \frac{1}{j\omega}\underbrace{\left(\frac{1}{C_A} + \frac{\omega^2 R_B{}^2 C_B{}^2}{C_B\left(1 + \omega^2 R_B{}^2 C_B{}^2\right)}\right)}_{C}$$

r – an effective equivalent dynamic resistance (note the dependence on ω)

C – an effective total overall capacitance

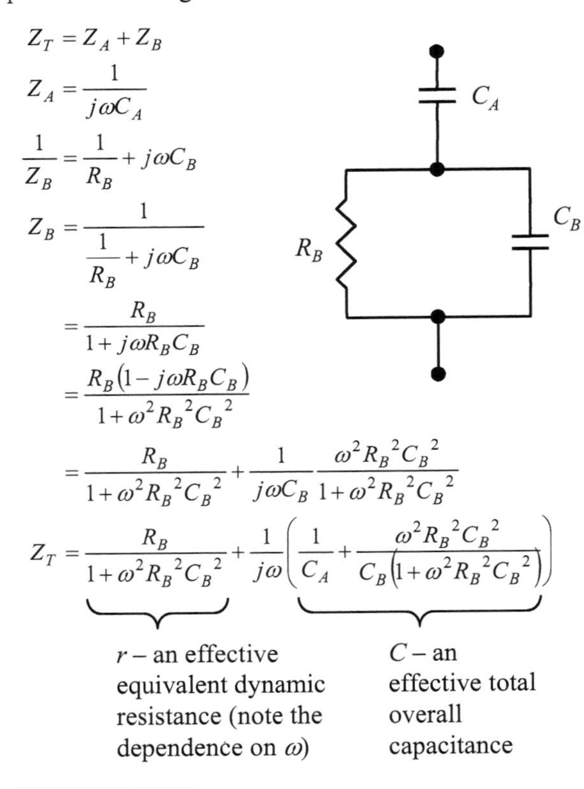

$$\boxed{Z_T = r + \frac{1}{j\omega C}}$$

Series R and C

$$Z = R + j\left(-X_C\right)$$

$$= R + \left(\frac{1}{j\omega C}\right)$$

Parallel R and C

$$\frac{1}{Z} = \frac{1}{R} + j\left(\frac{1}{X_C}\right)$$

$$= \frac{1}{R} + j\omega C$$

In general,

$$\frac{1}{a + bj} = \frac{a - bj}{a^2 + b^2}$$

3.1.7 Schering Bridge

The **Schering bridge** is used for the measurement of the equivalent dynamic resistance r and overall capacitance C of a capacitor.

$$Z_1 = r + \frac{1}{j\omega C}$$

$$Z_2 = R_2$$

$$Z_3 = \frac{1}{j\omega C_3}$$

$$Z_4 = \frac{R_4}{1 + j\omega C_4 R_4}$$

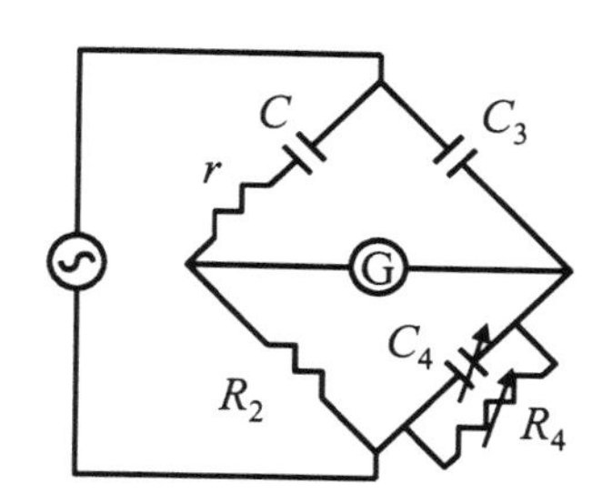

General bridge equation: $Z_1 Z_4 = Z_2 Z_3$ at balance condition.

$$\left(r - \frac{1}{\omega C} j\right)\left(\frac{R_4}{1 + j\omega C_4 R_4}\right) = \left(\frac{-1}{\omega C_3} j\right) R_2$$

$$\left(r - \frac{1}{\omega C} j\right)\left(\frac{1}{1 + j\omega C_4 R_4}\right) = -\frac{R_2}{R_4}\left(\frac{1}{\omega C_3} j\right)$$

$$\left(r - \frac{1}{\omega C} j\right)\frac{R_4}{R_2} = -\frac{1}{\omega C_3} j\left(1 + j\omega C_4 R_4\right)$$

$$\frac{rR_4}{R_2} - \frac{R_4}{R_2 \omega C} j = -\frac{1}{\omega C_3} j + \frac{\omega C_4 R_4}{\omega C_3}$$

$$\frac{rR_4}{R_2} = \frac{C_4 R_4}{C_3}$$

$$\frac{r}{R_2} = \frac{C_4}{C_3}$$

Therefore: $r = \frac{C_4}{C_3} R_2$

and: $\frac{-R_4}{R_2 \omega C} = -\frac{1}{\omega C_3}$

$$R_4 C_3 = R_2 C$$

$$C = \frac{R_4}{R_2} C_3$$

3.1.8 Measurement of Relative Permittivity

To measure the relative permittivity of a dielectric, the dielectric material is inserted between the plates of a parallel plate capacitor whose plate area and distance of separation are known, and then connected to a **Schering bridge**. The bridge measures the overall resistance r and capacitance C by adjusting R_3 and C_3 for a balance condition at each frequency.

$$r = \frac{R_B}{1+\omega^2 R_B{}^2 C_B{}^2}$$

$$\frac{1}{r} = \frac{1+\omega^2 R_B{}^2 C_B{}^2}{R_B}$$

$$= \frac{1}{R_B} + \omega^2 R_B C_B{}^2$$

$$= \left(R_B C_B{}^2\right)\omega^2 + \frac{1}{R_B}$$

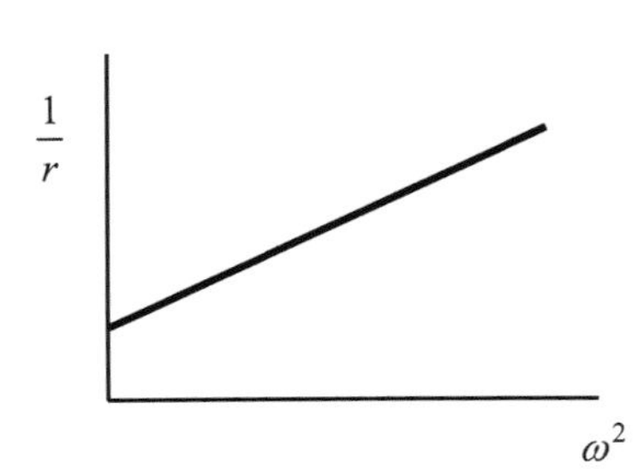

Plot of $1/r$ vs ω^2 should be linear with slope $C_B{}^2 R_B$ and intercept $1/R_B$, thus allowing R_B and C_B to be determined.

$$C = \left(\frac{1}{C_A} + \frac{\omega^2 R_B{}^2 C_B{}^2}{C_B\left(1+\omega^2 R_B{}^2 C_B{}^2\right)}\right)$$

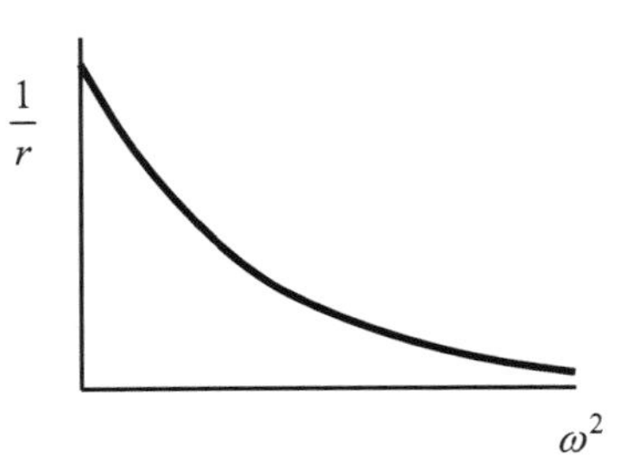

Plot of C vs ω^2 evaluated where $\omega^2 = 0$ enables C_A to be determined.

Once R_B and C_B have been determined, the loss angle δ can be found from:

$$\tan\delta = \frac{1}{\omega C_B R_B}$$

And knowing the dimensions of the capacitor (area of each plate A separation d), the relative permittivity ε_r of the dielectric material only can thus be found from:

$$C_B = \varepsilon_r \varepsilon_o \frac{A}{d}$$

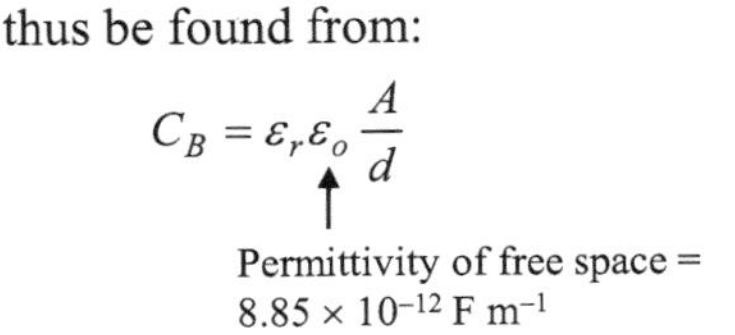

3.2 Solid Mechanics

Summary

$$F = kx$$ Hooke's law

$$\sigma = \frac{F}{A}$$ Stress

$$\varepsilon = \frac{\Delta l}{l}$$ Strain

$$E = \frac{\sigma}{\varepsilon}$$ Young's modulus

$$\varepsilon_x = \frac{1}{E}\left[\sigma_x - \nu\left(\sigma_y + \sigma_z\right)\right]$$ Triaxial stresses

$$\sigma_{1,2} = \frac{\sigma_x + \sigma_y}{2} \pm \sqrt{\left(\frac{\sigma_x - \sigma_y}{2}\right)^2 + \tau_{xy}{}^2}$$ Principal stresses

$$\tau_{\max} = \pm\frac{1}{2}\left(\sigma_1 - \sigma_2\right)$$

$$\frac{\partial^2 \sigma_x}{\partial x^2} + \frac{\partial^2 \tau_{xy}}{\partial x \partial y} = 0$$

$$\frac{\partial^2 \sigma_y}{\partial y^2} + \frac{\partial^2 \tau_{xy}}{\partial x \partial y} = 0$$

Equations of equilibrium and compatibility in 2D

$$\left(\frac{\partial^2}{\partial x^2} + \frac{\partial^2}{\partial y^2}\right)\left(\sigma_x + \sigma_y\right) = 0$$

$$M = \frac{E}{r} I$$ Bending of beams

3.2.1 Hooke's Law

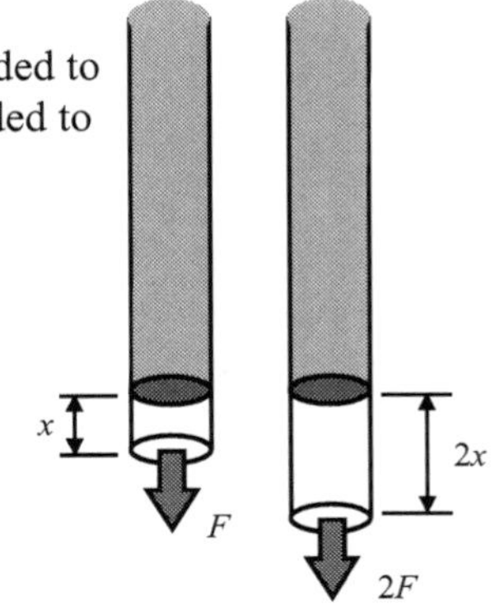

Experiments show that if a certain force was needed to stretch a bar by x, then double the force was needed to stretch the same bar by $2x$:

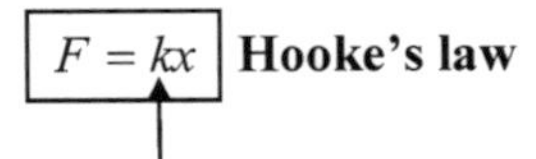

k depends on the type of material and the dimensions of the specimen.

Thomas Young (in 1807) described Hooke's relationship in a way which did not rely on the dimensions of a particular specimen.

$$F = kx$$

$$\frac{F}{A} = \frac{kx}{A} = \frac{kl}{A}\frac{x}{l}$$

$$E = \frac{kl}{A}$$

Let $\frac{F}{A} = E\frac{x}{l}$

$$\sigma = E\varepsilon$$

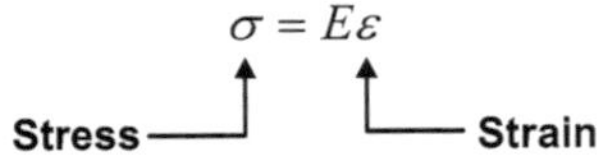

E is a material property which describes the elasticity, or **stiffness** of a material.

Similarly for loads applied in shear,

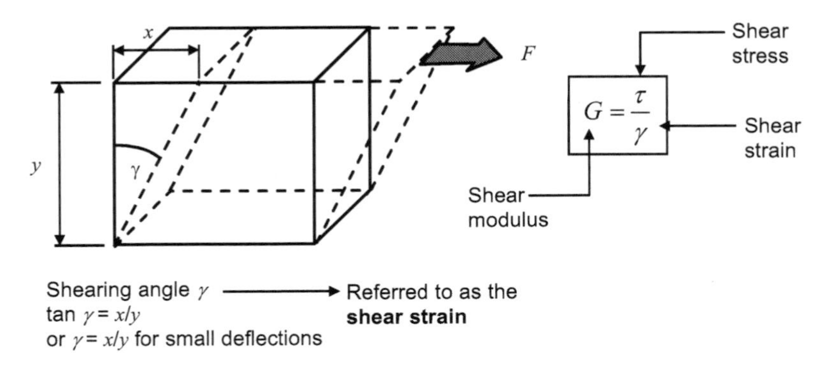

3.2.2 Stress

Tensile stress

When forces tend to pull on a body and thus stretch or elongate it, **tensile stresses** are produced within the material.

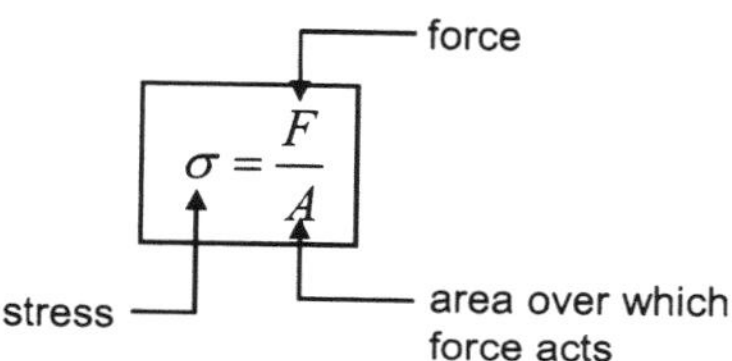

The units of stress are Pa (same as pressure).

Compression and tension are called **normal stresses**.

↓

Because the force producing the stress acts **normal** to the planes under consideration. The symbol σ is used for normal stresses.

Compressive stress

When forces tend to push on a body and thus shorten or compress it, **compressive stresses** are produced within the material.

Shear stress

Force acting parallel to area produces **shear stress** τ.

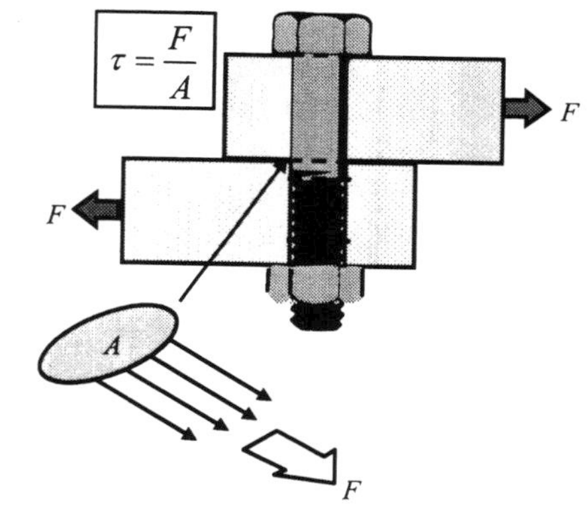

Stresses are labeled with subscripts. The first subscript indicates the direction of the normal to the plane over which the force is applied. The second subscript indicates the direction of the force. **Normal stresses** act normal to the plane, whereas **shear stresses** act parallel to the plane.

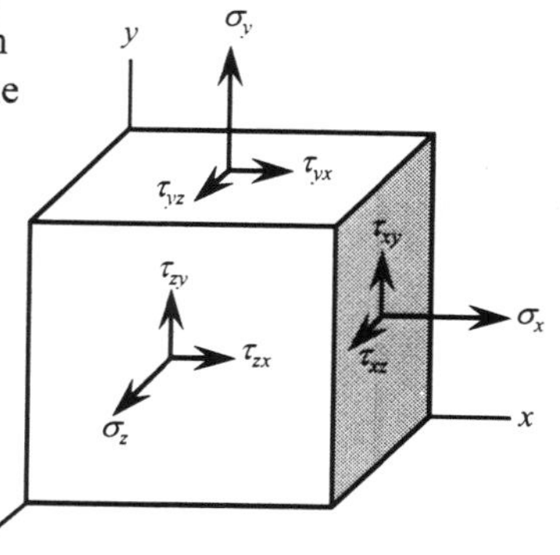

It is generally agreed that a positive shear stress results when the direction of the line of action of the force producing the stress and the direction of the outward normal to the surface are of the same sign; thus, the shear stresses τ_{xy} and τ_{xz} shown here are positive.

3.2.3 Strain

Linear strain

Application of a deforming force causes atoms within the body to be shifted away or displaced from their equilibrium positions. The net effect of this is a measurable change in dimensions of the body.

Strain is the fractional change in length of a body subjected to a deforming force.

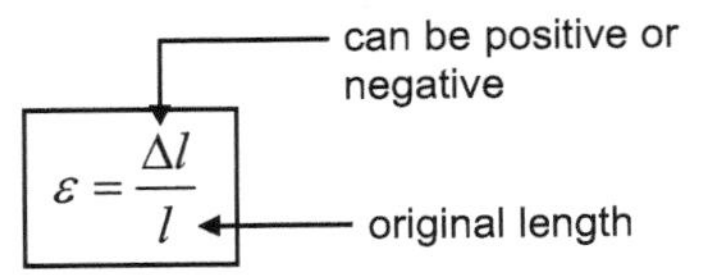

Experiments show that if Δl is plotted against l, (i.e., the displacement of the point is measured at several points along the body), the displacement Δl of the point varies linearly with the distance l.

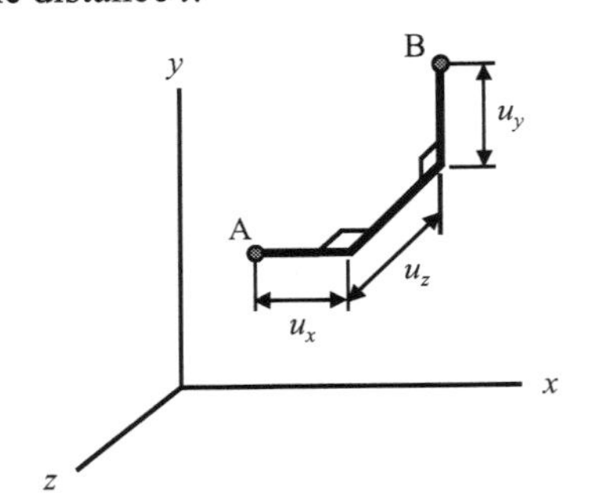

Thus, when a point within the solid undergoes displacements u_x, u_y and u_z, so as to move from A to B, the unit elongations, or strains, are defined as:

$$\varepsilon_x = \frac{\partial u_x}{\partial x}; \varepsilon_y = \frac{\partial u_y}{\partial y}; \varepsilon_z = \frac{\partial u_z}{\partial z}$$

Shear strain

When a point within a solid undergoes displacements as a result of applied stresses, angles between these points may change, leading to internal shear strains.

$$\gamma_{xy} = \frac{\partial u_x}{\partial y} + \frac{\partial u_y}{\partial x}; \gamma_{xz} = \frac{\partial u_x}{\partial z} + \frac{\partial u_z}{\partial x}; \gamma_{yz} = \frac{\partial u_y}{\partial z} + \frac{\partial u_z}{\partial y}$$

3.2.4 Poisson's Ratio

It is observed that for many materials, when stretched or compressed along the length within the elastic limit, there is a contraction or expansion of the sides as well as an extension or compression of the length. **Poisson's ratio** is the ratio of the fractional change in one dimension to the fractional change of the other dimension.

$$\nu = \frac{\frac{\Delta w}{w}}{\frac{\Delta l}{l}}$$

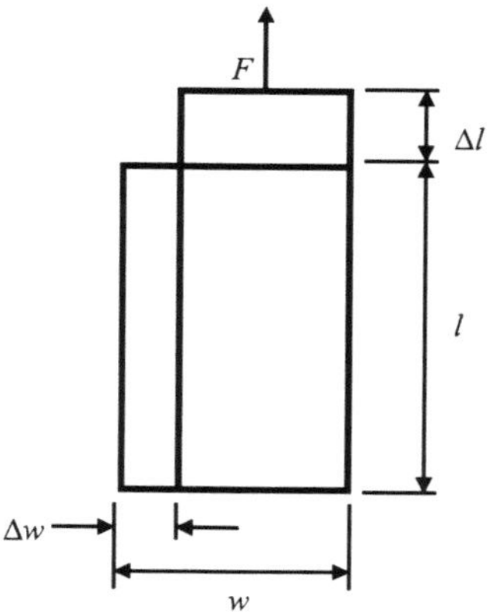

Poisson's ratio is a measure of how much a material tries to maintain a constant volume under compression or tension.

Consider a bar of square cross section $w \times w$ placed in tension under an applied force F. The initial total volume of the bar is:

$$V_1 = A_1 l$$

where $A_1 = w^2$. After the application of load, the length of the bar increases by Δl. The width of the bar decreases by Δw. The volume of the bar is now calculated from:

$$\begin{aligned} V_2 &= (l + \Delta l)(w - \Delta w)^2 \\ &= l(1+\varepsilon)w^2\left(1 - \frac{\Delta w}{w}\right)^2 \\ &= l(1+\varepsilon)A_1(1-\nu\varepsilon)^2 \\ &\approx A_1 l(1 + \varepsilon - 2\nu\varepsilon) \quad \text{since } \varepsilon^2 << 1 \end{aligned}$$

The change in volume is thus:

$$\begin{aligned} V_2 - V_1 &\approx A_1 l - A_1 l(1+\varepsilon - 2\nu\varepsilon) \\ &= A_1 l\varepsilon(1 - 2\nu) \end{aligned}$$

For there to be no volume change, ν has to be less than 0.5. $\nu > 0.5$ implies that the volume decreases with tension, an unlikely event. When $\nu = 0.5$, there is no volume change and the contraction in width is quite pronounced (e.g., rubber). When $\nu = 0$, the volume change is the largest and there is no perceptible contraction in width. Most materials have a value of ν within the range 0.2 to 0.4.

When the material contracts inwards (**plane stress** condition) under an applied tensile stress σ_T, there is no sideways stress induced in the material. If the sides of the material are held in position by external forces or restraints (**plane strain**), then there is a stress σ induced, the value of which is given by $\sigma = \nu\sigma_T$.

In terms of stresses and strains, in plane strain conditions (sides held in position), there is an effective increase in the stiffness of the specimen due to the induced sideways stresses. **Hooke's law** becomes:

$$\sigma = \frac{E}{1-\nu^2}\varepsilon$$

3.2.5 Stress Tensor

A **scalar field** (e.g., temperature) is represented by a single value, which is a function of x, y, z. In matrix form, we write:

$$\mathbf{T} = f(x, y, z)$$
$$U = [\mathbf{T}]$$

A **vector field** (e.g., the electric field) is represented by three components, $E = f(E_x, E_y, E_z)$, where each of these components may be a function of position x, y, z. In matrix form:

$$\mathbf{E} = \begin{bmatrix} E_x \\ E_y \\ E_z \end{bmatrix} \quad \text{where } \begin{aligned} E_x &= f(x, y, z) \\ E_y &= g(x, y, z) \\ E_z &= h(x, y, z) \end{aligned}$$

In solid mechanics, the stresses acting on a solid are given in terms of the normal and shear components acting on each surface. Arranged in matrix format, we write:

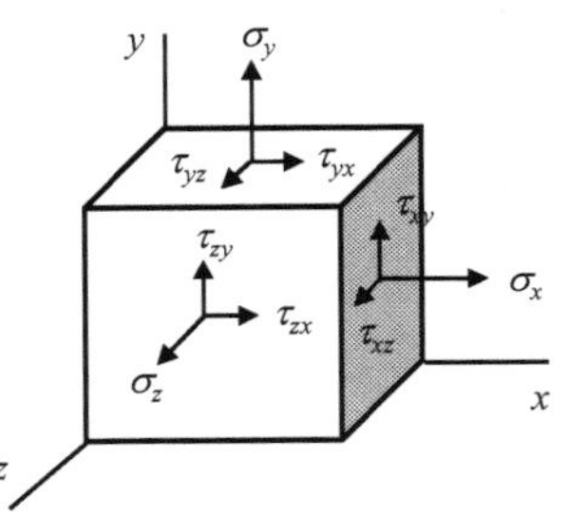

$$\begin{bmatrix} \sigma_{xx} & \tau_{xy} & \tau_{xz} \\ \tau_{yx} & \sigma_{yy} & \tau_{yz} \\ \tau_{zx} & \tau_{zy} & \sigma_{zz} \end{bmatrix}$$

The nine components of the stress matrix are referred to as the **stress tensor**.

A **tensor field**, such as the **stress tensor**, consists of nine components, each of which is a function of x, y and z. The tensor nature of stress arises from the ability of a material to support shear. Any applied force generally produces both normal (i.e., tensile and compressive) stresses and shear stresses.

It is often convenient to consider the total stress at a point within a body as the sum of the average, or **mean, stress** and the **stress deviations**.

$$\begin{bmatrix} \sigma_x & \tau_{xy} & \tau_{xz} \\ \tau_{yx} & \sigma_y & \tau_{yz} \\ \tau_{zx} & \tau_{zy} & \sigma_z \end{bmatrix} = \begin{bmatrix} \sigma_m & 0 & 0 \\ 0 & \sigma_m & 0 \\ 0 & 0 & \sigma_m \end{bmatrix} + \begin{bmatrix} \sigma_{xx} - \sigma_m & \tau_{xy} & \tau_{xz} \\ \tau_{yx} & \sigma_{yy} - \sigma_m & \tau_{yz} \\ \tau_{zx} & \tau_{zy} & \sigma_{zz} - \sigma_m \end{bmatrix}$$

The mean stress is given by: $\sigma_m = \frac{1}{3}(\sigma_{xx} + \sigma_{yy} + \sigma_{zz})$

The remaining stresses, are termed the **deviatoric stresses**. The mean stress is associated with the change in volume of the specimen, and the deviatoric stresses are responsible for any change in shape.

The stress tensor is written with two indices. Vectors require only one index and may be called tensors of the first **rank**. The stress tensor is of rank 2. Scalars are tensors of rank zero.

3.2.6 Triaxial Stresses and Strains

In the general case, stress and strain are related by a matrix of constants E_{ijkl} such that

$$\sigma_{ij} = E_{ijkl}\varepsilon_{kl}$$

For an isotropic solid (i.e., one having the same elastic properties in all directions), the constants E_{ijkl} reduce to two, the so-called Lamé constants μ, λ, and can be expressed in terms of two material properties: **Poisson's ratio**, ν, and **Young's modulus**, E, where

$$E = \frac{\mu(3\lambda + 2\mu)}{\lambda + \mu}; \; \nu = \frac{\lambda}{2(\lambda + \mu)}$$

For a condition of uniaxial **tension** or **compression**, the relationship between stress and strain, for small deformations, is given by:

$$\sigma = E\varepsilon$$

However, for the general state of **triaxial stresses**, one must take into account the strain arising from lateral contraction. For normal stresses and strains:

$$\varepsilon_x = \frac{1}{E}\left[\sigma_x - \nu(\sigma_y + \sigma_z)\right]$$

$$\varepsilon_y = \frac{1}{E}\left[\sigma_y - \nu(\sigma_x + \sigma_z)\right]$$

$$\varepsilon_z = \frac{1}{E}\left[\sigma_z - \nu(\sigma_x + \sigma_y)\right]$$

For shear stresses and strains:

$$\gamma_{xy} = \frac{1}{G}\tau_{xy}$$

$$\gamma_{yz} = \frac{1}{G}\tau_{yz}$$

$$\gamma_{xz} = \frac{1}{G}\tau_{xz}$$

where G is the **shear modulus**, a high value indicating a larger resistance to shear, given by

$$G = \frac{E}{2(1+\nu)}$$

3.2.7 Principal Stresses and Strains

The stress at some particular point in a body depends upon the angle of the plane over which the direction of stress is taken.

Consider the stress acting on a plane dA, which is tilted at an angle θ to the x axis, but whose normal is perpendicular to the z axis. It can be shown that the normal stress acting on dA is:

$$\sigma_\theta = \sigma_x \cos^2\theta + \sigma_y \sin^2\theta + 2\tau_{xy}\sin\theta\cos\theta$$

$$= \frac{1}{2}\left(\sigma_x + \sigma_y\right) + \frac{1}{2}\left(\sigma_x - \sigma_y\right)\cos 2\theta + \tau_{xy}\sin 2\theta$$

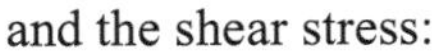

and the shear stress:

$$\tau_\theta = \left(\sigma_x - \sigma_y\right)\sin\theta\cos\theta + \tau_{xy}\left(\sin^2\theta - \cos^2\theta\right)$$

$$= \frac{1}{2}\left(\sigma_x - \sigma_y\right)\sin 2\theta - \tau_{xy}\cos 2\theta$$

It is possible to find three stresses, σ_1, σ_2, σ_3, which act in a direction normal to three orthogonal planes oriented in such a way that there is no shear stress across those planes. Only normal stresses act on these planes and they are called the **principal planes of stress**. The normal stresses acting on the principal planes are called the **principal stresses**. In the two dimensional case, the principal stresses are found from:

$$\sigma_{1,2} = \frac{\sigma_x + \sigma_y}{2} \pm \sqrt{\left(\frac{\sigma_x - \sigma_y}{2}\right)^2 + \tau_{xy}{}^2}$$

τ_{xy} is the shear stress across a plane perpendicular to the x axis in the direction of the y axis.

Shear stresses are a maximum at planes oriented at 45° to the principal planes. The maximum shear stresses are found from:

$$\tau_{\max} = \pm\sqrt{\left(\frac{\sigma_x - \sigma_y}{2}\right)^2 + \tau_{xy}{}^2} = \pm\frac{1}{2}\left(\sigma_1 - \sigma_2\right)$$

Fracture of brittle materials is most often due to the value of the maximum principal stress, while yield in ductile materials is often due to the value of the maximum shear stress.

3.2.8 Equilibrium

For a specimen whose applied loads are in equilibrium, the state of internal stress at points within the specimen must satisfy conditions of **static equilibrium**. Consider a two dimensional element of material within a solid body where the stresses vary from point to point throughout the body.

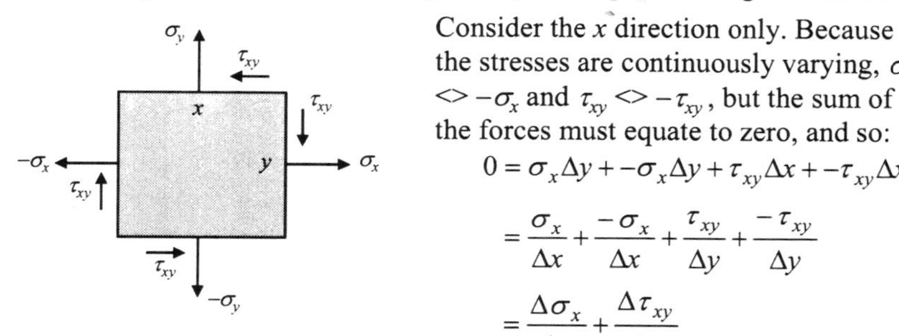

Consider the x direction only. Because the stresses are continuously varying, $\sigma_x <> -\sigma_x$ and $\tau_{xy} <> -\tau_{xy}$, but the sum of the forces must equate to zero, and so:

$$0 = \sigma_x \Delta y + -\sigma_x \Delta y + \tau_{xy} \Delta x + -\tau_{xy} \Delta x$$

$$= \frac{\sigma_x}{\Delta x} + \frac{-\sigma_x}{\Delta x} + \frac{\tau_{xy}}{\Delta y} + \frac{-\tau_{xy}}{\Delta y}$$

$$= \frac{\Delta \sigma_x}{\Delta x} + \frac{\Delta \tau_{xy}}{\Delta y}$$

In the limit of an infinitesimal small volume element in three dimensions, the condition for static **stress equilibrium** thus becomes:

$$\frac{\partial \sigma_x}{\partial x} + \frac{\partial \tau_{xy}}{\partial y} + \frac{\partial \tau_{xz}}{\partial z} = 0$$

$$\frac{\partial \tau_{yx}}{\partial x} + \frac{\partial \sigma_y}{\partial y} + \frac{\partial \tau_{yz}}{\partial z} = 0$$

$$\frac{\partial \tau_{zx}}{\partial x} + \frac{\partial \tau_{zy}}{\partial y} + \frac{\partial \sigma_z}{\partial z} = 0$$

Displacements of points within the solid are required to be smoothly varying from point to point within the specimen. Mathematically, this condition is expressed:

$$\frac{\partial^2 \varepsilon_x}{\partial y^2} + \frac{\partial^2 \varepsilon_y}{\partial x^2} = \frac{\partial^2 \gamma_{xy}}{\partial x \partial y}$$

$$\frac{\partial^2 \varepsilon_y}{\partial z^2} + \frac{\partial^2 \varepsilon_z}{\partial y^2} = \frac{\partial^2 \gamma_{yz}}{\partial y \partial z}$$

$$\frac{\partial^2 \varepsilon_z}{\partial x^2} + \frac{\partial^2 \varepsilon_x}{\partial z^2} = \frac{\partial^2 \gamma_{zx}}{\partial z \partial x}$$

Solutions to problems in elasticity generally require expressions for stress components which satisfy both **equilibrium** and **compatibility** conditions subject to the boundary conditions appropriate to the problem.

3.2.9 Calculation of Stresses and Displacements

Consider a two dimensional body placed under stress, and where the weight of the body is not significant. For static equilibrium and compatibility of displacements, we have:

$$\frac{\partial \sigma_x}{\partial x}+\frac{\partial \tau_{xy}}{\partial y}=0 \qquad \frac{\partial \sigma_y}{\partial y}+\frac{\partial \tau_{yx}}{\partial x}=0 \qquad \text{and} \qquad \frac{\partial^2 \varepsilon_x}{\partial y^2}+\frac{\partial^2 \varepsilon_y}{\partial x^2}=\frac{\partial^2 \gamma_{xy}}{\partial x \partial y}$$

and the relationship between stress and strain is:

$$\varepsilon_x=\frac{1}{E}\left[\sigma_x-\nu\sigma_y\right]; \quad \varepsilon_y=\frac{1}{E}\left[\sigma_y-\nu\sigma_x\right]; \quad \gamma_{xy}=\frac{1}{G}\tau_{xy}=\frac{2(1+\nu)}{E}\tau_{xy}$$

The compatibility equation becomes:

$$\frac{\partial^2}{\partial y^2}\left(\sigma_x-\nu\sigma_y\right)+\frac{\partial^2}{\partial x^2}\left(\sigma_y-\nu\sigma_x\right)=2(1+\nu)\frac{\partial^2}{\partial x \partial y}\tau_{xy}$$

Differentiating the equilibrium equations with respect to x and y, we obtain:

$$\boxed{\begin{aligned}\frac{\partial^2 \sigma_x}{\partial x^2}+\frac{\partial^2 \tau_{xy}}{\partial x \partial y}=0\\ \frac{\partial^2 \sigma_y}{\partial y^2}+\frac{\partial^2 \tau_{xy}}{\partial x \partial y}=0\end{aligned}}$$

$$2\frac{\partial^2 \tau_{xy}}{\partial x \partial y}=-\left(\frac{\partial^2 \sigma_x}{\partial x^2}+\frac{\partial^2 \sigma_y}{\partial y^2}\right)$$

The compatibility equation then becomes:

$$\frac{\partial^2 \sigma_x}{\partial y^2}-\nu\frac{\partial^2 \sigma_y}{\partial y^2}+\frac{\partial^2 \sigma_y}{\partial x^2}-\nu\frac{\partial^2 \sigma_x}{\partial x^2}=-(1+\nu)\left(\frac{\partial^2 \sigma_x}{\partial x^2}+\frac{\partial^2 \sigma_y}{\partial y^2}\right)$$

$$\frac{\partial^2 \sigma_x}{\partial y^2}+\frac{\partial^2 \sigma_y}{\partial x^2}+\frac{\partial^2 \sigma_x}{\partial x^2}+\frac{\partial^2 \sigma_y}{\partial y^2}=0$$

$$\boxed{\left(\frac{\partial^2}{\partial x^2}+\frac{\partial^2}{\partial y^2}\right)\left(\sigma_x+\sigma_y\right)=0}$$

Calculations of stresses and displacements for points within a body reduce to the solution of the above differential equations with the appropriate boundary conditions that apply to the problem at hand.

3.2.10 Moment of Inertia

The case of **bending moments** M applied to the ends of a beam results in the beam being bent into a circular arc – a condition of **pure bending**.

Consider a line drawn parallel to the axis at some position dx that represents the original position of material before bending. The distance $dx = rd\theta$. The displacement y at some point P along the line is $yd\theta$.

The strain at P is:

$$\varepsilon = \frac{yd\theta}{dx} = \frac{y}{r}$$

The stress at P is:

$$\sigma_x = \varepsilon E = \frac{E}{r}y$$

Now, consider an elemental area of area dA in the cross section. The force acting over dA is:

$$dF = \sigma_x dA = \frac{E}{r} ydA$$

The total bending moment due to this force is the sum of the moments for the forces dF:

$$M = \int_A \sigma_x ydA = \frac{E}{r}\int_A y^2 dA$$

The integral term is called the 2nd moment of area, or **moment of inertia**, and given the symbol I.

$$M = \frac{E}{r}I$$

$$\frac{1}{r} = \frac{M}{EI}$$

For example, the moment of inertia for a beam of width w and thickness t is given by a summation of products of areas and distances from the axis X_C of the cross-section. Dividing the beam into elements dy, we have:

$$I_{X_C} = AY^2 = \sum A_i y^2$$

$$= w\int_{-t/2}^{t/2} y^2 dy$$

$$= \left[\frac{wy^3}{3}\right]_{-t/2}^{t/2} = \left[2w\frac{t^3}{24}\right] = \frac{wt^3}{12}$$

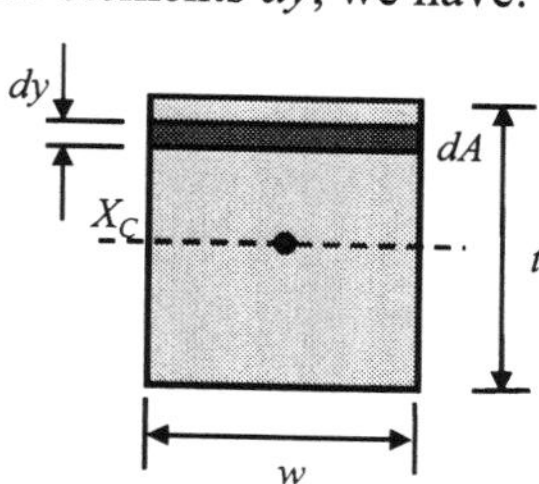

3.2.11 Stresses and Displacements in a Beam

The longitudinal stresses, σ_x, in a beam in **pure bending** may be expressed as a function of position throughout its thickness, y.

$$\sigma_x = \varepsilon E = \frac{E}{r} y$$

$$= \frac{M}{I} y \quad \text{since} \quad \frac{1}{r} = \frac{M}{EI}$$

For pure bending, $\sigma_y = \sigma_z = \tau_{yz} = \tau_{yx} = \tau_{zx} = 0$.

The maximum longitudinal stress occurs at $y = t/2$ and so,

$$\sigma_{max} = \frac{M}{I}\frac{t}{2}$$

$$\frac{M}{I} = \frac{2\sigma_{max}}{t}$$

$$\sigma_x = \sigma_{max}\frac{2y}{t}$$

From the definition of strain at a point, the strain at x is $\varepsilon_x = \frac{y}{r} = \frac{\partial \mu_x}{\partial x}$.

Taking into account the three dimensional deformation and Poisson's ratio, we have:

$$\varepsilon_y = \frac{1}{E}\left[\sigma_y - \nu\left(\sigma_x + \sigma_z\right)\right]$$

$$= \frac{1}{E}\left[0 - \nu\sigma_x\right] \quad \text{for pure bending where } \sigma_x \text{ and } \sigma_z = 0$$

$$= -\nu\frac{\sigma_x}{E}$$

$$= -\nu\frac{E}{r}y\frac{1}{E}$$

$$\varepsilon_y = \frac{\partial u_y}{\partial y} = -\nu\frac{y}{r} = -\nu\varepsilon_x$$

Similarly, $\varepsilon_z = \frac{\partial u_z}{\partial z} = -\nu\frac{y}{r} = -\nu\varepsilon_x$

In pure bending, all the shear strains are zero, and so:

$$0 = \frac{\partial u_x}{\partial y} + \frac{\partial u_y}{\partial x} = \frac{\partial u_x}{\partial z} + \frac{\partial u_z}{\partial x} = \frac{\partial u_y}{\partial z} + \frac{\partial u_z}{\partial y}$$

The displacement in the x direction is the integral:

$$u_x = \int \frac{y}{r} dx \qquad \text{since} \quad \frac{y}{r} = \frac{\partial u_x}{\partial x}$$

$$= \frac{yx}{r} + C \qquad u_x = 0 \;@\; x = 0 \therefore C = 0$$

From the shear strains, in pure bending:

$$0 = \frac{\partial u_x}{\partial y} + \frac{\partial u_y}{\partial x}$$

$$\frac{\partial u_x}{\partial y} = -\frac{\partial u_y}{\partial x}$$

$$-\frac{\partial u_y}{\partial x} = \frac{x}{r}$$

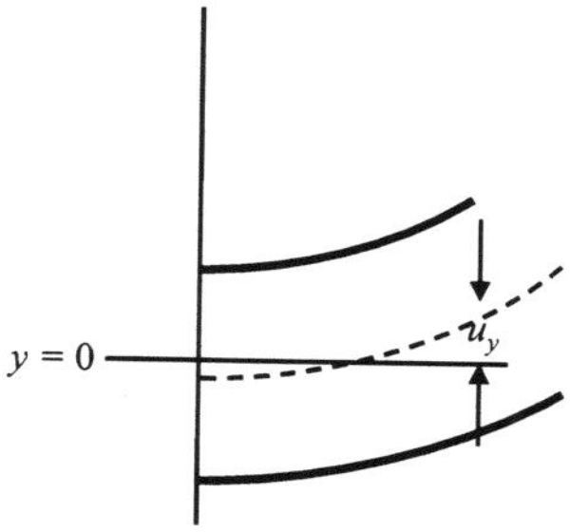

$$u_y = -\int \frac{x}{r} dx$$

$y = 0$

u_y

$$= -\frac{x^2}{2r} + C \qquad u_y = 0 \;@\; x = 0 \therefore C = 0$$

The deflection of the beam u_y is expressed as a function of x:

$$\frac{1}{r} = \frac{M}{EI}$$

$$u_y = -\frac{x^2 M}{2EI}$$

This idealised case of **pure bending** occurs due to the constant bending moment M being applied to the ends of the beam and M is constant along the length of the beam. In general, beams may carry distributed loads and/or a variety of point loads and be supported as either fixed or free at one or both ends. In most cases, the bending moment is not constant along the length of the beam. In these cases, the **elastic curve**, or **deflection curve**, is found from the solution of a second order differential equation:

$$\frac{d^2 u_y}{dx^2} = \frac{M(x)}{EI}$$

Particular solutions are found by imposing boundary conditions. For example, for a cantilevered beam, $dy/dx = 0$ and $y = 0$ at $x = 0$.

3.3 Signal Processing

Summary

$$V_{out} = \frac{1}{RC}\int V_{in}\,dt$$ Integrator (low pass filter)

$$\frac{V_{out}}{V_{in}} = \frac{1}{1 + R\omega Cj}$$ Frequency domain

$$\frac{V_{out}}{V_{in}} = \frac{1}{1 + RCs}$$ s domain

$$V_{out} = RC\frac{dV_{in}}{dt}$$ Differentiator (high pass filter)

$$\frac{V_{out}}{V_{in}} = \frac{R\omega C}{\sqrt{R^2\omega^2C^2 + 1}}$$ Frequency domain

$$\frac{V_{out}}{V_{in}} = \frac{RCs}{1 + RCs}$$ s domain

3.3.1 Transfer Function

Many measurements of physical phenomena are done using a transducer to convert the physical process into an electrical signal, which is typically filtered and amplified before being stored or sent to an output display device. The formal relationship between the input signal and the output response of an electrical circuit is called the **transfer function**.

A simple transfer function is the ratio of the output voltage over the input voltage. For example, for the simple *RC* **low pass filter** shown below, the transfer function is:

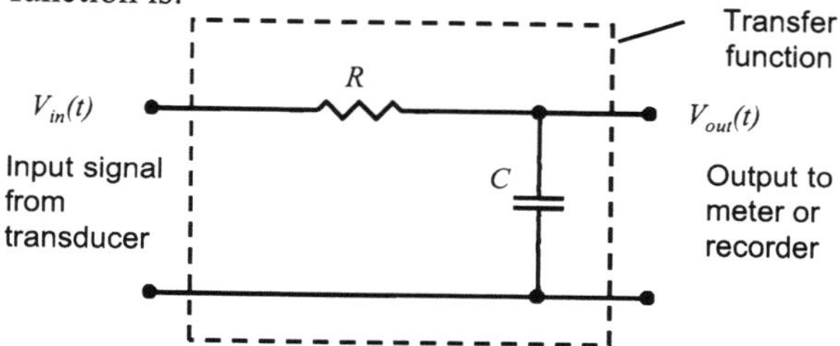

The mathematical expression for the transfer function is given by the ratio of the output impedance to the input impedance.

In simple form:

$$V_{in} = IZ = I\sqrt{R^2 + X_C^{\ 2}}$$

$$V_{out} = IX_C$$

$$\left|\frac{V_{out}}{V_{in}}\right| = \frac{1}{\omega C\sqrt{R^2 + \frac{1}{\omega^2 C^2}}} = \frac{1}{\sqrt{1 + R^2\omega^2 C^2}}$$

In complex form:

$$V_{in} = I(R - X_C j) = I\left(R - \frac{1}{\omega C}j\right)$$

$$V_{out} = I(-X_C j) = I\left(-\frac{1}{\omega C}j\right)$$

$$\frac{V_{out}}{V_{in}} = \frac{\frac{-1}{\omega C}j}{R - \frac{1}{\omega C}j} = \frac{1 - R\omega Cj}{1 + R^2\omega^2 C^2} = \frac{1}{1 + R\omega Cj}$$

As shown here, the transfer function is not necessarily a constant. In the case of a filter, the transfer function H is a function of ω.

3.3.2 Transforms and Operators

Many physical phenomena can be described by **differential equations**. A **differential operator** is a short-hand way of writing "the derivative with respect to." For example:

$$\frac{dy}{dx} = Dy \quad \text{and} \quad \frac{d^2y}{dx^2} = D^2y$$

After the operator D has been applied to the original function y, a new function is formed. That is, the original function has been **transformed** into another function.

original function → $y*D = Y$ ← new function (operator ↓ D)

$$y*D = Y$$

Another example of an operator is the **integral transform operator** T which has the form:

$$T[f(t)] = \int_{-\infty}^{\infty} f(t)K(s,t)\,dt = F(s)$$

$F(s)$ is the transform of $f(t)$.

Here, f is a function of t which is transformed by the operator T to give a new function $F(s)$. K is a function of the variables s and t. The function $F(s)$ is the **integral transform** of the original function $f(t)$.

Consider some interesting forms of the function $K(s,t)$:

$$K(s,t) = 0 \qquad t < 0$$
$$= e^{-st} \qquad t \geq 0$$

In this example, the resulting integral transform is called the **Laplace transform** $L[f(t)]$ of the function $f(t)$. The Laplace transform is defined as:

$$L[f(t)] = \int_0^{\infty} f(t)e^{-st}\,dt$$

The resulting integral, that is, $L[f(t)]$, is a function of s only.

$$L[f(t)] = F(s)$$

$F(s)$ is the Laplace transform of $f(t)$. The symbol L is the **Laplace operator** which acts on $f(t)$ to give the transformed function $F(s)$.

3.3.3 Low Pass Filter – Integrator

Consider an **integrator** circuit:

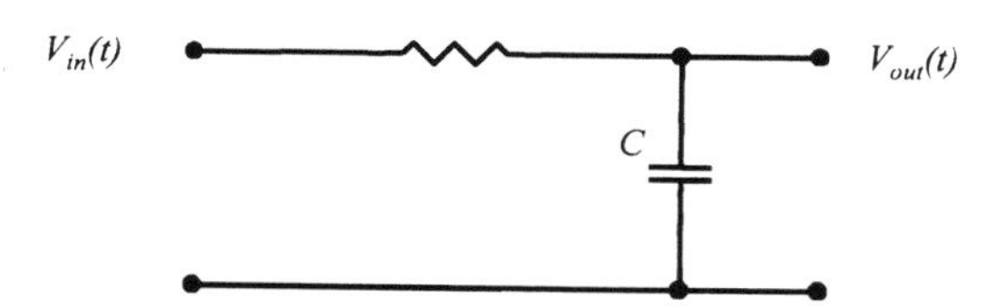

The output voltage is the time integral of the input voltage:

$$V_{out} = \frac{1}{RC}\int V_{in}dt$$

when RC is large.

If the input signal is a sine wave, then the output signal is a cosine wave (whose amplitude decreases with increasing frequency of the input signal).

In complex notation, we have:

$$V_{out} = \frac{1}{1+RCj\omega}V_{in}$$

When RC is large, then:

$$V_{out} \approx \frac{1}{RCj\omega}V_{in}$$

But,

$$V_{out} = \frac{1}{RC}\int V_{in}dt$$

Let $\frac{d}{dt} = s$

and $\int dt = \frac{1}{s}$

s is a **differential operator** since the application of s to a function takes the time derivative of that function.

Thus,

$$V_{out} = \frac{1}{RC}\int V_{in}dt$$

$$= \frac{1}{RCs}V_{in}$$

It appears therefore that $s = j\omega$.

Is this s the same as that in the Laplace transform?

Consider the function $y = e^{-st}$

Thus, $\frac{dy}{dt} = -se^{-st}$

$= -sy$

That is, s *is* a differential operator for this function. Thus:

$$s = \frac{d}{dt}$$

In the **Laplace transform**, s can be considered a differential operator and not just an ordinary variable. This powerful notion enables us to convert transfer functions from the **time domain** to the **s domain**, where the algebra may be more readily solved, and then convert back to the time domain for the final output.

3.3.4 Integrator

t-domain analysis

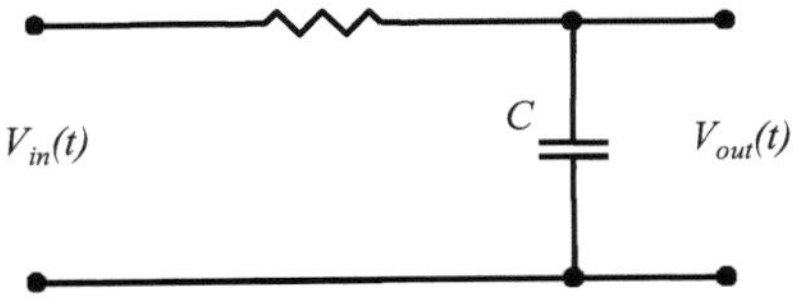

$$V_{out} = \frac{q}{C}$$

$$\frac{dQ}{dt} = I \therefore Q = \int I dt$$

Thus $V_{out} = \frac{1}{C}\int I dt$

Now, $V_R = IR$

and thus $\therefore V_{out} = \frac{1}{RC}\int V_R dt$

Now $V_R \approx V_{in}$ when RC is large

or $V_c << V_r$

thus $V_{out} = \frac{1}{RC}\int V_{in} dt$ Output voltage signal is the integral of the input voltage signal.

$V_{in}(t)$ C $V_{out}(t)$

ω-domain analysis

$$V_{in} = I(R - X_C j)$$

$$= I\left(R - \frac{1}{\omega C} j\right)$$

$$V_{out} = I(-X_C j)$$

$$= I\left(-\frac{1}{\omega C} j\right)$$

$$\frac{V_{out}}{V_{in}} = \frac{\frac{-1}{\omega C} j}{R - \frac{1}{\omega C} j}$$

$$= \frac{1 - R\omega C j}{1 + R^2\omega^2 C^2}$$

$$= \frac{1}{1 + R\omega C j} = H(\omega)$$

s-domain analysis

$$\frac{d}{dt} = s$$

$$V_{in} = RsQ + \frac{1}{C}Q$$

$$= Q\left(Rs + \tfrac{1}{C}\right)$$

Now, $V_{out} = \frac{Q}{C}$

thus $= \frac{1}{C}\left(\frac{V_{in}}{Rs + \frac{1}{C}}\right)$

$$\frac{V_{out}}{V_{in}} = \frac{1}{1 + RCs} = H(s)$$

This equation is a **transfer function** in s. This is not an integral equation in the time domain, nor a complex algebraic expression in the frequency domain, but a simple algebraic expression involving the differential operator s.

3.3.5 Differentiator

t-domain analysis

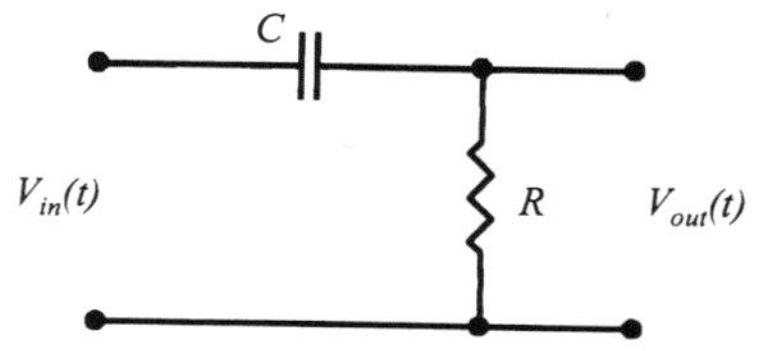

$$V_{out} = IR$$

$$I = \frac{dQ}{dt}$$

$$= C\frac{dV_C}{dt}$$

$$V_{out} = RC\frac{dV_C}{dt}$$

$V_C >> V_R \therefore V_C \approx V_{in}$ when RC is small.

$$V_{out} = RC\frac{dV_{in}}{dt}$$

Output voltage is the derivative of the input voltage.

ω-domain analysis

$$V_{in} = I(R + -X_C j)$$

$$V_{out} = IR$$

$$\frac{V_{out}}{V_{in}} = \frac{R}{R - X_c j}$$

$$\left|\frac{V_{in}}{V_{out}}\right| = \frac{R}{\sqrt{R^2 + \frac{1}{\omega^2 C^2}}}$$

$$= \frac{R\omega C}{\sqrt{R^2\omega^2 C^2 + 1}}$$

$$= H(\omega)$$

s-domain analysis

$$\frac{d}{dt} = s$$

$$V_{in} = RsQ + \frac{1}{C}Q$$

$$= Q\left(Rs + \tfrac{1}{C}\right)$$

Now, $V_{out} = RsQ$

Thus $V_{out} = Rs\left(\dfrac{V_{in}}{Rs + \frac{1}{C}}\right)$

$$= \frac{V_{in}}{1 + \frac{1}{RsC}}$$

$$\frac{V_{out}}{V_{in}} = \frac{RCs}{1 + RCs} = H(s)$$

3.3.6 Mechanical Property Measurements

An interesting application of signal processing and analysis is in the testing of the **mechanical properties of materials**. Many tests on materials involve the application of a load or a force, and the measurement of a deflection.

In an ideal case, materials behave as linear springs and obey **Hooke's law**, or are **Newtonian fluids** with a time-dependent response.

In practice, materials fall somewhere in between these two extremes and are often modelled by a collection of springs and dashpots connected in series or parallel combinations.

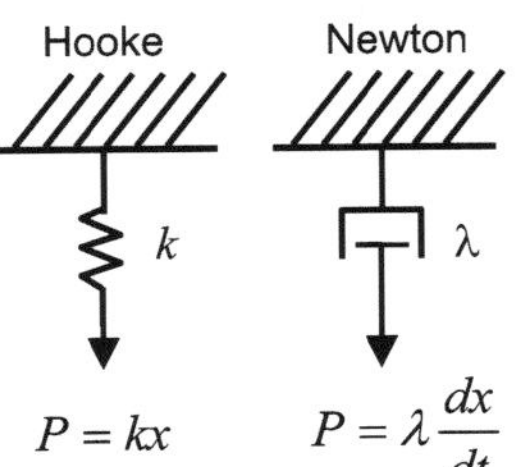

There is, therefore, a connection between the load and the displacement of a specimen which depends upon its material properties. More generally, we can say that the transfer function between load and displacement is given by the ratio of the **Fourier transform** of the component signals. For example, for a dynamic application of a load P, the resulting displacement h is:

$$P = P_o e^{i(\omega t+\phi)} \quad h = h_o e^{i(\omega t)}$$

For example:

$$p = h\underbrace{\left[\left(\frac{1}{k_1 + i\omega\lambda} + \frac{1}{k_2}\right)^{-1}\right]}_{\text{Transfer function}}$$

$$TF = G' + iG''$$

Storage modulus $G' = \dfrac{|P_o|}{|h_o|}\cos\phi$ $\qquad$ $G'' = \dfrac{|P_o|}{|h_o|}\sin\phi$ Loss modulus

$$TF = \frac{F(P(t))}{F(h(t))}$$

The Fourier transforms $F(P(t))$ and $F(h(t))$ of the experimental quantities P and h are themselves complex quantities, the real part of the ratio of them thus gives the storage and the imaginary part the loss modulus of the system.

One advantage of approaching this type of measurement using Fourier transforms is that the load need not be a single frequency. If a superposition of frequencies is applied, a Fourier transform of the signals provides a way to extract out the single-frequency response of the material thus allowing properties to be measured over a range of frequencies in a short time.

3.4 Fourier Optics

Summary

$m\lambda = a\sin\theta$ Diffraction minima

$$f(x) = A_o + \sum_{n=1}^{\infty}\left[A_n \cos nkx + B_n \sin nkx\right]$$ Fourier series

$$A_o = \frac{1}{l}\int_0^l f(x)\,dx$$

$$A_n = \frac{2}{l}\int_0^l f(x)\cos nkx\,dx$$

$$B_n = \frac{2}{l}\int_0^l f(x)\sin nkx\,dx$$

$$f(x) = \int_0^{\infty}\left(A(k)\cos kx + B(k)\sin kx\right)dk$$ Fourier integral

$$A(k) = \frac{1}{\pi}\int_{-\infty}^{\infty} f(x)\cos kx\,dx$$

$$B(k) = \frac{1}{\pi}\int_{-\infty}^{\infty} f(x)\sin kx\,dx$$

$\text{N.A.} = n_o \sin\theta$ Numerical aperture

3.4.1 Diffraction

In **geometrical optics**, the path of light through lenses and prisms can be calculated using the laws of reflection and refraction where light rays travel in straight lines. In **physical optics**, more attention is given to the wave nature of light as it passes through materials and accounts for effects like diffraction and interference.

The phenomenon of diffraction can be readily demonstrated by illuminating a narrow vertical slit with plane waves of light.

Slit of width a being illuminated by plane waves of light. The wave front can be thought of as the superposition of an infinite number of spherical waves from point sources.

"parallel" rays

P

$90-\theta$

θ

slit a

Δx

Viewing screen

D

At a large distance D from slit, rays on the screen are virtually parallel (not shown parallel here) and arrive at P at some angle $90-\theta$.

Based on the superposition of waves from the different point sources, the angular position of the first, and all subsequent, minima is found from:

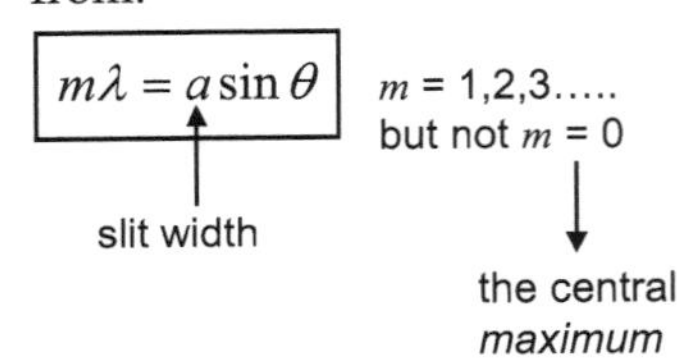

When the viewing screen is placed close to the object (in this case, the slit) the diffraction image on the screen resembles the outline of the slit, with fringing around the edges. This is Fresnel diffraction. When the viewing screen is placed a large distance from the object, the image on the screen becomes less recognisable as the object and spreads out horizontally in a series of bright and dark fringes – a **Fraunhofer diffraction pattern**.

As a decreases (slit gets narrower), the angle θ increases (pattern spreads out further). As the slit gets wider (a increases) the angle θ decreases (conditions approach that of geometrical optics – no spreading out of rays).

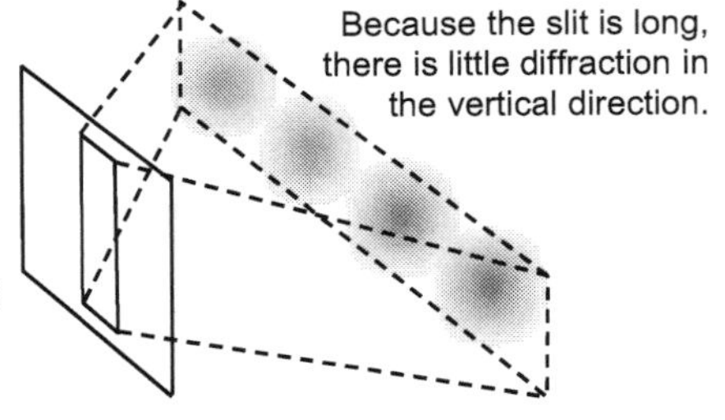

3.4.2 Fourier Transform – Graphical Approach

Consider an object, such a photographic film, of width x which is semi-transparent with bands of dark and light stripes of spacing d, so that the amplitude of the intensity of transmitted light coming through it varies in a smooth sinusoidal manner.

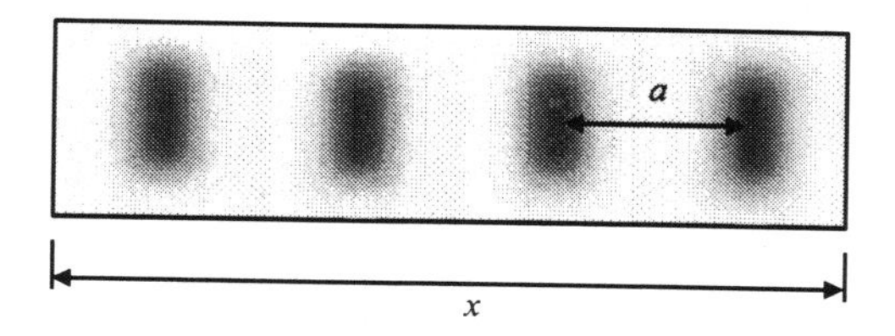

Note, amplitude is different from intensity. The eye responds to intensity, which is the amplitude squared.

The amplitude of the transmitted light is plotted:

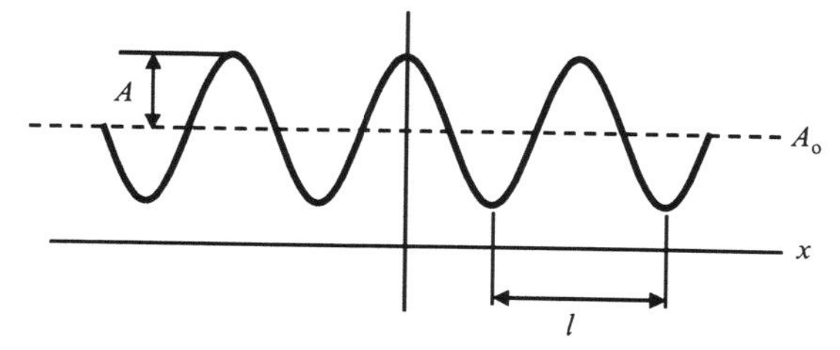

and contains a steady "DC" term upon A_o which is superimposed a cosine function of amplitude A. Now here, l is the *period* of the wave (since time quantities are now being replaced by spatial quantities).

Now, let us define a quantity k so that $k = \dfrac{2\pi}{l}$.

k is actually the **spatial frequency** in radians/metres (equivalent to ω radians/second in the time domain).

Now in general, the equation for a **travelling wave** is $u(t) = A\cos(\omega t)$.

For **spatial frequencies**, we thus write $A(x) = A\cos\left(\dfrac{2\pi}{l}x\right) = A\cos kx$.

where $A(x)$, in this case, is the transmitted amplitude of the light at a position x.

The **amplitude function** $A(x)$ therefore is the sum of a "DC" term, with spatial frequency zero, and a single cosine component with spatial frequency k.

3.4.3 Fourier Transform – Single Frequency

The **Fourier transform** of our amplitude function is a graphical description of the component frequencies. We plot k on the horizontal axis, and "amplitude" on the vertical axis. That is, the Fourier transform is plotted in the k domain, or, as it is often called, ***k* space**. For our example amplitude function, the Fourier transform is:

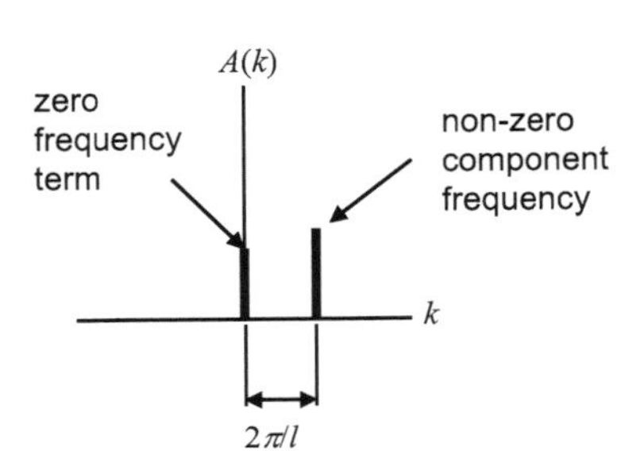

Remember l is not wavelength but frequency – a spatial frequency. Decreasing l (bands in object closer together) means an increase in spatial frequency.

The line at $k = 2\pi/l$ indicates the component of the single frequency term in the original function, where as the single line at $k = 0$ is the "DC" zero frequency term. If the spacing of the actual bands of intensity in the original object were to decrease, the spatial frequency increases and the line would move outwards along the k axis. That is, the spacing of the lines in the **Fourier transform** is an inverse measure of the magnitude of the **spatial frequencies** in the original object.

The height of the lines in the Fourier transform indicate the relative contribution to the total **amplitude spectrum**. For example, a low **zero frequency** term and high outer frequency terms would result in a high contrast object. If the height of the zero frequency terms is a lot higher than the outer frequency terms, then we will most likely see a washed out low contrast object.

The present example is for a very simple, sinusoidal, transmission function. More realistic objects have more complicated Fourier transforms.

3.4.4 Diffraction Grating

A more complicated, but still somewhat simple, object worthy of discussion is an object with a transmission function like a square wave – that is, a series of parallel slits being illuminated uniformly from behind – a **diffraction grating**. The amplitude function can be plotted:

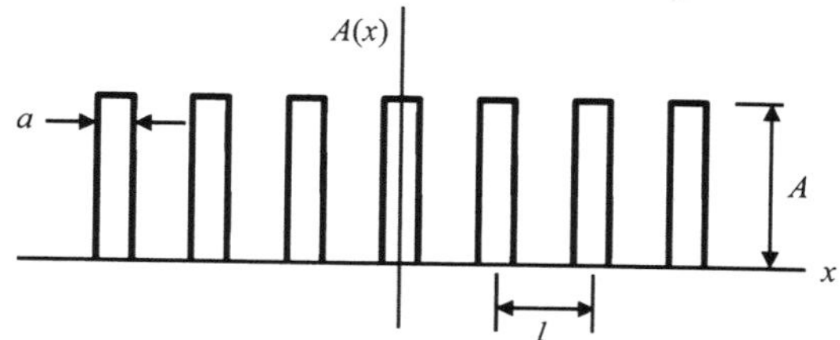

Because each aperture in the original object has sharp edges, we need many Fourier components to construct the original object.

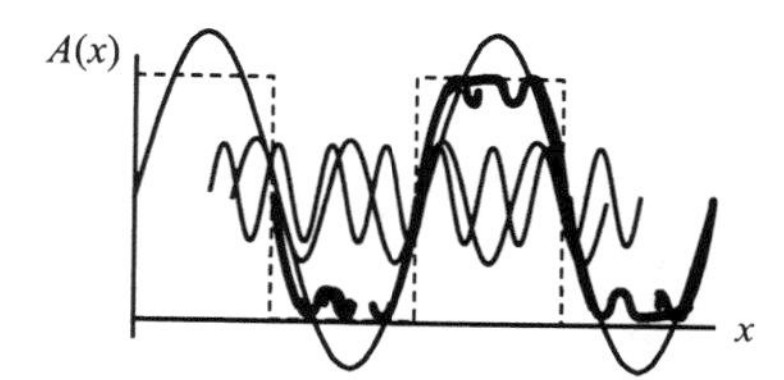

The **Fourier transform** has the form:

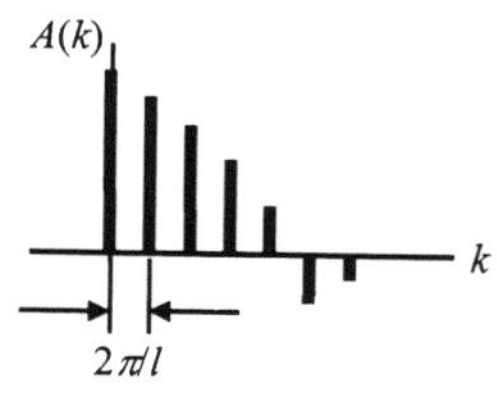

Each vertical line represents a **spatial frequency** component which is needed to reconstruct the original object. The close spacing of the components in k space indicates that original slits are relatively far apart. If the original slits in the object were closer together (small l) then the spacing of the Fourier components would be larger.

That some spatial frequency components are negative in amplitude indicates a **phase shift** – and that the amplitude of those components is subtracted in order to reconstruct the original function.

The component at the **zero frequency** position indicates the amplitude of the non-periodic features of the original object – in this case, this would represent the area of the slits (each slit passing an equal amount of light, they being uniformly illuminated).

3.4.5 Single Slit

It is interesting to determine the **Fourier transform** of an illuminated **single slit**. The amplitude transmission function is drawn:

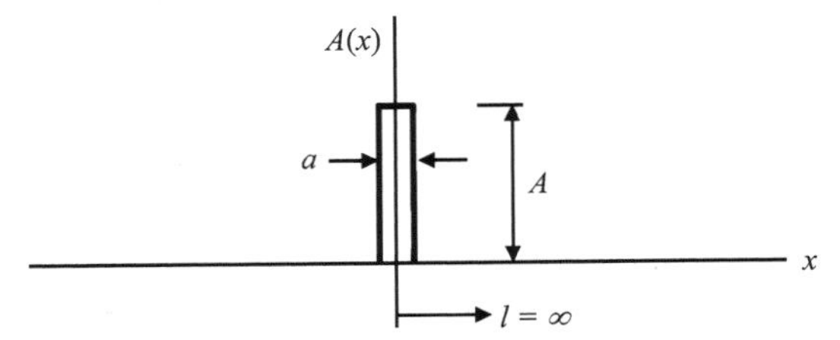

This object is a limiting case of the previous diffraction grating (multiple slit) object where $l = \infty$.

Again, this object contains sharp edges, and so we expect many Fourier components would be needed to reconstruct it. However, notice that in the previous example, as l increases, the spacing of the bars $(2\pi/l)$ decreases. Here, l goes to infinity and so the bars collapse to zero spacing, thus forming a continuous curve. The Fourier transform thus becomes:

The amplitude function is the amplitude of the electric field of the light wave. Optical detectors respond to intensity rather than field strength. These plots show the **intensity distribution**, which is what we would actually see or record if the transform were measured or viewed. The intensity distribution is the amplitude squared.

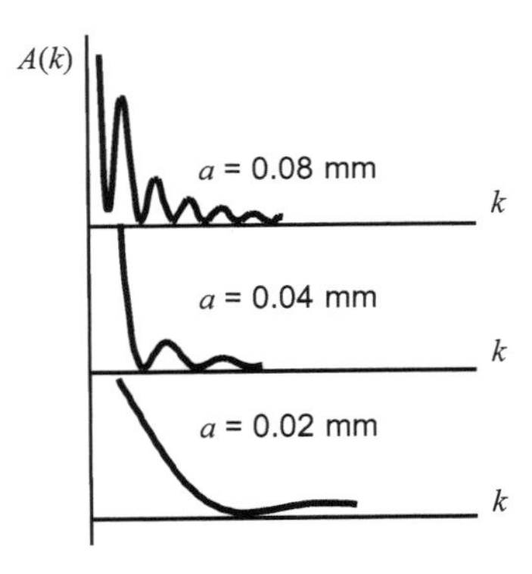

Decreasing the width of the slit a makes the component spatial frequencies spread out in k space.

It can be seen that the Fourier transform, in the case of a single-slit object, is the envelope of the vertical bars for the discrete frequencies of the multiple slit **diffraction grating** pattern.

3.4.6 Double Slit

A very important object in physical optics is a **double slit** object. The amplitude transmission function is drawn:

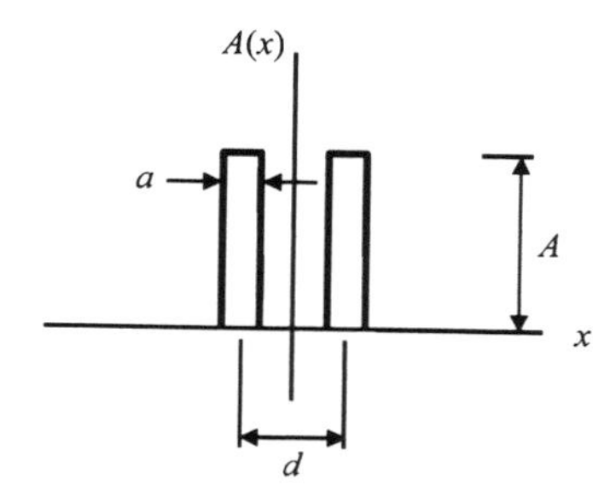

In this case, the double slit transform is **modulated** by the single slit transform.

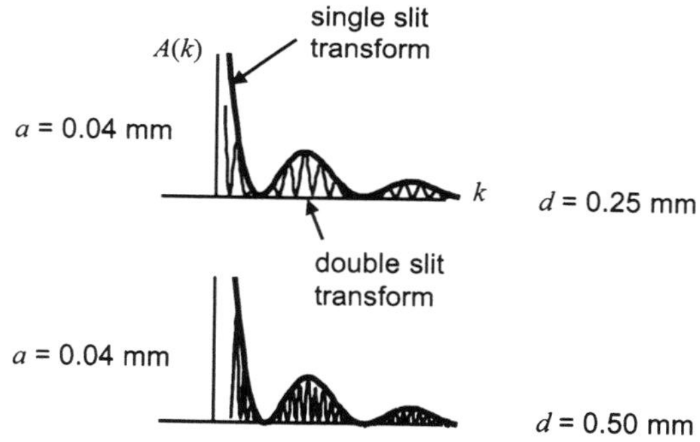

If we were to add more slits, then this modulation would be such that we approach that of a series of vertical bars as shown earlier – that is, the peaks in the outer envelope would become sharper until they became single frequency bars in k space.

As d gets larger, k becomes smaller and so the frequency components in the Fourier transform become closer together.

Decreasing the width of the slit a spreads out the modulation (distance between maxima in the outer envelope) of the pattern in k space.

3.4.7 Fourier Transform – Mathematical Approach

For a spatially periodic function $f(x)$ of period l with spatial frequency k, the function may be represented by a **Fourier series** which is written:

$$f(x) = A_o + \sum_{n=1}^{\infty}\left[A_n \cos nkx + B_n \sin nkx\right] \qquad l = \frac{2\pi}{k}$$

with $A_o = \dfrac{1}{l}\displaystyle\int_0^l f(x)\,dx$ and $A_n = \dfrac{2}{l}\displaystyle\int_0^l f(x)\cos nkx\,dx$

the "DC" term

$$B_n = \frac{2}{l}\int_0^l f(x)\sin nkx\,dx$$

Amplitude terms for component frequency nk

Consider a simple cosine function with a "DC" offset $A(x) = A_o + A\cos kx$.

We wish to compute the **Fourier transform** of this function. The function can be expressed as a Fourier series thus:

$$A(x) = A_o + \sum_{n=1}^{\infty}\left[A_n \cos nkx\right]$$

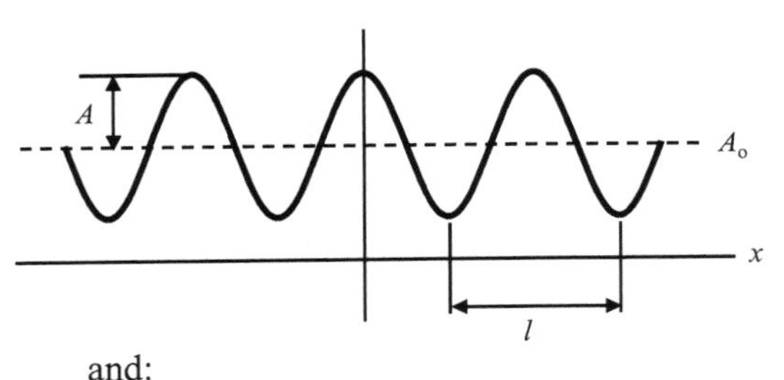

So, we have:

$$A_o = \frac{1}{l}\int_0^l (A_o + A\cos kx)\,dx$$

$$= \frac{1}{l}\int_0^l A_o\,dx + \frac{1}{l}\int_0^l A\cos kx\,dx$$

$$= A_o \quad \text{(as expected)}$$

and:

$$A_1 = \frac{2}{l}\int_0^l (A_o + A\cos kx)\cos kx\,dx$$

$$= \frac{2}{l}\int_0^l A_o \cos kx dx + \frac{2}{l}\int_0^l A\cos^2 kx dx$$

$$= \frac{2A}{l}\int_0^l \cos^2 kx dx$$

$$= \frac{2A}{kl}\left[\frac{1}{2}kx + \frac{1}{4}\sin 2kx\right]_0^l$$

$$= A \quad \text{(as expected)}$$

That is, the Fourier components of this function can be plotted:

A_o A

k

$2\pi/l$

3.4.8 Fourier Transform – Continuous Function

If the period l increases to a large value, then in the limit as l goes to infinity, we write the Fourier series as an integral sum of the continuous variable k:

$$f(x)=\int_0^{\infty}(A(k)\cos kx+B(k)\sin kx)dk \quad \text{where} \quad A(k)=\frac{1}{\pi}\int_{-\infty}^{\infty}f(x)\cos kx dx$$

$$B(k)=\frac{1}{\pi}\int_{-\infty}^{\infty}f(x)\sin kx dx$$

The functions $A(k)$ and $B(k)$ are called the Fourier cosine and Fourier sine transforms of $f(x)$. The **Fourier transforms** are a continuous function of k.

Euler's formula allows us to express trigonometric quantities in terms of complex exponentials. This often leads to simplifications in the handling of these equations. Thus, we can write:

$$F(k)=\int_{-\infty}^{\infty}f(x)e^{-jkx}dx$$

$$f(x)=\frac{1}{2\pi}\int_{-\infty}^{\infty}F(k)e^{jkx}dx$$

The function $f(x)$ is can be thought of as the **inverse Fourier transform** of $F(k)$:

$$f(x)=F^{-1}[F(k)]$$

$F(k)$ is a complex function with real and imaginary components. The real part of $F(k)$ gives the amplitude spectrum of the cosine components, and the imaginary part gives the amplitude spectrum of the sine components of the original function.

3.4.9 Lenses

A **Fraunhofer diffraction pattern** of an object such as a single slit is formed at a very large distance from the object. The diffraction pattern can be more conveniently viewed if a lens (L_2) is placed in the optical path so that the diffraction pattern is focused at the secondary focal point where we might position a viewing screen.

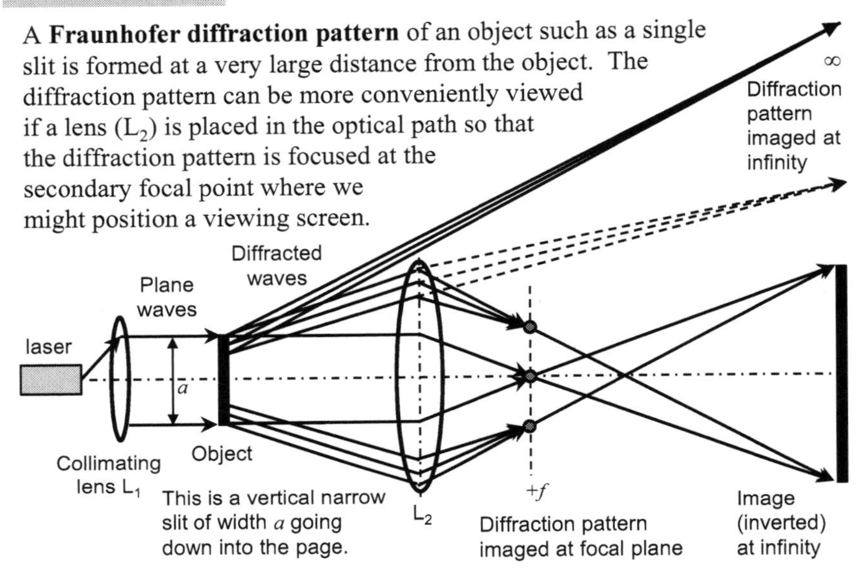

The central maximum in the diffraction pattern represents the non-periodic "DC" terms in the object. The outlying fringes or maxima arise from the periodic features of the object. In the case of a slit, the outlying maxima represent the high spatial frequencies of the component sine waves needed to reconstruct the sharp corners of the geometry of the slit.

Another lens L_3 can be positioned so that the diffraction pattern from L_2 becomes the object for L_3, and the **inverse transform** of the object is formed at a more conveniently located image plane.

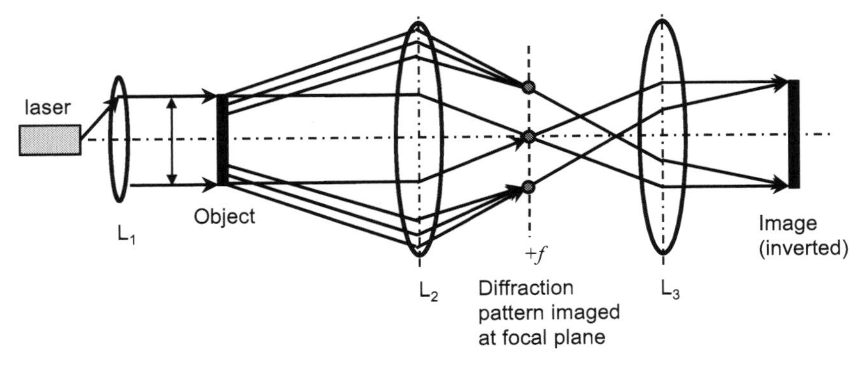

3.4.10 Spatial Filtering

Because the **diffraction pattern** from L_2 is a **Fourier transform**, then we can block out selected spatial frequencies using a filter. For example, to smooth the image (by removing the high frequency components) we block the outer points in the diffraction pattern from being seen by the inverse transform lens L_3.

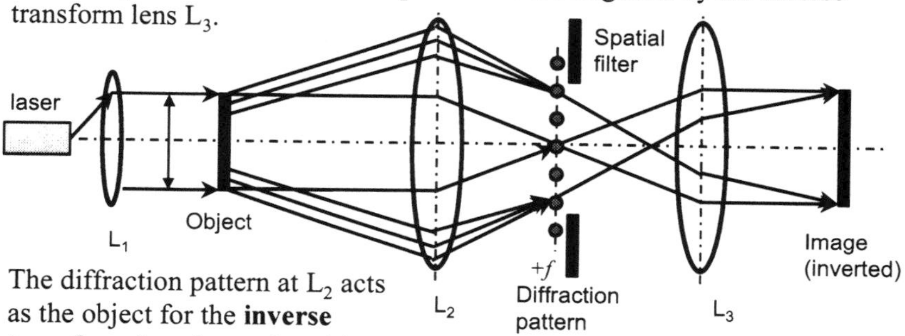

The diffraction pattern at L_2 acts as the object for the **inverse transform** lens L_3, and so this now only sees the low frequency terms. The diffraction pattern of L_3 forms at the back focal plane of L_3 and is the reconstructed image of the object. But, since the high spatial frequencies have been removed by the **spatial filter**, this reconstructed image of the original object is now smoothed.

Conversely, it can be now appreciated that non-periodic objects (such as the majority that are seen) can be represented by an infinite number of continuous components and, as such, a real physical lens will not capture those high frequency terms whose diffraction pattern lies outside the width of the lens. Thus, the objective lens itself, in a **microscope** or a **telescope**, acts as a spatial filter by virtue of its finite size, thus limiting the resolving power of an optical instrument. To increase the resolving power of such an instrument, one must increase the aperture to capture more of the high frequency terms in the diffraction pattern. This is **Abbe's theory of image formation** and was employed by Carl Zeiss in the manufacture of microscopes in the 19th century.

Should one wish to see very small features in an object, it is not only the magnification of the instrument that is important, but the **numerical aperture**. A small, sharp image from a large aperture lens is more desirable than a large blurry image.

Refractive index of medium (e.g., $n_{air} = 1$)

$$\text{N.A.} = n_o \sin\theta$$

Cone half-angle of ray at edge of lens

θ

3.5 Quantum Mechanics

Summary

$-\frac{\hbar^2}{2m}\frac{\partial^2\Psi}{\partial x^2}+V(x,t)\Psi = i\hbar\frac{\partial\Psi}{\partial t}$	Schrödinger equation
$\Psi(x,t)=\psi(x)\phi(t)$	Wave equation
$-\frac{\hbar^2}{2m}\frac{\partial^2\psi}{\partial x^2}+V\psi = E\psi$	Time-independent Schrödinger equation
$\phi(t)=e^{-i\frac{E}{\hbar}t}$	Time-dependent Schrödinger equation
$E = hf$	Planck's equation
$\Delta x\Delta p \geq 2\pi\hbar$	Uncertainty principle
$\langle x\rangle = \int_{-\infty}^{\infty} xP(x,t)dx = \int_{-\infty}^{\infty}\Psi^* x\Psi dx$	Normalisation
$\int_{-\infty}^{\infty}\Psi^*\Psi dx = 1$	Expectation value
$\langle\psi_m\vert\psi_n\rangle = \int\psi_m{}^*\psi_n dx$	Overlap Integral
$\hat{H}\Psi(x,t)=i\hbar\frac{\partial\Psi(x,t)}{\partial t}$	Hamiltonian
$\left\langle\Psi_1\mid\hat{O}\mid\Psi_2\right\rangle = \int\left(\hat{O}\Psi_1\right)^*\Psi_2 dx = \int\Psi_1{}^*\left(\hat{O}\Psi_2\right)dx$	
$\left\langle E_n{}^{(1)}\right\rangle = \left\langle\psi_n\mid\hat{H}_1\mid\psi_n\right\rangle = \int_{-\infty}^{\infty}\psi_n{}^*\hat{H}_1\psi_n dx$	Perturbation theory
$\frac{d\hat{H}}{d\alpha}=0$	Variational method

3.5.1 Quantum Mechanics

The total energy of a particle (e.g., an **electron** in an atom, a free electron, an electron in an electric field, a conduction electron in a solid) is the sum of the potential and kinetic energies. Expressed in terms of **momentum** p, and mass m, this is stated:

$$E = \frac{1}{2}mv^2 + V$$

$$= \frac{1}{2}pv + V \quad \text{since } p = mv$$

$$E = \frac{p^2}{2m} + V \quad \text{Energy equation}$$

let $\omega = 2\pi f$

and $k = \dfrac{2\pi}{\lambda}$

and $\hbar = \dfrac{h}{2\pi}$

thus $E = hf = \hbar\omega$

Considering movement in one dimension only,

$$\text{let } \hat{p} = -i\hbar\frac{\partial}{\partial x}$$

$$\hat{E} = i\hbar\frac{\partial}{\partial t}$$

$$i^2 = -1$$

The $\partial/\partial x$ and $\partial/\partial t$ terms are **differential operators**. For example, when the differential operator $\partial/\partial t$ acts on the displacement variable x, we obtain $\partial x/\partial t$ = velocity. We use the symbol ∂ instead of d here because we will apply these operators to a function which is dependent on both x and t, and so the partial derivatives must be used. The differential operator can also act on a function. For example, if $f(x,t) = 2x + 3t$, then $\partial f(x,t)/\partial t = 3$.

Now, the **potential energy**, V, may depend upon the position and time, and so in general, $V = V(x,t)$.

The quantities $\hat{p}$ and $\hat{E}$ above are **differential operators**. When they operate on a function involving x and t, the result is the momentum p and energy E, respectively. The energy equation becomes a differential operator equation.

$$i\hbar\frac{\partial}{\partial t} = -\frac{\hbar^2}{2m}\frac{\partial^2}{\partial x^2} + V(x,t)$$

Note: $(-i)^2 = -1$

We now let this differential operator equation operate on a function Ψ. Ψ is called a **wave function**, and may itself be a function of x and t.

We can now write the total energy as:

$$-\frac{\hbar^2}{2m}\frac{\partial^2\Psi}{\partial x^2} + V(x,t)\Psi = i\hbar\frac{\partial\Psi}{\partial t}$$

This is the **Schrödinger equation**. It is a differential equation . The solution to this equation is the function Ψ. Thus, we solve the Schrödinger equation by finding the form of Ψ for various forms of the potential function V.

A simple example is $\Psi(x,t) = A \sin(kx - \omega t)$, the wave function for a sinusoidal travelling wave. When dealing with more complicated functions like matter waves, the wave function is a complex function. For example, one solution of the wave equation (a free electron) has the form $\Psi(x,t) = (A\cos(kx) + B\sin(kx))e^{-i\omega t}$.

3.5.2 Solution to the Schrödinger Equation

The solution to the Schrödinger wave equation is the **wave function** Ψ. For many cases of interest, the potential function V is a function of x only, that is, the potential is static (independent of time). This allows the wave equation to be separated into time-independent and time-dependent equations that can be readily solved independently.

$$-\frac{\hbar^2}{2m}\frac{\partial^2\Psi}{\partial x^2}+V\Psi=i\hbar\frac{\partial\Psi}{\partial t}$$

Schrödinger differential wave equation

$$\text{let } \Psi=\psi(x)\phi(t)$$

$$-\frac{\hbar^2}{2m}\frac{\partial^2\psi\phi}{\partial x^2}+V\psi\phi=i\hbar\frac{\partial\psi\phi}{\partial t}$$

$$-\frac{\hbar^2}{2m}\frac{\partial^2\psi\phi}{\partial x^2}\frac{1}{\psi\phi}+V=i\hbar\frac{\partial\psi\phi}{\partial t}\frac{1}{\psi\phi}$$

ϕ is a function of t.
ψ is a function of x.

$$\frac{1}{\psi}\left[-\frac{\hbar^2}{2m}\frac{\partial^2\psi}{\partial x^2}+V\psi\right]=i\hbar\frac{\partial\phi}{\partial t}\frac{1}{\phi}$$

G is a constant that just connects the two equations. It is termed the **separation constant** because it allows the variables to be separated into two equations. The physical significance of G shall be shown to be the energy E.

$$-\frac{\hbar^2}{2m}\frac{\partial^2\psi}{\partial x^2}+V\psi=G\psi$$

Time-independent Schrödinger equation

$$i\hbar\frac{\partial\phi}{\partial t}\frac{1}{\phi}=G$$

Time-dependent Schrödinger equation

The resulting solutions of these equations are functions, one a function of x, the other a function of t. When these two functions are multiplied together, we obtain the **wave function**. The wave function $\Psi(x,t)$ is the solution to the original Schrödinger differential wave equation.

$$\boxed{\Psi(x,t)=\psi(x)\phi(t)}$$

$\psi(x)$ is a solution to the **time-independent equation**. $\phi(t)$ is the solution to the **time-dependent equation**.

In general, there may be many solutions to the time-independent equation each differing by a multiplicative constant. The collection of functions $\psi(x)$ that are solutions are called **eigenfunctions**, or **characteristic functions**. The **eigenfunctions** for the time-independent equation $\psi(x)$ determine the space dependence of the wave function Ψ. The quantum state associated with an eigenfunction is called an **eigenstate**.

3.5.3 Interpretation of the Wave Function

We might ask, just what is the physical significance of the **wave function** Ψ? In classical physics, a wave function describes the displacement of a particle, or the amplitude of an electric field, or some other phenomenon, at a given coordinate x at some time t. For example, the amplitude of the electric field in an electromagnetic wave at some location x at time t can be expressed:

$$E(x,t) = E_o \sin(kx - \omega t)$$

The **energy density** is the energy contained within the electric field per unit volume (J m^{-3}). The **intensity** of the field is measurement of power (i.e., rate of energy transfer) transmitted per unit area by the field. The average (or rms) power of an electromagnetic wave is $I_{av} = \varepsilon_o c\left[\frac{1}{2}E_o{}^2\right]$.

The important feature here is that the energy carried by a wave is proportional to the *square* of the amplitude of the wave. In the case of electromagnetic waves, what we are really measuring as energy is the density of photons (i.e., number per unit volume) – since each photon carries its own quanta of energy.

By analogy to the case of photons, the wave function for an electron has a connection with the energy carried by it since **Schrödinger's equation** is an energy equation. **Born** postulated that the square of the amplitude of the wave function is a measure of the **probability density** of an electron. Since Ψ is complex, in order to obtain a real physical value for this probability, we use the product:

$$\boxed{P(x,t) = |\Psi|^2 = \Psi^* \Psi}$$

where Ψ^* is the complex conjugate of Ψ. $|\Psi|^2$ is interpreted as a **probability density function**. For example, the probability that an electron is located within a small increment Δx around x at time t is $P(x,t)\Delta x$.

When a small particle, such as an electron (or a proton, or a photon), travels from place to place, it does so using all possible paths that connect the two places. Some paths are more probable than others. The electron is not smeared out into some kind of wave – it retains its particle-like nature. It is the *probabilities* that are wave-like.

For example, say there is a spate of car thefts in the east part of a city. The next week, more than the usual number of thefts occur in the centre of the city. In the next week, it is found that a large number of thefts occur in the west. A probability wave is moving from east to west through the city – whereby the chances of finding an increased number of car thefts depend upon the time (i.e., which week) and the place (east, centre or west).

3.5.4 The Time-Dependent Equation

We shall use the example of an electron in a potential field to illustrate the nature of quantum mechanics, although it should be remembered that the principles also apply to other objects, such as protons, neutrons and photons.

The solution to the time-dependent equation involves the use of an **auxiliary equation**. We proceed as follows:

$$i\hbar\frac{\partial\phi}{\partial t}\frac{1}{\phi}=G$$

$$\frac{\partial\phi}{\partial t}=-G\phi\frac{i}{\hbar}$$

Auxiliary equation $\mathrm{m}=-G\frac{i}{\hbar}$

$$\therefore \phi=e^{-\frac{Gi}{\hbar}t}$$

$$=\cos\frac{Gt}{\hbar}-i\sin\frac{Gt}{\hbar}$$

Euler's formula
$e^{(a+bi)x}=e^{ax}(\cos bx+i\sin bx)$

i.e., $\phi(t)$ has frequency $\omega=\frac{G}{\hbar}$

but $\omega=\frac{E}{\hbar}$ where E is the total **energy** of the particle.

$$\therefore G=E$$

$$\boxed{\phi(t)=e^{-i\frac{Et}{\hbar}}}$$

Comparing with the general exponential form of the **wave equation**:

$$y(t)=Ae^{i\omega t}$$

we see that the time-dependent part of the wave function represents the **phase** of the probability wave and so the time-independent part represents the **amplitude** of the wave.

$$\Psi(x,t)=\psi(x)\phi(t)$$

That is, although the total wave function Ψ is a function of x and t, the amplitude of the wave function is independent of t. That is, the positional probability density is independent of time. Under these conditions, the electron is said to be in a **stationary state**. In this case, the probability amplitude is given by: $P(x)=|\Psi|^2=|\psi|^2$

3.5.5 Normalisation and Expectation

In order for the positional probability of an electron to have physical meaning, the electron must be somewhere within the range between $-\infty$ and $+\infty$. That is:

$$\int_{-\infty}^{\infty} \Psi^* \Psi \, dx = 1$$

The amplitude of the wave function is found from the solution to the time-independent equation. We shall see that the general solution $\psi(x)$ to this equation contains a constant of arbitrary value. For a particular solution, the value of the constant depends upon the boundary conditions of the problem. The most general situation is that the electron must be somewhere. That is, the total probability of finding the electron between $-\infty$ and $+\infty$ is 1. When the value of the constant has been found from the boundary conditions, the wave function is said to be **normalised**.

What then is the expected location of a particle, such as an electron? Since the electron must be somewhere between $-\infty$ and $+\infty$, the expected value is the weighted sum of the individual probabilities over all values of x.

$$\langle x \rangle = \int_{-\infty}^{\infty} x P(x,t) \, dx = \int_{-\infty}^{\infty} \Psi^* x \Psi \, dx$$

$\langle x \rangle$ is the **expectation value** of the electron's position. This is not necessarily the most likely value of x. The most likely value of x for a given measurement is given by the maximum (or maxima) in the probability density function. The expectation value is the average value of x that would be obtained if x were measured many times. When the probability density is symmetric about $x = 0$, the expectation value for $x = 0$.

Expectation values for **energy** and **momentum** may also be calculated from:

$$\langle E \rangle = \int_{-\infty}^{\infty} \Psi^* \hat{E} \Psi \, dx$$

$$\langle p \rangle = \int_{-\infty}^{\infty} \Psi^* \hat{p} \Psi \, dx$$

The terms inside the integral sign are customarily written in the order shown here to reflect a style of "**bra-ket**" operator notation introduced by **Dirac**.

3.5.6 Zero Potential

Note: Here, we are using the technique of separation of variables with complex roots to determine the general solution to the wave equation (see Section 2.7.3) .

Consider the case $V(x) = 0$

$$-\frac{\hbar^2}{2m}\frac{\partial^2\psi}{\partial x^2} = E\psi$$

$$\frac{\partial^2\psi}{\partial x^2} = \frac{2Em}{-\hbar^2}\psi$$

$$\frac{\partial^2\psi}{\partial x^2} + \frac{2Em}{-\hbar^2}\psi = 0$$

Euler's formula
$e^{ix} = \cos x + i\sin x$

Auxiliary equation: $\mathrm{m}^2 + \frac{2Em}{\hbar^2} = 0$ and so $\mathrm{m} = \pm i\frac{\sqrt{2Em}}{\hbar}$

Letting $k = \frac{\sqrt{2Em}}{\hbar}$ Note that E (and k) can take on any value > 0. The boundary conditions do not require any discreteness (n).

we obtain: $\psi(x) = C_1 e^{+ikx} + C_2 e^{-ikx}$.

Converting to trigonometric form using Euler's formula:

$$\begin{aligned}\psi(x) &= C_1\cos kx + C_1 i\sin kx + C_2\cos(-kx) + C_2 i\sin(-kx)\\ &= (C_1 + C_2)\cos kx + (C_1 - C_2)i\sin kx\end{aligned}$$

The **eigenfunctions** become $\psi(x) = A\cos kx + Bi\sin kx$ $\begin{cases} A = C_1 + C_2 \\ B = C_1 - C_2 \end{cases}$

The **wave function** is thus:

$$\begin{aligned}\Psi(x,t) &= \psi(x)\phi(t)\\ &= (A\cos kx + Bi\sin kx)e^{-i\omega t} \qquad \text{since } \frac{E}{\hbar} = \omega\end{aligned}$$

or
$$\begin{aligned}\Psi(x,t) &= \left(C_1 e^{ikx} + C_2 e^{-ikx}\right)e^{-i\omega t}\\ &= C_1 e^{i(kx-\omega t)} + C_2 e^{-i(kx+\omega t)} \text{ in exponential form.}\end{aligned}$$

wave travelling $+x$ wave travelling $-x$

This is a general solution that describes the **superposition** of a wave travelling to the right ($+kx$) and one travelling to the left ($-kx$) with amplitudes C_1 and C_2, respectively.

Note, we need to select one of the possible solutions since we can't have a free electron travelling to the left and then to the right. It has to be one or the other. If it changed direction, then some force would act upon it and it would not be "free."

A particular solution for the case of a probability wave travelling in the $+x$ direction can be found by setting $C_2 = 0$ and so $C_1 = A = B$ and hence:

$$\begin{aligned}\Psi(x,t) &= (A\cos kx + Ai\sin kx)e^{-i\omega t}\\ &= \left(Ae^{ikx}\right)e^{-i\omega t}\end{aligned}$$

$$\boxed{\Psi(x,t) = Ae^{i(kx-\omega t)} = A\cos(kx-\omega t) + Ai\sin(kx-\omega t)}$$

That this represents a **travelling wave** can be seen from the real part of $\Psi(x, t)$.

$$\Psi(x,t)_{re} = A\cos(kx - \omega t)$$

While the amplitude A of the wave might remain constant, its position is dependent on the time t. That is, whenever $kx - \omega t = \pi/2, 3\pi/2$, etc. we have a node (where the function $\Psi = 0$). If we fix a time t, then these nodes will appear at periodic intervals of x. As t increases, the positions x for the nodes must also increase and so the nodes "travel" along in the x direction – that is, the wave travels (and the electron travels along with it).

Expectation values:

$$\hat{p} = -i\hbar\frac{\partial}{\partial x}$$

$$\hat{E} = i\hbar\frac{\partial}{\partial t}$$

$$k = \frac{\sqrt{2Em}}{\hbar}$$

The expected, or average, value of **momentum** is found from:

$$\langle p\rangle = \int_{-\infty}^{\infty}\Psi^*\hat{p}\Psi dx$$

$$= \int_{-\infty}^{\infty}\Psi^*(-i\hbar)\frac{d\Psi}{dx}dx$$

$$= \int_{-\infty}^{\infty}\Psi^*(-i\hbar)(ik)Ae^{i(kx-\omega t)}dx$$

$$= \int_{-\infty}^{\infty}-i\hbar(ik)\Psi^*\Psi dx$$

$$= \hbar k\int_{-\infty}^{\infty}\Psi^*\Psi dx$$

$$= \hbar k$$

$$\boxed{\langle p\rangle = \sqrt{2Em}}$$ **de Broglie relation**

The expected, or average, value of **energy** E for a free electron is found from:

$$\langle E\rangle = \int_{-\infty}^{\infty}\Psi^*\hat{E}\Psi dx$$

$$= \int_{-\infty}^{\infty}\Psi^* i\hbar\frac{d\Psi}{dt}dx$$

$$= \int_{-\infty}^{\infty}Ae^{-i(kx-\omega t)}i\hbar(-i\omega)Ae^{i(kx-\omega t)}dx$$

$$= \int_{-\infty}^{\infty}-i^2\hbar\omega\Psi^*\Psi dx$$

$$= \hbar\omega$$

$$\boxed{\langle E\rangle = hf}$$ **Planck's equation**

Note that E can take on any value; there is no discreteness (n) in the expression for k (and hence E).

The positive value of momentum here is consistent with the sign convention we adopted for matter waves, that is $(kx - \omega t)$ for an electron and wave moving in the positive x direction. If the probability wave is moving in the $+x$ direction, then the electron is very likely also moving in the same direction – since the amplitude (squared) of the probability wave determines where the electron is most likely to be, as if it were carried along by the wave.

Normalisation:

$$1 = \int_{-\infty}^{\infty} \Psi^*\Psi dx = \int_{-\infty}^{\infty} Ae^{-i(kx-\omega t)}Ae^{i(kx-\omega t)}dx = A^2\int_{-\infty}^{\infty} dx \quad \text{(Divergent integral)}$$

The difficulty with this normalisation is that the limits of integration are far larger than that which would ordinarily apply in a real physical situation. If the electron is bound by a large, but finite boundary, then (as in the case of a square well potential) a non-zero value of A may be calculated while retaining an approximation to the ideal case of infinite range.

Expectation value of x:

We might well ask what the expectation value is for the position x of the electron?

$$|\Psi|^2 = \Psi^*\Psi = Ae^{-i(kx-\omega t)}Ae^{i(kx-\omega t)} = A^2 \quad \text{or} \quad |\psi|^2 = \psi^*\psi = Ae^{-ikx}Ae^{ikx} = A^2$$

That is, the positional probability density is a constant, which means that the electron has an equal probability of being located anywhere along x. This is in accordance with the uncertainty principle, since in this case, the momentum (and hence the velocity) can be precisely calculated, but the position is completely undetermined.

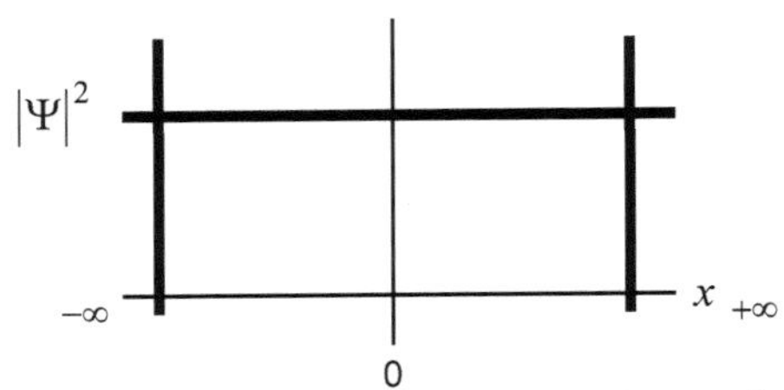

Note that for the equation $\psi(x)$ to be a valid solution to the time-independent part of the Schrödinger equation, it does not matter what the value of k (and hence E) is as long as it is a constant in time and independent of x. This means that for a given value of E, the *square* of the amplitude of the resulting wave function is a constant, independent of x (and t). That is, for a free electron, the electron can be equally likely to be anywhere in x with a velocity v (and momentum p). The uncertainty in x is infinite. The value of p is known precisely. The energy E is not quantised for a completely free electron.

3.5.7 Particle in a Box

A particularly important case which can be solved using the **Schrödinger equation** is that of a one dimensional motion of an electron between two rigid walls separated by a distance L. Such a scenario is called a **particle in a box**. We wish to compute the probability of finding the electron at any particular position between 0 and L according to the principles of quantum mechanics. The motion of the electron is assumed to consist of completely elastic collisions between the walls. At any position between the walls, it is assumed to have a constant velocity (and hence, momentum) independent of time. That is, the (kinetic) energy is a constant and expressed as:

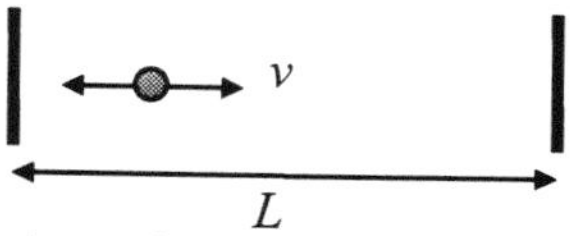

$$E = \frac{p^2}{2m}$$

Note, this is NOT simple harmonic motion between the walls. It is *constant velocity* between the walls.

Since the momentum of the electron is a constant, from the **de Broglie** relation $\lambda = h/p$ we have a characteristic single wavelength λ. Therefore, we can expect that the solution to the time-independent wave equation will be of the form of a travelling wave: $\psi(x) = A \sin kx$ where $k = 2\pi/\lambda$.

The boundary conditions associated with the walls are satisfied as long as $k = n\pi/L$, or, that the allowed wavelengths of the electron are $2L/n$ where $n = 1,2,3\ldots$. These are **standing wave** patterns between the walls of the box.

The energies for each standing wave are found from the allowed values of momentum and the de Broglie relation:

$$E_n = \frac{n^2 h^2}{8mL^2}$$

n = 1 is the **zero point energy**.

Unlike the case of a free electron, the presence of the walls imposes a restriction on the allowed values of E which in turn leads to a non-uniform probability of finding the electron at any particular location between them.

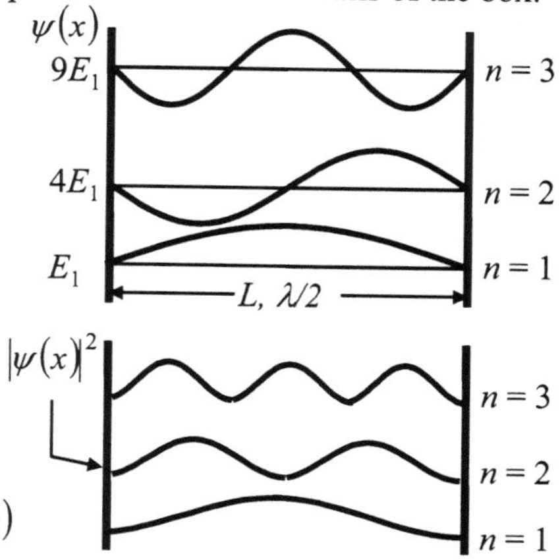

Now, $\frac{\partial^2 \psi(x)}{\partial x^2} = -k^2 A \sin kx = -k^2 \psi(x)$

Thus,

$$\frac{\hbar^2}{2m}\frac{n^2\pi^2}{L^2}\psi(x) = \frac{n^2h^2}{8mL^2}\psi(x) = E\psi(x)$$

which is in accordance with the Schrödinger equation: $-\frac{\hbar^2}{2m}\frac{\partial^2 \psi}{\partial x^2} = E\psi$

Probability that the electron will be found within dx of any position x between 0 and L

3.5.8 Infinite Square Well

Consider the case of an **infinite square well potential**:

$$V(x) = \infty \qquad x \le -\frac{a}{2}; x \ge \frac{a}{2}$$

$$= 0 \qquad -\frac{a}{2} < x < \frac{a}{2}$$

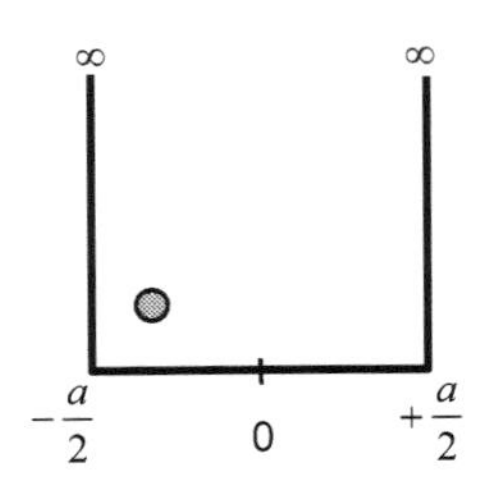

The particle, say an electron, is confined to the region inside the well.

where $V(x) = 0$, we have $-\frac{\hbar^2}{2m}\frac{\delta^2\psi}{\delta x^2} = E\psi$ which is the same as the zero potential case.

Therefore the **eigenfunctions** are $\psi(x) = C_1 e^{+ikx} + C_2 e^{-ikx}$.

This is a general solution to the wave equation for the case of $V(x) = 0$ which is the superposition of two travelling waves in opposite directions. In the present case, the electron might be free to travel within the walls of the container, but when it gets to one wall, it bounces back in the other direction. So, here, we do not select one or the other direction as we did in the zero potential case. Instead, both directions must be considered together and further, the travelling waves must have the same amplitude.

In trigonometric form, we obtain (as in the zero potential case):

$$\psi(x) = C_1 \cos kx + C_1 i \sin kx + C_2 \cos(-kx) + C_2 i \sin(-kx)$$
$$= (C_1 + C_2)\cos kx + (C_1 - C_2)i \sin kx$$

However, unlike the case of zero potential, the travelling waves, in opposite directions, have the same amplitude. Thus, $C_1 = C_2$ and so:

$$\psi(x) = A\cos kx \qquad A = 2C_1$$

Or, alternately, $C_2 = -C_1$ and so:

$$\psi(x) = Bi\sin kx \qquad B = 2C_1$$

For the first case, the full wave function is $\Psi(x,t) = (A\cos kx)(e^{-i\omega t})$

This is a **standing wave** because the amplitude term (A cos kx) depends only on x and not on t. That is, no matter what time we look, the amplitude of the matter wave at some value of x remains unchanged.

Therefore, the most general form of solution for the standing wave pattern is the superposition of these two solutions so that:

$$\boxed{\psi(x) = A\cos kx + Bi\sin kx} \qquad \left\{\begin{array}{l} A = 2C_1 \\ B = 2C_1 \end{array}\right.$$

Boundary conditions:

Let $x = \frac{a}{2}; \psi(x) = 0$

$$\psi(x) = A\cos\frac{ka}{2} + Bi\sin\frac{ka}{2}$$
$$= 0$$

Let $x = -\frac{a}{2}; \psi(x) = 0$

$$\psi(x) = A\cos\frac{-ka}{2} + Bi\sin\frac{-ka}{2}$$
$$= A\cos\frac{ka}{2} - Bi\sin\frac{ka}{2}$$
$$= 0$$

The boundary conditions here are the restriction that the electron has zero probability of being at the walls of the well. This is equivalent to saying that the eigenfunctions reduce to zero at these locations in x. The square of the amplitude of the eigenfunctions is equivalent to the probability density because we are dealing with a **standing wave**, or **stationary state**.

Thus: $A = 0;\ Bi\sin\frac{ka}{2} = 0$ or $B = 0;\ A\cos\frac{ka}{2} = 0$

Eigenfunctions:

$A = B = 0$ is a trivial solution – that is, the particle is not inside the well. Non-trivial eigenfunctions are found by letting, say, $A = 0$ (or $B = 0$) and letting k take on values such that:

(i) $A = 0;\ Bi\sin\frac{ka}{2} = 0$

$$\frac{ka}{2} = n\pi$$
$$k = \frac{2n\pi}{a} \quad n = 1,2,3,4\ldots$$
or
$$= \frac{n\pi}{a} \quad n = 2,4,6\ldots$$

(ii) $B = 0;\ A\cos\frac{ka}{2} = 0$

$$\frac{ka}{2} = n\frac{\pi}{2}$$
$$k = \frac{n\pi}{a} \quad n = 1,3,5,7\ldots$$

Thus: $\psi_n(x) = A_n\cos k_n x \quad n = 1,3,5,7\ldots$

$$k_n = \frac{n\pi}{a}$$

$\psi_n(x) = B_n i\sin k_n x \quad n = 2,4,6,8\ldots$

Normalisation:

To determine the values of the constants A and B, the eigenfunctions are normalised. For the odd n case:

$$1 = \int_{-a/2}^{a/2} \psi^* \psi dx$$

$$= A^2 \int_{-a/2}^{a/2} \cos^2 \frac{n\pi}{a} x dx$$

$$= A^2 \int_{-a/2}^{a/2} \frac{1}{2} + \frac{1}{2} \cos \frac{2n\pi}{a} x dx$$

$$= A^2 \left[\frac{1}{2} x - \frac{a}{4n\pi} \sin \frac{2n\pi}{a} x \right]_{-a/2}^{a/2}$$

$$= A^2 \left[\frac{a}{2} \right]$$

Now, $\cos^2\theta = 1 - \sin^2\theta$

and $\cos 2\theta = \cos^2\theta - \sin^2\theta$

$\sin^2\theta = \cos^2\theta - \cos 2\theta$

Thus $\cos^2\theta = 1 - \cos^2\theta + \cos 2\theta$

$= \dfrac{1 + \cos 2\theta}{2}$

$A^2 = \dfrac{2}{a}$ Similarly for the even case, $B^2 = \dfrac{2}{a}$

The normalised **eigenfunctions** are thus:

$\psi_n(x) = \sqrt{2/a} \cos k_n x \quad n = 1,3,5,7\ldots$

$\psi_n(x) = \sqrt{2/a} \sin k_n x \quad n = 2,4,6,8\ldots$

While the full **wave function** is written:

$\Psi_n(x,t) = \left(\sqrt{2/a} \cos k_n x\right)\left(e^{-i\omega t}\right) \quad n = 1,3,5,7\ldots$

$\Psi_n(x,t) = \left(\sqrt{2/a} \sin k_n x\right)\left(e^{-i\omega t}\right) \quad n = 2,4,6,8\ldots$

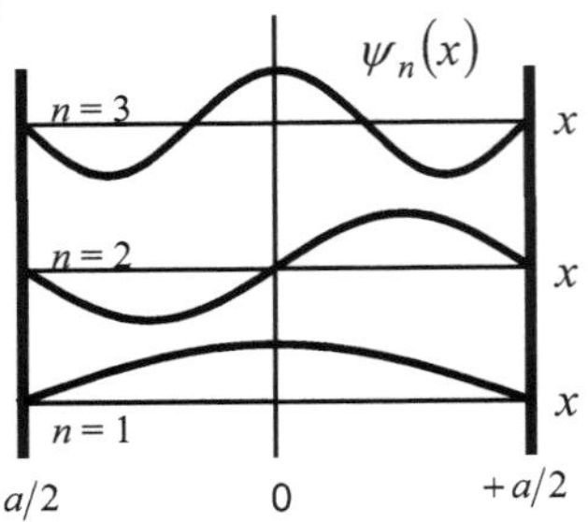

A plot of the probability distributions shows the positional probability of the location of the electron for each of the allowable energy levels. In all cases, the probability is of course zero at the walls where the energy barrier is infinitely high.

Note that because of symmetry around $x = 0$, the **expectation value**, or average value, of x is zero.

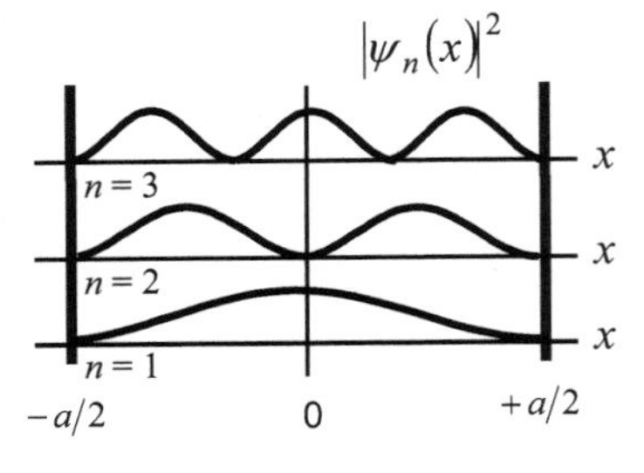

Eigenvalues:

$$k = \frac{n\pi}{a} = \frac{\sqrt{2Em}}{\hbar}$$

$$\frac{n^2\pi^2}{a^2} = \frac{2Em}{\hbar^2}$$

$$\boxed{E = \frac{n^2\pi^2\hbar^2}{2a^2m}} \quad n = 1,2,3,4\ldots$$

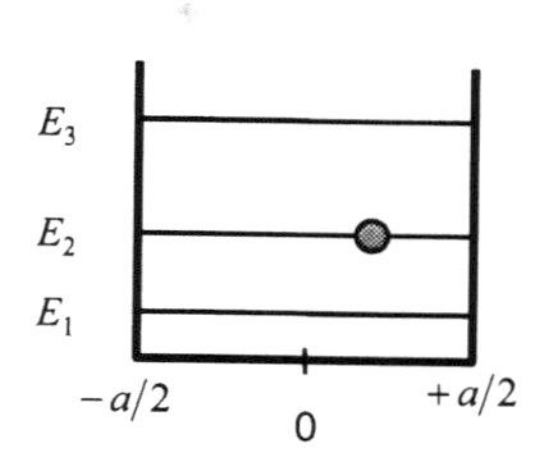

The energy $n = 1$ corresponds to the zero point energy:

$$E_1 = \frac{\pi^2\hbar^2}{2a^2m}$$

Energy is quantised. Each possible value of E is an **eigenvalue**. Unlike the case of a free electron (such as in the zero potential case), energy is quantised in this case as a consequence of it being *bound* (between the walls of the well).

When the particle is in the well, $\Delta x \approx a$. The uncertainty in the momentum is thus:

$$\Delta p = \frac{\hbar}{2\Delta x} = \frac{\hbar}{2a} \quad \text{for } E_1$$

$$p_1 = \sqrt{2mE_1} = \frac{\pi\hbar}{a}$$

$$\Delta p = 2p_1 = \frac{2\pi\hbar}{a}$$

$$\Delta x\Delta p = \frac{2\pi\hbar}{a}a = 2\pi\hbar$$

which is consistent with **Heisenberg's uncertainty principle**.

3.5.9 Step Potential

A particularly important potential with practical consequences is the case of a **step potential** V_o. The case of the energy of the electron $E < V_o$ is considered here.

V_o

x

0

$$V(x) = V_o \quad x > 0$$
$$= 0 \quad x < 0$$

For the case of $x < 0$, the general solution to the wave equation is the same as that developed for a free particle (the zero potential). Expressed in exponential form:

$$\psi(x) = C_1 e^{+ikx} + C_2 e^{-ikx} \quad \text{where } k_1 = \frac{\sqrt{2mE}}{\hbar}$$

For the case of $x > 0$, we have $V(x) = V_o$ and so:

$$-\frac{\hbar^2}{2m}\frac{\partial\psi}{\partial x^2} + V_o\psi = E\psi$$

$$\frac{\partial\psi}{\partial x^2} + \frac{2m}{-\hbar^2}(V_o - E)\psi = 0$$

$$m^2 + \frac{2m}{-\hbar^2}(V_o - E) = 0 \qquad \text{auxiliary equation}$$

It is necessary that eigenfunctions ψ must be single-valued, finite and continuous. This enables us to match the two solutions together at the step ($x = 0$).

$$m = \pm\frac{\sqrt{2m(V_o - E)}}{\hbar} \qquad k_2 = \frac{\sqrt{2m(V_o - E)}}{\hbar}$$

$$\psi(x) = C_3 e^{+k_2 x} + C_4 e^{-k_2 x}$$

As x approaches infinity, $\psi(x)$ must be finite and so $C_3 = 0$. At $x = 0$, the value of $\psi(x)$ and $\partial\psi/\partial x$ must match for $x < 0$ and $x > 0$. In evaluating $\psi(x)$ and $\partial\psi/\partial x$ at $x = 0$ we find that:

$$\psi(x) = \frac{C_4}{2}\left(1 + i\frac{k_2}{k_1}\right)e^{ik_1 x} + \frac{C_4}{2}\left(1 - i\frac{k_2}{k_1}\right)e^{-ik_1 x} \qquad x < 0$$

$$= C_4 e^{-k_2 x} \qquad x > 0$$

That is, unlike the classical Newtonian treatment, quantum mechanics predicts an exponential decrease in the eigenfunction (and also the probability amplitude) on the right hand side of the step. This has important implications for the phenomenon of **tunnelling**.

The first term in the solution for $x < 0$ represents the wave function for the electron approaching the step. The second term represents the electron being reflected from the step. The combined waveforms and associated probability amplitudes are a **standing wave** which represents the probability of the electron being at any point to the left of the step.

3.5.10 Finite Square Well

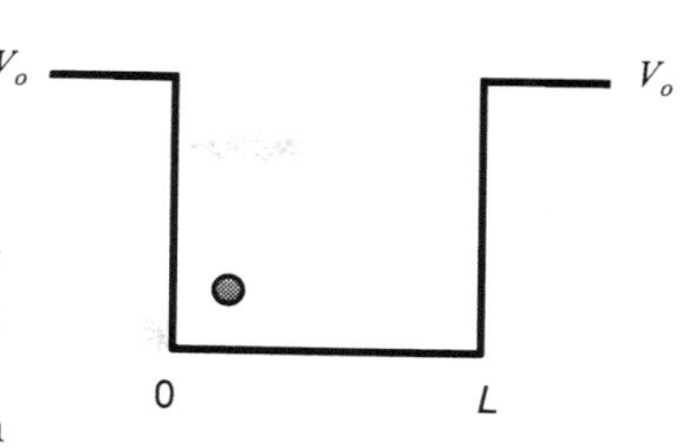

The electron is confined to the region inside the well unless it has sufficient energy to overcome the work function V_o.

$$V(x) = V_o \quad x \leq 0; x \geq L$$
$$= 0 \quad 0 < x < L$$

Take, for example, a free electron in a metal. Should an electron near the surface of a metal acquire sufficient kinetic energy to leave the surface, the surface is left with a net positive charge – and so the electron is immediately attracted back towards it. In classical Newtonian mechanics, the electron can only escape the surface completely if it has enough energy to overcome this Coulomb attraction – the **work function** – of the metal. This is an example of a **square well potential**.

Inside the well, $V(x) = 0$ and so the solution to the zero potential form of the Schrödinger equation can be used:

$$\psi(x) = C_1 e^{ik_1 x} + C_2 e^{-ik_1 x} \text{ where } k_1 = \frac{\sqrt{2mE}}{\hbar}$$

Outside the well, the solution for the step potential can be used:

$$\psi(x) = C_3 e^{k_2 x} + C_4 e^{-k_2 x} \text{ where } k_2 = \frac{\sqrt{2m(V_o - E)}}{\hbar}$$

For a finite solution, $C_3 = 0$ when $x > L$ and $C_4 = 0$ for $x < 0$. The eigenfunctions must also match in slope at the boundary walls of the well (because the **Schrödinger equation** shows that the second derivative of ψ must be finite if $(E - V)$ is finite). The solution to the Schrödinger equation shows that this can only happen at certain values of E.

There are only a finite number of states which can exist where the electron energy is less than V_o. These are called **bound states**. When the electron energy E is greater than this, the electron escapes the bound state of the well and is free (and can have any energy E).

Note that there is a finite probability of the electron being located outside the well even if its energy E is less than V_o.

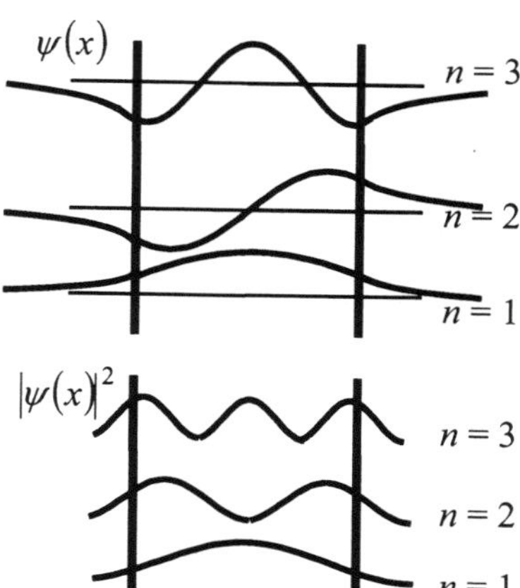

3.5.11 Potential Barrier

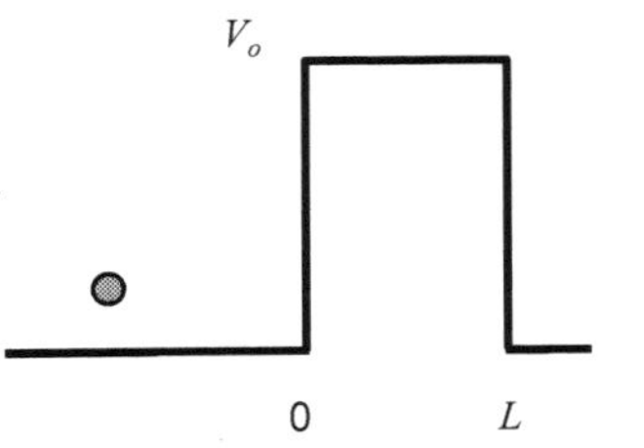

The sides of the finite square well potential V_o can be considered a **potential barrier**. In many physical situations, the width L of the barrier is finite. In classical Newtonian mechanics, an electron can only find itself on the other side of the barrier if it acquires sufficient potential energy to surmount the barrier (i.e., by being given a sufficient amount of an initial kinetic energy, for example, from, say, heating). In quantum mechanics, the solution to the Schrödinger equation for this potential allows for the possibility of the **electron tunnelling** through the barrier and appearing on the other side even when the electron energy is insufficient to surmount the barrier.

On either side of the barrier, the solution to the Schrödinger equation is sinusoidal in accordance with the solution for the zero potential. Within the barrier (for the case of the electron energy E being less than V_o), the solution is an exponential (as in the step potential). As before, we require the solutions to be continuous and finite for all values of x. In matching the **eigenfunctions** in the three regions $x < 0$, $0 < x < L$ and $x > L$, one possible solution has the form:

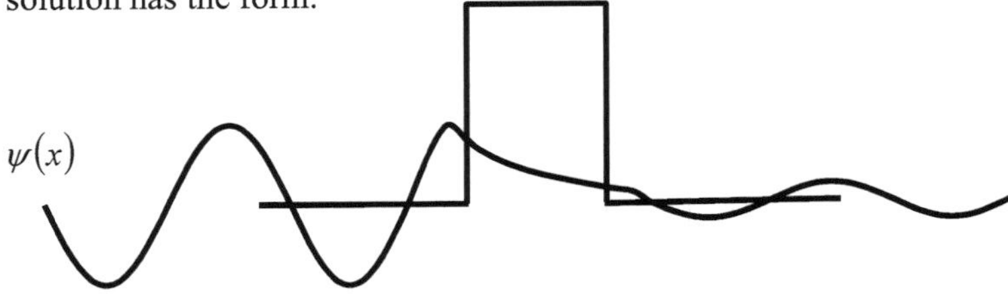

In general, the form of the solution depends upon the width L of the barrier and the ratio of the electron energy to the barrier potential, for example, whether the energy E is greater or less than V_o. For the case of $E > V_o$, the wave function is partly reflected at the barrier, a circumstance which has particular significance for the formation of energy gaps in the band structure of solids.

Tunnelling through barriers, whether we are considering electrons or any other atomic particle, cannot be explained by conventional classical Newtonian physics, yet has considerable practical importance from the conduction of electricity through contacts and junctions to the processes involved in nuclear decay. It is an everyday occurrence.

3.5.12 Harmonic Oscillator

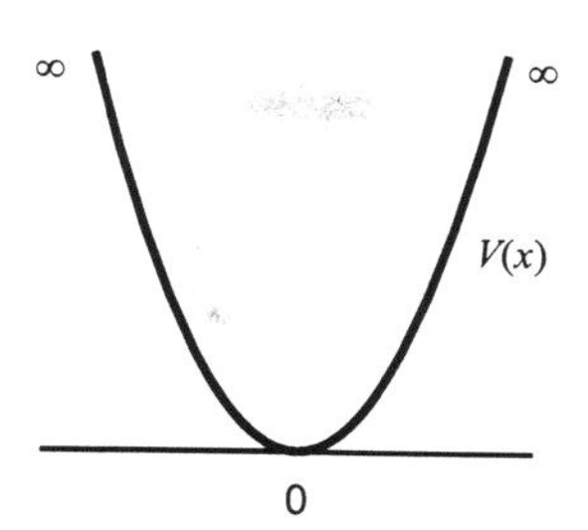

Consider the case of a **simple harmonic oscillator potential**:

$$V(x) = \frac{1}{2}Cx^2$$

The time-independent equation becomes:

$$-\frac{\hbar^2}{2m}\frac{\partial^2 \psi}{\partial x^2} + \frac{Cx^2}{2}\psi = E\psi$$

A **power series** solution (not derived here) yields:

n	**Eigenfunctions**
0	$\psi_o = A_o e^{-u^2/2}$
1	$\psi_1 = A_1 u e^{-u^2/2}$
2	$\psi_2 = A_2 (2u^2 - 1) e^{-u^2/2}$
3	$\psi_3 = A_3 (2u^3 - 3u) e^{-u^2/2}$
4	$\psi_4 = A_4 (4u^4 - 12u^2 + 3) e^{-u^2/2}$

$$\psi_n(u) = A_n e^{-u^2/2} H_n(u)$$

$H_n(u)$: **Hermite polynomial**

where $u = \left(\frac{(Cm)^{1/4}}{\hbar^{1/2}}\right)x$

The allowed energies, or **eigenvalues**, are:

$$\boxed{E_n = \left(n + \frac{1}{2}\right)hf} \quad n = 0,1,2,3\ldots$$

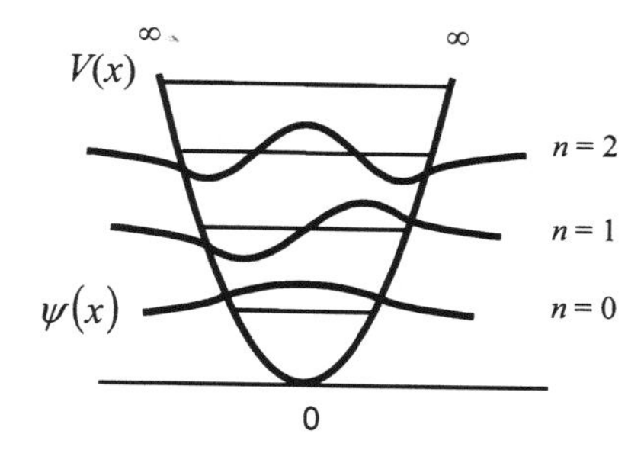

Note that, compared with the infinite square well potential, the energy levels for the harmonic oscillator are equally spaced hf. Note also the existence of a zero point energy at $n = 0$.

The harmonic oscillator potential has particular importance in describing the state of any system that exhibits small oscillations about a central position (e.g., vibrations of atoms, acoustic and thermal properties of solids, and the response of solids to electromagnetic waves).

3.5.13 Coulomb Potential – Bohr Atom

The single-electron **Bohr atom** can be analysed using the Schrödinger equation with the Coulomb potential. Since the potential function is three dimensional, $V(x,y,z)$, it is easier to work with spherical polar coordinates since then, the potential function becomes a function of one variable (r) only.

$$V(x,y,z)=\frac{-Zq_e}{4\pi\varepsilon_o\sqrt{x^2+y^2+z^2}}$$

$$V(r)=\frac{-Zq_e{}^2}{4\pi\varepsilon_o r}$$ in spherical polar coordinates, r, θ, ϕ

The **Schrödinger equation** becomes:

$$-\frac{\hbar^2}{2m}\left[\frac{1}{r^2}\frac{\partial}{\partial r}\left(r^2\frac{\partial}{\partial r}\right)+\frac{1}{r^2\sin^2\theta}\frac{\partial}{\partial\phi^2}+\frac{1}{r^2\sin\theta}\frac{\partial}{\partial\theta}\left(\sin\theta\frac{\partial}{\partial\theta}\right)\right]\Psi+V(r)=E\Psi$$

Let: $\Psi(r,\theta,\phi)=R(r)\Theta(\theta)\Phi(\phi)$

$$-\frac{\hbar^2}{2m}\left[\frac{1}{r^2}\Theta\Phi\frac{\partial}{\partial r}\left(r^2\frac{\partial R}{\partial r}\right)+\frac{R\Theta}{r^2\sin^2\theta}\frac{\partial^2\Phi}{\partial\phi^2}+\frac{R\Phi}{r^2\sin\theta}\frac{\partial}{\partial\theta}\left(\sin\theta\frac{\partial\Theta}{\partial\theta}\right)\right]\Psi$$
$$+V(r)R\Theta\Phi=ER\Theta\Phi$$

Multiply by: $\dfrac{-2mr^2\sin^2\theta}{\hbar^2 R\Theta\Phi}$

$$-\frac{\hbar^2}{2m}\left[\frac{\sin^2\theta}{R}\frac{\partial}{\partial r}\left(r^2\frac{\partial R}{\partial r}\right)+\frac{1}{\Phi}\frac{\partial^2\Phi}{\partial\phi^2}+\frac{\sin\theta}{\Theta}\frac{\partial}{\partial\theta}\left(\sin\theta\frac{\partial\Theta}{\partial\theta}\right)\right]$$
$$+\frac{-2m}{\hbar^2}V(r)r^2\sin^2\theta=\frac{-2mEr^2\sin^2\theta}{\hbar^2}$$

$$\frac{1}{\Phi}\frac{\partial^2\Phi}{\partial\phi^2}=\frac{-\sin^2\theta}{R}\frac{\partial}{\partial r}\left(r^2\frac{\partial R}{\partial r}\right)-\frac{\sin\theta}{\Theta}\frac{\partial}{\partial\theta}\left(\sin\theta\frac{\partial\Theta}{\partial\theta}\right)-\frac{2m}{\hbar^2}r^2\sin^2\theta(E-V(r))$$
$$=-m_l^2 \text{ a constant}$$

$$\frac{\partial^2\Phi}{\partial\phi^2}=-m_l^2\Phi$$

$$-m_l^2 = \frac{-\sin^2\theta}{R}\frac{\partial}{\partial r}\left(r^2\frac{\partial R}{\partial r}\right) - \frac{\sin\theta}{\Theta}\frac{\partial}{\partial\theta}\left(\sin\theta\frac{\partial\Theta}{\partial\theta}\right) - \frac{2mr^2\sin^2\theta}{\hbar^2}(E - V(r))$$

$$\frac{-m_l^2}{\sin^2\theta} = -\frac{1}{R}\frac{\partial}{\partial r}\left(r^2\frac{\partial R}{\partial r}\right) - \frac{1}{\Theta\sin\theta}\frac{\partial}{\partial\theta}\left(\sin\theta\frac{\partial\Theta}{\partial\theta}\right) - \frac{2mr^2}{\hbar^2}(E - V(r))$$

$$\frac{1}{R}\frac{\partial}{\partial r}\left(r^2\frac{\partial R}{\partial r}\right) + \frac{2mr^2}{\hbar^2}(E - V(r)) = \frac{m_l^2}{\sin^2\theta} - \frac{1}{\Theta\sin\theta}\frac{\partial}{\partial\theta}\left(\sin\theta\frac{\partial\Theta}{\partial\theta}\right)$$

$$= l(l+1) \quad \text{a constant}$$

- m_l is called the **magnetic quantum number**
- l is called the **angular momentum quantum number**

$$\frac{\partial^2\Phi}{\partial\phi^2} = -m_l^2\Phi \quad \text{function of } \phi$$

$$\frac{1}{R}\frac{\partial}{\partial r}\left(r^2\frac{\partial R}{\partial r}\right) + \frac{2mr^2}{\hbar^2}(E - V(r)) = l(l+1) \quad \text{function of } r$$

$$\frac{m_l^2}{\sin^2\theta} - \frac{1}{\Theta\sin\theta}\frac{\partial}{\partial\theta}\left(\sin\theta\frac{\partial\Theta}{\partial\theta}\right) = l(l+1) \quad \text{function of } \theta$$

The value of this approach is that the original partial differential equation involving three variables is broken down into three ordinary differential equations of one variable each.

The solutions in this case exist only for certain values of the quantum numbers m for $\Phi(\varphi)$, l for $\Theta(\theta)$ and n for $R(r)$ such that:

$$n = 1,2,3...$$

$$l = 0,1,2...n-1$$

$$m = -l, -l+1, ...0.... +l-1, l$$

The **eigenfunctions** $\psi_{n,l,m}$ provide information about the probability amplitudes, or probability density functions, of the electron for each allowed state. The **eigenvalues**, or allowed energies E, are:

$$\boxed{E = -\frac{Z^2 q_e^{\,4} m}{(4\pi\varepsilon_o)^2 2\hbar^2 n^2}} \quad n = 1,2,3...$$

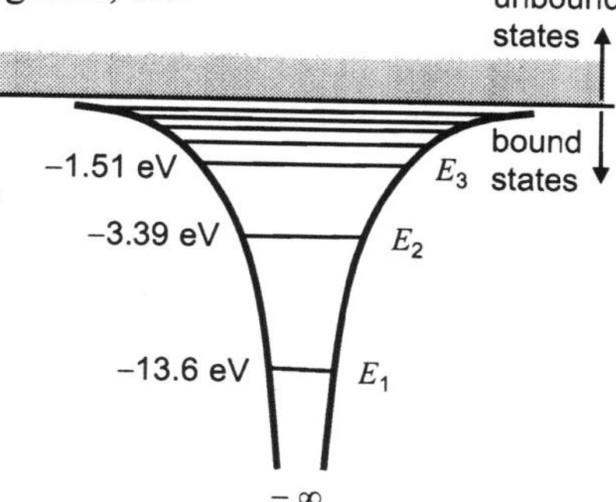

For a given value of principal quantum number n, there are several different possible values of l and m. When there are two or more eigenfunctions (i.e., combinations of n, m and l) that result in the same eigenvalue (or energy), these states are said to be **degenerate**.

3.5.14 Superposition

This **superposition** of solutions is a general feature of solutions to linear differential equations. Thus, if the function $\Psi_n(x, t)$ are solutions to the Schrödinger wave equation, then so is any combination of these.

$$\Psi(x,t) = c_1\Psi_1 + c_2\Psi_2 + ... + c_n\Psi_n$$

The constants c_n allow us to adjust the weighting or proportion of each component wave function if desired.

Linear combinations of eigenfunctions $\psi(x)$ are also solutions to the time-independent equation as long as they correspond to the same value of E.

Consider a *special case* whereby an electron can be in either of one or two separate regions x_1 and x_2. In general, the probability of finding the electron at some position x is given by:

$$P(x,t) = |\Psi(x,t)|^2 = |c_1\Psi_1(x,t) + c_2\Psi_2(x,t)|^2$$

The normalisation condition requires that the total probability, when added over all values of x, is equal to unity, so:

$$\int_x P(x,t)dx = \int_x |\Psi|^2 dx = \int_x |c_1\Psi_1 + c_2\Psi_2|^2 dx = \int_x \Psi^*\Psi dx = 1$$

$$= \int_x (c_1\Psi_1^* + c_2\Psi_2^*)(c_1\Psi_1 + c_2\Psi_2)dx$$

$$1 = c_1^2 \int_x \Psi_1^*\Psi_1 dx + \int_x c_1c_2\Psi_1^*\Psi_2 + c_1c_2\Psi_2^*\Psi_1 dx + c_2^2 \int_x \Psi_2^*\Psi_2 dx$$

If we take the real part of the wave function, this simplifies to:

$$1 = c_1^2 + c_2^2 + 2c_1c_2 \int_x \Psi_1^*\Psi_2 dx$$

The probabilities attached to each wave function individually are:

$$P_1 = c_1^2 \int_{x_1} \Psi_1^*\Psi_1 dx = c_1^2 \qquad P_2 = c_2^2 \int_{x_2} \Psi_2^*\Psi_2 dx = c_2^2$$

In this special case, each wave function is separately normalised (that is, the electron can only be in either one of two regions x_1 or x_2). The total probability of being in either x_1 or x_2 is thus: $1 = c_1^2 + c_2^2$,

which means that $2c_1c_2 \int_x \Psi_1^*\Psi_2 dx = 0$.

This integral, the **overlap integral**, equals zero in this case because we specified that the electron can only be in one region, either x_1 or x_2. This of course, is not generally the case. For example, in the double slit experiment there is an appreciable overlap of probabilities from each slit for the electron striking the distant screen midway between the slits.

3.5.15 Orthogonal Eigenfunctions

Consider the two eigenfunctions associated with the **infinite square well potential**:

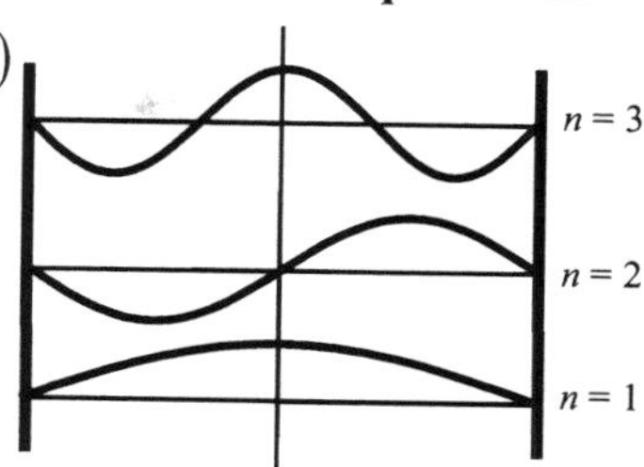

$$\psi_1 = \sqrt{2/a}\cos\frac{\pi x}{a}$$

$$\psi_3 = \sqrt{2/a}\cos\frac{3\pi x}{a} \qquad k_n = \frac{n\pi}{a}$$

To determine if there is any overlap in these eigenfunctions, the overlap integral is evaluated:

$$\int \psi_1^* \psi_3 dx = \int_{-a/2}^{a/2} \frac{2}{\sqrt{a}}\cos\frac{\pi x}{a}\frac{2}{\sqrt{a}}\cos\frac{3\pi x}{a}dx$$

$$= \frac{4}{a}\int_{-a/2}^{a/2} \cos\frac{\pi x}{a}\cos\frac{3\pi x}{a}dx \qquad \text{Integration by parts}$$

$$= \frac{4}{a}\left[\frac{1}{3\pi}\cos\frac{\pi x}{a}\sin\frac{3\pi x}{a} + \frac{1}{12}\left(\sin\frac{2\pi x}{a} - \frac{1}{2}\sin\frac{4\pi x}{a}\right)\right]_{-a/2}^{a/2}$$

$$= \frac{4}{a}\left[\frac{1}{3\pi}\cos\frac{\pi}{2}\sin\frac{3\pi}{2} + \frac{1}{12}\left(\sin\pi - \frac{1}{2}\sin 2\pi\right)\right]$$

$$- \frac{4}{a}\left[\frac{1}{3\pi}\cos\frac{-\pi}{2}\sin\frac{-3\pi}{2} + \frac{1}{12}\left(\sin(-\pi) - \frac{1}{2}\sin -2\pi\right)\right]$$

$$= 0$$

A similar result occurs when the overlap integral for ψ_1 and ψ_2 is evaluated. When the overlap integral for two eigenfunctions evaluates to zero, the eigenfunctions are said to be **orthogonal**. In this case, these eigenfunctions are orthogonal because they have different eigenvalues (i.e., energies) even though they may share the same space $-a/2$ to $+a/2$.

Orthogonality between functions is similar to that found when dealing with vectors. Indeed, the overlap integral is equivalent to the **dot product** between two vectors. In general,

$$\boxed{\langle\psi_m|\psi_n\rangle = \int \psi_m^* \psi_n dx}$$

where the bracket notation (known as **bra-ket**, introduced by **Dirac**) is used in place of the • symbol used for conventional vectors. Eigenfunctions are thus also referred to as **eigenvectors**.

3.5.16 Operator Notation

In three dimensions, the vector differential operator is written using the notation "**del**":
$$\nabla = \frac{\partial}{\partial x}\mathbf{i} + \frac{\partial}{\partial y}\mathbf{j} + \frac{\partial}{\partial z}\mathbf{k}$$

and so the **Schrödinger equation** can be written: $-\frac{\hbar^2}{2m}\nabla^2\Psi + V\Psi = i\hbar\frac{\partial\Psi}{\partial t}$.

We can form a new operator $\hat{H}$, so that $\hat{H} = -\frac{\hbar^2}{2m}\nabla^2 + V$.

$\hat{H}$ is called the **energy operator**, or the **Hamiltonian operator**. The Schrödinger equation is written, in operator notation, as:
$$\boxed{\hat{H}\Psi(x,t) = i\hbar\frac{\partial\Psi(x,t)}{\partial t}}$$

Note, for a particle moving in one dimension, we can write:
$$\hat{H} = \frac{p^2}{2m} + V(x)$$

The time-independent Schrödinger equation becomes $\hat{H}\psi(x) = E\psi(x)$.

We say that $\psi(x)$ are eigenfunctions of the operator $\hat{H}$ with eigenvalues E. For the most part, eigenfunctions associated with different eigenvalues are orthogonal (the exceptions are cases of **degeneracy**, where it is possible to have the same energy for different eigenstates). Other quantities (like position, momentum, etc.) can also be represented by the eigenvalues of operators.

In general, the expectation value of an observable quantity (eigenvalue) O associated with an operator $\hat{O}$ for a system with wave functions $\Psi(x,t)$ is found from:
$$\langle O\rangle = \int_{-\infty}^{\infty}\Psi^*\hat{O}\Psi\,dx$$

For physically observable quantities, the eigenvalues must be real numbers and the wave functions must be **orthogonal**. An operator with this property is called **Hermitian**. It can be shown that for Hermitian operators:
$$\int_{-\infty}^{\infty}\left(\hat{O}\Psi_1\right)^*\Psi_2\,dx = \int_{-\infty}^{\infty}\Psi_1^*\left(\hat{O}\Psi_2\right)dx$$

which in Dirac's notation is: $\left\langle\hat{O}\Psi_1 \mid \Psi_2\right\rangle = \left\langle\Psi_1 \mid \hat{O}\Psi_2\right\rangle$.

Both expressions are usually combined to an even more shortened form:
$$\boxed{\left\langle\Psi_1 \mid \hat{O} \mid \Psi_2\right\rangle =}\int\left(\hat{O}\Psi_1\right)^*\Psi_2\,dx \text{ or } \int\Psi_1^*\left(\hat{O}\Psi_2\right)dx$$

For example, in a quantum mechanical system:
$$\left\langle\Psi \mid \hat{E} \mid \Psi\right\rangle = \langle E\rangle = \int_{-\infty}^{\infty}\Psi^*\left(\hat{E}\Psi\right)dx = \int_{-\infty}^{\infty}\left(\hat{E}\Psi\right)^*\Psi\,dx$$

3.5.17 Commutators

Let the **expectation value** of an observable be expressed $\langle O\rangle = \int \Psi^* \hat{O}\Psi dx$.

We are interested in how the value of $\langle O\rangle$ changes with time:

$$\frac{d}{dt}\langle O\rangle = \int \left(\Psi^* \hat{O}\right)\frac{d}{dt}\Psi + \left(\hat{O}\Psi\right)\frac{d}{dt}\Psi^* dx$$

$$\text{But } \hat{H}\Psi = i\hbar\frac{\partial}{\partial t}\Psi$$

$$\text{Thus } \frac{d}{dt}\Psi = \frac{1}{i\hbar}\hat{H}\Psi;\quad \frac{d}{dt}\Psi^* = -\frac{1}{i\hbar}\left(\hat{H}\Psi\right)^*$$

$$\frac{d}{dt}\langle O\rangle = \frac{1}{i\hbar}\int \left(\Psi^* \hat{O}\right)\hat{H}\Psi + -\hat{O}\Psi\left(\hat{H}\Psi\right)^* dx$$

$$= \frac{i}{\hbar}\int \hat{O}\Psi\left(\hat{H}\Psi\right)^* - \Psi^*\hat{O}\hat{H}\Psi dx$$

$$\text{But} \int \hat{O}\Psi\left(\hat{H}\Psi\right)^* dx = \int \Psi^*\hat{H}\hat{O}\Psi dx \quad \text{Hermitian operator}$$

$$\text{Therefore } \frac{d}{dt}\langle O\rangle = \frac{i}{\hbar}\int \Psi^*\hat{H}\hat{O}\Psi - \Psi^*\hat{O}\hat{H}\Psi dx$$

$$= \frac{i}{\hbar}\int \Psi^*\left(\hat{H}\hat{O} - \hat{O}\hat{H}\right)\Psi dx$$

$\left(\hat{H}\hat{O} - \hat{O}\hat{H}\right)$ is called the **commutator** and is written: $\left[\hat{H},\hat{Q}\right]$

$$\frac{d}{dt}\langle O\rangle = \frac{d}{dt}\left\langle \Psi \mid \hat{O} \mid \Psi\right\rangle = \frac{i}{\hbar}\int \Psi^*\left[\hat{H},\hat{O}\right]\Psi dx = \frac{i}{\hbar}\left\langle \Psi \mid \left[\hat{H},\hat{O}\right] \mid \Psi\right\rangle$$

$$= \frac{i}{\hbar}\left\langle\left[\hat{H},\hat{O}\right]\right\rangle$$

When the commutator is zero, the two operators are said to commute. The physical observables associated with the operators may be measured precisely at the same time. For example, the energy and momentum of a free electron (zero potential) are not quantised and may be measured independently – they are compatible, and their operators commute, whereas the position x, and the momentum may not be simultaneously measured.

Another aspect of the commutator is that when the commutator is zero, the observable quantity O is unchanging in time. Put another way, the observable quantity is said to be conserved. An example is angular momentum in the spherically symmetric hydrogen atom.

3.5.18 Perturbation Theory

Consider the time-independent Schrödinger equation. Using operator notation:

$$\hat{H}\psi_n = E_n\psi_n$$

In many cases of practical interest, the ideal quantum system is perturbed by some small effect – for example, the application of an electric field from a nearby particle or some external source. How may the eigenfunctions and eigenvalues of the perturbed system be represented? The **Hamiltonian** is written as the sum of the original, plus the **perturbation**:

$$\hat{H} = \hat{H}_o + \lambda\hat{H}_1$$

where the factor λ quantifies the strength of the perturbation.

When the perturbation λ is small, the eigenfunctions for the system as a whole can be represented by a convergent series:

$$\psi_n = \psi_n^{(0)} + \lambda\psi_n^{(1)} + \lambda^2\psi_n^{(2)}...$$

and the eigenvalues:

$$E_n = E_n^{(0)} + \lambda E_n^{(1)} + \lambda^2 E_n^{(2)}...$$

An approximation for the expectation value of $E^{(1)}$ for the first term of the perturbation is:

$$\left\langle E_n^{(1)} \right\rangle = \left\langle \psi_n \mid \hat{H}_1 \mid \psi_n \right\rangle$$

$$= \int_{-\infty}^{\infty} \psi_n^* \hat{H}_1 \psi_n dx$$

Let us take a first order approximation to a solution for the perturbed system, that is, the 0th and 1st terms in the series only. The Schrödinger equation is:

$$\hat{H}_0\psi_n^{(0)} + \lambda\hat{H}_1\psi_n^{(1)} = E_n^{(0)}\psi_n^{(0)} + \lambda E_n^{(1)}\psi_n^{(1)}$$

Often, but not always, a first order approximation is reasonable for our purposes and the effect on the energies of the perturbation can be obtained analytically.

For a second order approximation, the expression for the perturbed component of the energy (the derivation of which is not shown here) is:

$$E_n^{(2)} = \sum_{p \neq n} \frac{\left| \left\langle \psi_p \mid \hat{H}_1 \mid \psi_n \right\rangle \right|^2}{E_n^0 - E_p^0}$$

3.5.19 Perturbation Theory – Infinite Square Well

Consider the time-independent equation for a one dimensional **infinite square well potential**. The perturbed system has a small potential in the middle of the well, as shown. We will determine the energy of the perturbed system, to a first order approximation, for the lowest energy level $n = 1$:

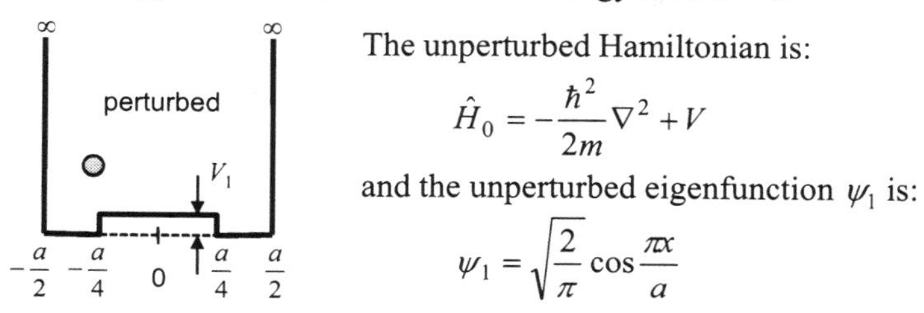

The unperturbed Hamiltonian is:

$$\hat{H}_0 = -\frac{\hbar^2}{2m}\nabla^2 + V$$

and the unperturbed eigenfunction ψ_1 is:

$$\psi_1 = \sqrt{\frac{2}{\pi}}\cos\frac{\pi x}{a}$$

The perturbation is a constant V_1 within the limits shown.

$$V(x) = V_1 \qquad -\frac{a}{4} \le x \le \frac{a}{4}; x = 0 \text{ otherwise}$$

For the unperturbed system, we have eigenvalues:

$$E = \frac{n^2\pi^2\hbar^2}{2a^2 m} \qquad n = 1,2,3,4\ldots$$

and eigenfunctions:

$$\psi_n(x) = A_n \cos k_n x \qquad n = 1,3,5,7\ldots$$

$$\psi_n(x) = B_n i \sin k_n x \qquad n = 2,4,6,8\ldots \qquad k_n = \frac{n\pi}{a}$$

We may wish to estimate the lowest energy level of the perturbed system. Thus:

$$\left\langle E_1^{(1)} \right\rangle = \left\langle \psi_1 \,|\, \hat{H}_1 \,|\, \psi_1 \right\rangle \quad \text{where } \hat{H}_1 = V_1$$

$$= \int_{-a/4}^{a/4} A_1^* \cos\frac{\pi x}{a} V_1\, A_1 \cos\frac{\pi x}{a} dx$$

We only need to integrate between $-a/4$ and $a/4$ because the perturbation is zero otherwise.

$$= |A_1|^2 V_1 \int_{-a/4}^{a/4} \cos^2\frac{\pi x}{a} dx$$

$$\int \cos^2 x dx = \frac{1}{2}x + \frac{1}{4}\sin 2x$$

$$= \left(\frac{2}{a}\right) V_1 \left[\frac{a}{4} - \frac{a}{2\pi}\right]$$

$$A^2 = \frac{2}{a} \quad \text{from normalisation condition}$$

$$= V_1\left(\frac{1}{2} + \frac{1}{\pi}\right)$$

$$E_1 = \frac{\pi^2\hbar^2}{2ma^2} + \lambda V_1\left(\frac{1}{2} + \frac{1}{\pi}\right)$$

to a first order approximation since $E_1 = E_1^{(0)} + \lambda E_1^{(1)}$

3.5.20 Harmonic Oscillator – First Order Perturbation

Consider a one dimensional **harmonic oscillator** with charge $-q$ that is perturbed by an electric field E in the positive x direction where the perturbation is small.

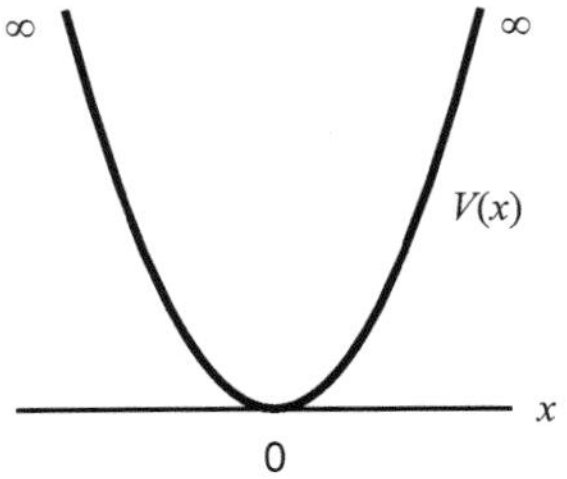

For the unperturbed case, we have:
$$V(x) = \frac{1}{2}Cx^2$$

The **Hamiltonian** takes the form:

$$\hat{H}_0 = -\frac{\hbar^2}{2m}\nabla^2 + \frac{1}{2}Cx^2$$

The eigenfunctions are $\psi_n(u) = A_n e^{-u^2/2} H_n(u)$.

where $u = \left(\frac{(Cm)^{1/4}}{\hbar^{1/2}}\right)x$

The allowed energies, or **eigenvalues**, are:

$$E_n = \left(n + \frac{1}{2}\right)hf \quad n = 0,1,2,3\ldots$$

Now, the perturbation is an electric field E acting on a charge q. The potential energy associated with this depends on the position x and is given by the product $U = qEx = \lambda x$ since E and q are assumed to be constant. The product $\lambda = qE$ thus quantifies the strength of the perturbation.

The Hamiltonian for the perturbed system becomes:

$$\hat{H} = \hat{H}_0 + \lambda\hat{H}_1$$

$$= -\frac{\hbar^2}{2m}\nabla^2 + \frac{1}{2}Cx^2 + \lambda x \quad \text{that is, } \hat{H}_1 = x \text{ in this case.}$$

In this example, we use the same symbol E for both electric field and energy. The context should make the use clear.

A first order approximation for the energy levels E for the perturbed system is found from:

$$E_n = E_n^{(0)} + \lambda E_n^{(1)}$$

$$\left\langle E_n^{(1)}\right\rangle = \left\langle \psi_n \mid \hat{H}_1 \mid \psi_n \right\rangle = \int \psi_n^* \hat{H}_1 \psi_n dx$$

Here, $\hat{H}_1 = x$, and so:

$$\left\langle E_n^{(1)}\right\rangle = \int \psi_n^* x \psi_n dx$$

$$= 0$$

This integral evaluates to zero because in this case, we are essentially finding the expectation value of x, which, by symmetry, is at $x = 0$.

The conclusion therefore is that to a first order approximation, a perturbation of the form λx does not affect the spacing of the energy levels of the simple harmonic oscillator from the original unperturbed energies.

3.5.21 Harmonic Oscillator – Second Order Perturbation

A second order representation of the energy levels for the perturbed **simple harmonic oscillator** is: $E_n = E_n^{(0)} + \lambda E_n^{(1)} + \lambda^2 E_n^{(2)}$

where $E_n^{(2)} = \sum_{p \neq n} \frac{\left|\langle \psi_p \mid \hat{H}_1 \mid \psi_n \rangle\right|^2}{E_n^0 - E_p^0}$ and here, $\hat{H}_1 = x$

Now, $\langle \psi_p \mid \hat{H}_1 \mid \psi_n \rangle = \langle \psi_p \mid x \mid \psi_n \rangle$

$$= \left(\frac{\hbar}{m\omega}\right)^{1/2}\left(\frac{n}{2}\right)^{1/2} \qquad p = n-1$$

$$= \left(\frac{\hbar}{m\omega}\right)^{1/2}\left(\frac{n+1}{2}\right)^{1/2} \qquad p = n+1$$

$$= 0 \qquad \text{otherwise}$$

This result is not derived here, but only stated. For justification, see 2.8.24.

Thus:

$$E_n^{(2)} = \left(\frac{\hbar}{m\omega}\right)\frac{n}{2}\left(\frac{1}{E_n^0 - E_{n-1}^0}\right) + \left(\frac{\hbar}{m\omega}\right)\frac{n+1}{2}\left(\frac{1}{E_n^0 - E_{n+1}^0}\right)$$

$$E_n^0 = \left(n + \frac{1}{2}\right)\hbar\omega; E_{n-1}^0 = \left(n - \frac{1}{2}\right)\hbar\omega; E_{n+1}^0 = \left(n + \frac{3}{2}\right)\hbar\omega$$

$$E_n^{(2)} = \left(\frac{\hbar}{m\omega}\right)\frac{n}{2\hbar\omega}\frac{1}{\left(n+\frac{1}{2}\right) - \left(n - \frac{1}{2}\right)} + \left(\frac{\hbar}{m\omega}\right)\frac{n+1}{2\hbar\omega}\frac{1}{\left(n+\frac{1}{2}\right) - \left(n+\frac{3}{2}\right)}$$

$$= -\frac{1}{2m\omega^2}$$

The perturbed energy levels can now be expressed:

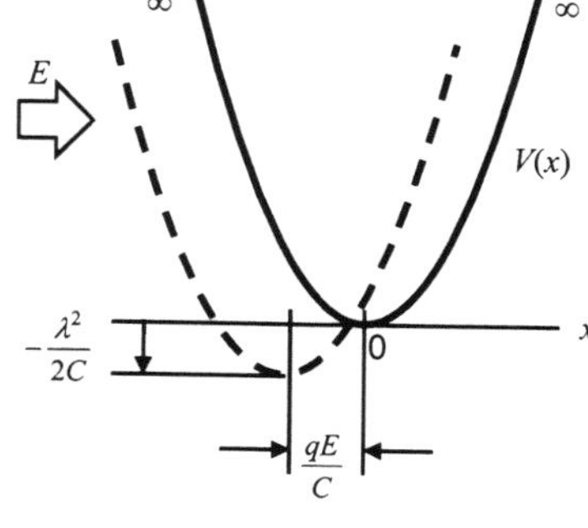

$$E_n = E_n^{(0)} + \lambda E_n^{(1)} + \lambda^2 E_n^{(2)}$$

$$= \left(n + \frac{1}{2}\right)\hbar\omega + \lambda(0) - \lambda\frac{1}{2m\omega^2}$$

$$= \left(n + \frac{1}{2}\right)\hbar\omega - \frac{\lambda^2}{2C} \qquad \text{where } \omega = \sqrt{\frac{C}{m}}$$

$$= \left(n + \frac{1}{2}\right)\hbar\omega - \frac{q^2E^2}{2C} \qquad \text{and } \lambda = qE$$

for a mass m moving in simply harmonic motion

3.5.22 Harmonic Oscillator – Eigenfunctions

For the case of $n = 1, p = 0$, we are interested in the first two eigenfunctions:

$$\psi_o = A_o e^{-u^2/2} \qquad \text{where } u = \left(\frac{(Cm)^{1/4}}{\hbar^{1/2}}\right)x$$

$$\psi_1 = A_1 u e^{-u^2/2}$$

Now: $C = m\omega^2$

Thus: $u = \left(\frac{\left(m^2\omega^2\right)^{1/4}}{\hbar^{1/2}}\right)x = \left(\frac{m\omega}{\hbar}\right)^{1/2} x$ and $du = \left(\frac{m\omega}{\hbar}\right)^{1/2} dx$

Applying the normalisation condition:

$$\psi_o = A_o e^{-u^2/2} \qquad\qquad \psi_1 = A_1 u e^{-u^2/2}$$

$$1 = \int_{-\infty}^{\infty} A_o^2 e^{-u^2}\left(\frac{\hbar}{m\omega}\right)^{1/2} du \qquad 1 = \int_{-\infty}^{\infty} A_1^2 u^2 e^{-u^2}\left(\frac{\hbar}{m\omega}\right)^{1/2} du$$

$$= \left(\frac{\hbar}{m\omega}\right)^{1/2} A_o^2\sqrt{\pi} \qquad\qquad = \left(\frac{\hbar}{m\omega}\right)^{1/2} A_1^2\frac{\sqrt{\pi}}{2}$$

$$A_o = \left(\frac{m\omega}{\hbar}\right)^{1/4}\left(\frac{1}{\pi}\right)^{1/4} \qquad\qquad A_1 = \left(\frac{m\omega}{\hbar}\right)^{1/4}\sqrt{2}\left(\frac{1}{\pi}\right)^{1/4}$$

Standard integrals

$$\int_{-\infty}^{\infty} x^2 e^{-x^2} dx = \frac{\sqrt{\pi}}{2}$$

$$\int_{-\infty}^{\infty} e^{-x^2} dx = \sqrt{\pi}$$

Thus: $\left\langle \psi_p \mid \hat{H}_1 \mid \psi_n \right\rangle = \left\langle \psi_p \mid x \mid \psi_n \right\rangle$

$$= \int_{-\infty}^{\infty} A_o e^{-u^2/2} x A_1 u e^{-u^2/2} dx$$

$$= \int_{-\infty}^{\infty} A_o e^{-u^2/2} u\left(\frac{\hbar}{m\omega}\right)^{1/2} A_1 u e^{-u^2/2}\left(\frac{\hbar}{m\omega}\right)^{1/2} du$$

$$= A_0 A_1\left(\frac{\hbar}{m\omega}\right)\int_{-\infty}^{\infty} u^2 e^{-u^2} du$$

$$= \left(\frac{m\omega}{\hbar}\right)^{1/4}\left(\frac{1}{\pi}\right)^{1/4}\left(\frac{m\omega}{\hbar}\right)^{1/4}\sqrt{2}\left(\frac{1}{\pi}\right)^{1/4}\left(\frac{\hbar}{m\omega}\right)\left[\frac{1}{2}\sqrt{\pi}\right]$$

$$= \left(\frac{\hbar}{m\omega}\right)^{1/2}\left(\frac{1}{2}\right)^{1/2}$$

3.5.23 Harmonic Oscillator – Exact Solution

A second order representation of the energy levels for the perturbed **simple harmonic oscillator** can be determined exactly (as distinct from the approximate method using perturbation methods).

$$\hat{H} = \frac{p^2}{2m} + \frac{1}{2}Cx^2 + \lambda x$$

The potential term is expressed:

$$\begin{aligned} V(x) &= \frac{1}{2}Cx^2 + \lambda x \\ &= \frac{1}{2}C\left(x^2 + \frac{2\lambda}{C}x\right) \\ &= \frac{1}{2}C\left(x^2 + \frac{2\lambda}{C}x + \frac{\lambda^2}{C^2} - \frac{\lambda^2}{C^2}\right) \\ &= \frac{1}{2}C\left(x^2 + \frac{2\lambda}{C}x + \frac{\lambda^2}{C^2}\right) - \frac{\lambda^2}{2C} \end{aligned}$$

$$V(x) = \underbrace{\frac{1}{2}C\left(x + \frac{\lambda}{C}\right)^2}_{} - \frac{\lambda^2}{2C}$$

This term has the form $1/2CX^2$.

Note, $X = x+\lambda/C$ shifts the centre of the potential function to the left (for a negative q and positive E) since $\lambda = qE$. The $\lambda^2/2C$ term shifts the potential function down.

That is, if we express the potential function in terms of a new variable $X = x+\lambda/C$, then the form of the eigenvalues, in this case, is the same as that obtained using second order approximation given by the perturbation method.

$$E_n = \left(n + \frac{1}{2}\right)\hbar\omega - \frac{\lambda^2}{2C} \qquad n = 0,1,2,3...$$

In this case (harmonic oscillator perturbed by a constant electric field), the second order approximation to the problem is an exact, complete solution. Higher order approximations are zero.

3.5.24 Variational Method

The **variational method** is used to approximate the lowest, or ground state, energy level of a quantum system. The method relies on (i) a guess of the form of the wave function ψ_t and (ii) the fact that the expectation value for the Hamiltonian will never be lower than the exact value – this last condition relying on the normalisation condition of the wave function.

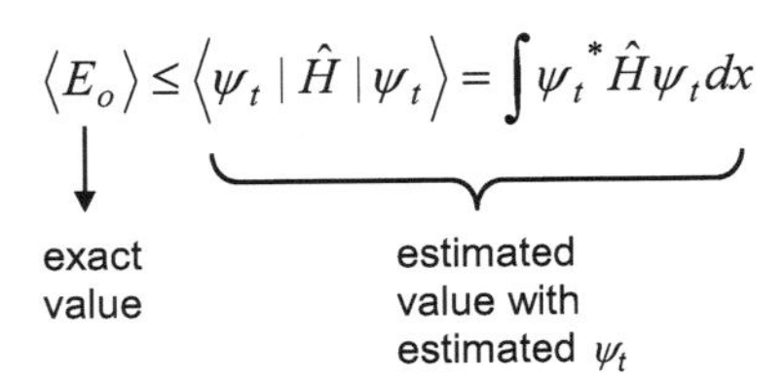

The estimated or **trial wave function** ψ_t is expressed in terms of an adjustable parameter α. The value of α for which the integral is minimised gives the best estimate of E_0.

The procedure is:

- Settle on a trial eigenfunction – this can be very similar in form to the known unperturbed case if the actual quantum system is not too far different.
- Evaluate the integral to determine the Hamiltonian for the quantum system – this will be expressed in terms of x (one dimensional case) and the adjustable parameter α.
- Determine the value of α for which $\dfrac{d\hat{H}}{d\alpha} = 0$.
- Substitute this value for α back into the Hamiltonian to obtain E_o.
- If an exact solution is known, verify the procedure by checking that: $\langle \psi_t | \hat{H} | \psi_t \rangle \geq E_o$

exact value

3.5.25 Variational Method – Example

Consider the **harmonic oscillator potential** with the Hamiltonian:

$$\hat{H} = \frac{p^2}{2m} + \lambda x^4 = -\frac{\hbar^2}{2m}\frac{\partial^2}{\partial x^2} + \lambda x^4$$

Let the trial function be $\psi(x) = \left(\frac{2\alpha}{\pi}\right)^{1/4} e^{-\alpha x^2}$ where α is the adjustable parameter.

We take each term individually:

$$\left\langle \psi_t \,|\, \frac{p^2}{2m} \,|\, \psi_t \right\rangle = \left(\frac{2\alpha}{\pi}\right)^{1/2} \int_{-\infty}^{\infty} e^{-\alpha x^2} \frac{-\hbar^2}{2m}\frac{\partial^2}{\partial x^2} e^{-\alpha x^2} dx$$

$$= \left(\frac{2\alpha}{\pi}\right)^{1/2} \frac{-\hbar^2}{2m} \int_{-\infty}^{\infty} e^{-\alpha x^2}\left((-2\alpha x)^2 e^{-\alpha x^2} + -2\alpha e^{-\alpha x^2}\right) dx$$

$$= \left(\frac{2\alpha}{\pi}\right)^{1/2} \frac{-\hbar^2}{2m} \int_{-\infty}^{\infty} 4\alpha^2 x^2 e^{-2\alpha x^2} + -2\alpha e^{-2\alpha x^2} dx$$

$$= \left(\frac{2\alpha}{\pi}\right)^{1/2} \frac{-\hbar^2}{2m} \left(4\alpha^2 \int_{-\infty}^{\infty} x^2 e^{-2\alpha x^2} dx + \int_{-\infty}^{\infty} -2\alpha e^{-2\alpha x^2} dx\right)$$

$$= \left(\frac{2\alpha}{\pi}\right)^{1/2} \frac{-\hbar^2}{2m} \left(\frac{4\alpha^2}{2}\sqrt{\pi}(2\alpha)^{-3/2} + -2\alpha \frac{\sqrt{\pi}}{\sqrt{2\alpha}}\right)$$

$$= \frac{\hbar^2}{2m}\alpha$$

and:

$$\left\langle \psi_t \,|\, x^4 \,|\, \psi_t \right\rangle = \left(\frac{2\alpha}{\pi}\right)^{1/2} \int_{-\infty}^{\infty} e^{-2\alpha x^2} x^4 dx$$

$$= \left(\frac{2\alpha}{\pi}\right)^{1/2} \left(\frac{3}{4}\sqrt{\pi}(2\alpha)^{-5/2}\right)$$

$$= \frac{3}{16\alpha^2}$$

$$\left\langle \psi_t \,|\, \lambda x^4 \,|\, \psi_t \right\rangle = \frac{3\lambda}{16\alpha^2}$$

Standard integrals

$$\int_{-\infty}^{\infty} x^2 e^{-\alpha x^2} dx = \frac{1}{2}\sqrt{\pi}\alpha^{-3/2}$$

$$\int_{-\infty}^{\infty} x^4 e^{-\alpha x^2} dx = \frac{3}{4}\sqrt{\pi}\alpha^{-5/2}$$

$$\int_{-\infty}^{\infty} e^{-\alpha x^2} dx = \frac{1}{\alpha}\sqrt{\pi}$$

The linear combination is thus:

$$\left\langle \psi_t \mid \hat{H} \mid \psi_t \right\rangle = \frac{\hbar^2}{2m}\alpha + \frac{3\lambda}{16\alpha^2}$$

The equation now has to be minimised with respect to α.

$$\frac{d\hat{H}}{d\alpha} = 0$$

$$0 = \frac{\hbar^2}{2m} + \frac{3\lambda(-2)}{16\alpha^3}$$

$$\frac{\hbar^2}{2m} = \frac{3}{8}\frac{\lambda}{\alpha^3}$$

$$\alpha = \left(\frac{3}{4}\frac{\lambda m}{\hbar^2}\right)^{1/3}$$

Our estimate for E_o is thus:

$$\begin{aligned} E_o &= \left\langle \psi_t \mid \hat{H} \mid \psi_t \right\rangle \\ &= \frac{\hbar^2}{2m}\alpha + \frac{3\lambda}{16\alpha^2} \\ &= \frac{\hbar^2}{2m}\left(\frac{3}{4}\frac{\lambda m}{\hbar^2}\right)^{1/3} + \frac{3\lambda}{16}\left(\frac{3}{4}\frac{\lambda m}{\hbar^2}\right)^{-2/3} \\ &= 1.082\lambda^{1/3}\left(\frac{\hbar^2}{2m}\right)^{-2/3} \end{aligned}$$

An exact solution for this problem yields $E_o = 1.060\lambda^{1/3}\left(\frac{\hbar^2}{2m}\right)^{-2/3}$.

Note that $1.082 > 1.060$ in accordance with our requirement that

$$\left\langle \psi_t \mid \hat{H} \mid \psi_t \right\rangle \geq E_o$$

↓

Exact value

3.5.26 Helium Atom

A **hydrogen atom** consists of one electron orbiting a central nucleus. The energy levels can be found from the wave function for a spherically symmetric potential – the Bohr atom. A **helium atom** consists of two electrons orbiting a nucleus of charge $+2q_e$. The electrons are close enough to have a mutual influence on their wave functions. That is, the wave function for a single electron is perturbed by the presence of the other electron.

For the case of no interaction, the potential function, the **Hamiltonian** and the ground state energy level ($n = 1$) are:

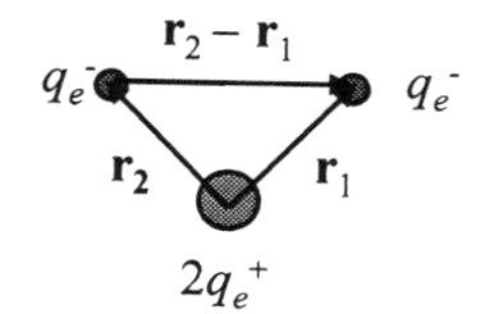

$$V(r) = -\frac{2q_e^2}{4\pi\varepsilon_o r}$$

$$\hat{H} = \frac{p_1^2}{2m} + \frac{p_1^2}{2m} - \frac{2q_e^2}{4\pi\varepsilon_o r_1} - \frac{2q_e^2}{4\pi\varepsilon_o r_2}$$

$$E = -\frac{2^2 q_e^4 m}{(4\pi\varepsilon_o)^2 2\hbar^2}$$
$$= -108\,\text{eV}$$

The experimental value is –78.975 eV. To obtain a better estimate from quantum mechanics, the perturbation, or interaction, has to be included. We let the perturbation be of the form:

$$E = \frac{q_e^2}{|\mathbf{r}_1 - \mathbf{r}_2|}$$

Using perturbation theory, the predicted perturbation to the ground state energies is:

$$E_1 = \left\langle \psi(r_1)\psi(r_2) \middle| \frac{q_e^2}{|\mathbf{r}_2 - \mathbf{r}_1|} \middle| \psi(r_1)\psi(r_2) \right\rangle$$

Calculations show that this evaluates to +34 eV (for $Z = 2$).

Allowing for two electrons, the predicted ground state energy of each electron is:

$$E = E_o + E_1$$
$$= -108 + (2)(17)$$
$$= -74.8\,\text{eV}$$

3.5.27 Transitions

Stationary quantum states occur when the potential V is a function of position only. The complete set of solutions to the Schrödinger equation for a potential $V(x)$ is, by the principle of superposition:

$$\Psi(x,t) = C_1\Psi_1 + C_2\Psi_2 + ... + C_n\Psi_n$$

$$\text{or} \quad \Psi(x,t) = \sum_n C_n(t)\psi_n(x)\phi_n(t) \qquad \text{since } \Psi = \psi(x)\phi(t)$$

$$= \sum_n C_n(t)\psi_n(x)e^{-i\frac{E_n}{\hbar}t} \qquad \text{and } \phi(t) = e^{-i\frac{Et}{\hbar}}$$

C_n are the weightings for each quantum state n and are expressed as a function of time to take into account the changing probability density of states in a system when transitions occur. Note, the quantum *states* themselves are still stationary, and so we can retain the procedure for the separation of variables, but the probability density of the states may be time dependent. This may happen, for example, during the excitation of an atom where an electron is promoted from a lower energy level, say the ground state, to a higher energy level by the absorption of a photon.

When an electron is in the **ground state**, the probability function is simply:

$$|\Psi|^2 = \Psi^*\Psi = \psi^* e^{+i\frac{E_n}{\hbar}t} \psi e^{-i\frac{E_n}{\hbar}t} = \psi^*\psi$$

That is, $|\Psi|^2$ is independent of time. When an electron makes a transition from state n to m, (say the ground state to an excited state), the transition involves a mixed state wave function.

$$\Psi(x,t) = C_n\Psi_n + C_m\Psi_m = C_n(t)\psi_n(x)e^{-i\frac{E_n}{\hbar}t} + C_m(t)\psi_m(x)e^{-i\frac{E_m}{\hbar}t}$$

$$\Psi^*\Psi = \left[C_n\psi_n^* e^{+i\frac{E_n}{\hbar}t} + C_m\psi_m^* e^{+i\frac{E_m}{\hbar}t}\right]\left[C_n\psi_n e^{-i\frac{E_n}{\hbar}t} + C_m\psi_m e^{-i\frac{E_m}{\hbar}t}\right]$$

$$= C_n^2\psi_n^*\psi_n + C_n\psi_n^* C_m\psi_m e^{i\frac{E_n - E_m}{\hbar}t} + C_m\psi_m^* C_n\psi_n e^{-i\frac{E_n - E_m}{\hbar}t} + C_m^2\psi_m^*\psi_m$$

It is sufficient to note from the above that the magnitude of the wave function contains oscillatory terms involving

$$\omega = \frac{E_n - E_m}{\hbar}$$

That is, the wave function oscillates with a frequency ω – precisely that of the photon frequency predicted by **Bohr**.

3.5.28 Transition Rate

Quantum mechanics provides an explanation of not only why a transition involves the emission or absorption of a photon in terms of the frequency of the mixed-state wave function, but also the rate at which a transition occurs.

Let $\hat{H} = \hat{H}_o + \hat{H}_1$

$\hat{H}_o$ → unperturbed Hamiltonian

$\hat{H}_1$ → perturbed Hamiltonian (e.g., time-varying potential from electromagnetic radiation)

$$\hat{H} = -\frac{\hbar^2}{2m}\frac{\partial^2}{\partial x^2} + V$$

$$\hat{H}\psi(x) = E\psi(x)$$

$$V_1(x,t) = qE(x,t) = qE_o \sin(kx - \omega t)$$

$\sin(kx-\omega t)$ → time-varying electric field component of an electromagnetic wave

The Schrödinger equation is written:

$$\hat{H}\Psi(x,t) = \left(\hat{H}_o + \hat{H}_1\right)\Psi = i\hbar\frac{\partial \Psi}{\partial t}$$

$$\left(\hat{H}_o + V_1(x,t)\right)\Psi = i\hbar\frac{\partial \Psi}{\partial t}$$

So, in general, starting from state n, we have:

$$\left(\hat{H}_o + V(x,t)\right)\Psi = \underbrace{\hat{H}_o\left(\sum_n C_n\psi_n e^{-i\frac{E_n}{\hbar}t}\right)} + V_1(x,t)\sum_n C_n\psi_n e^{-i\frac{E_n}{\hbar}t} = i\hbar\frac{\partial \Psi}{\partial t}$$

$$\psi(x) = \sum_n C_n\psi_n(x)$$

$$\hat{H}_o\psi(x) = E\psi(x) = \sum_n E_n C_n\psi_n(x)$$

$$\hat{H}_o\Psi(x,t) = \sum_n E_n C_n\psi_n(x)e^{-\frac{En}{\hbar}t}$$

Thus:

$$\left(\hat{H}_o + V_1(x,t)\right)\Psi = \sum_n E_n C_n\psi_n e^{-i\frac{E_n}{\hbar}t} + V_1(x,t)\sum_n C_n\psi_n e^{-i\frac{E_n}{\hbar}t} = i\hbar\frac{\partial \Psi}{\partial t}$$

where $i\hbar\frac{\partial \Psi}{\partial t} = i\hbar\sum_n C_n\psi_n \frac{-i}{\hbar}E_n e^{-i\frac{E_n}{\hbar}t} + \frac{\partial C_n(t)}{\partial t}\psi_n e^{-i\frac{E_n}{\hbar}t}$ by the chain rule

$$= \sum_n E_n C_n\psi_n e^{-i\frac{E_n}{\hbar}t} + i\hbar\sum_n \frac{\partial C_n(t)}{\partial t}\psi_n e^{-i\frac{E_n}{\hbar}t}$$

and so, cancelling like terms, we obtain:

$$i\hbar\sum_n \frac{\partial C_n(t)}{\partial t}\psi_n e^{-i\frac{E_n}{\hbar}t} = \sum_n C_n V_1(x,t)\psi_n e^{-i\frac{E_n}{\hbar}t}$$

Since the state n is acted upon by the time-varying function $V(x,t)$, this causes the energy of the electron to change from E_n to E_m. As far as the left hand side of the above equation is concerned, we obtain:

$$\left\langle \psi_m \mid i\hbar\sum_n \frac{\partial C_n}{\partial t}\psi_n e^{-i\frac{E_n}{\hbar}t}\right\rangle = i\hbar\sum_n \frac{\partial C_n}{\partial t}\left\langle \psi_m^* \mid \psi_n\right\rangle e^{-i\frac{E_n}{\hbar}t}$$

Since ψ_m and ψ_n are orthogonal, all the terms in the sum reduce to zero except the case where $m = n$ and so:

$$i\hbar\sum_n \frac{\partial C_n}{\partial t}\left\langle \psi_m^* \mid \psi_n\right\rangle e^{-i\frac{E_n}{\hbar}t} = i\hbar\frac{\partial C_m}{\partial t}e^{-i\frac{E_m}{\hbar}t}$$

On the right hand side, we obtain:

$$\left\langle \psi_m \mid \sum_n C_n V_1(x,t)\psi_n e^{-i\frac{E_n}{\hbar}t}\right\rangle = \sum_n C_n\left\langle \psi_m \mid V_1(x,t) \mid \psi_n\right\rangle e^{-i\frac{E_n}{\hbar}t}$$

and equating the two:

$$i\hbar\frac{\partial C_m}{\partial t}e^{-i\frac{E_m}{\hbar}t} = \sum_n C_n\left\langle \psi_m \mid V_1(x,t) \mid \psi_n\right\rangle e^{-i\frac{E_n}{\hbar}t}$$

$$\frac{\partial C_m}{\partial t} = \frac{1}{i\hbar}\sum_n C_n\left\langle \psi_m \mid V_1(x,t) \mid \psi_n\right\rangle e^{i\frac{E_m-E_n}{\hbar}t}$$

The physical significance of the terms is:

$\dfrac{\partial C_m}{\partial t} =$ the rate of change of probability density of state m

$\dfrac{E_m - E_n}{\hbar} = \omega$ the frequency of the transition.

A **quantum transition** is accompanied by the energy change from the energy of one vibrational mode of the matter wave to the energy of another. During a transition, electrons do not "jump" from one level to another like people going to different floors in a building, but change their vibrational mode (i.e., their probability density distribution) from one pattern to another.

Appendix

A.1 Useful Information
A.2 Some Standard Integrals
A.3 Special Functions

A.1 Useful Information

$A = \frac{1}{2}bh$	Area of a triangle
$A = \pi r^2$	Area of a circle
$C = 2\pi r$	Circumference of a circle
$A = bh$	Area of a parallelogram
$V = \frac{4}{3}\pi r^3$	Volume of a sphere
$S = 4\pi r^2$	Surface area a sphere

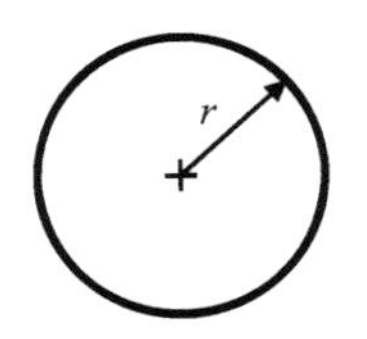

$$\pi = 3.14159265...$$
$$e = 2.7182818...$$
$$\sqrt{2} = 1.41421356...$$
$$1\,\text{rad} = 360/2\pi = 57.29578...^{\circ}$$

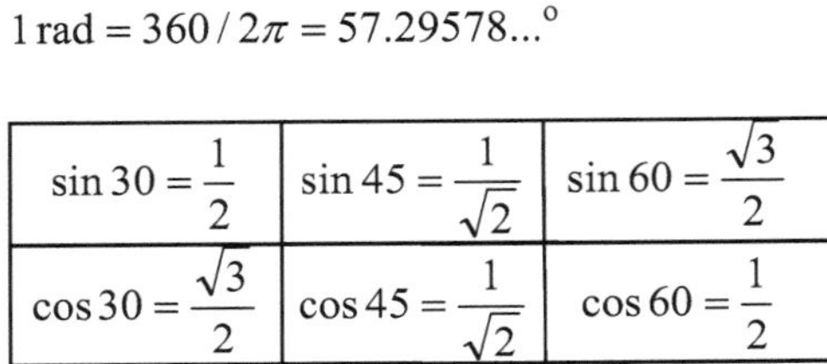

$\sin 30 = \frac{1}{2}$	$\sin 45 = \frac{1}{\sqrt{2}}$	$\sin 60 = \frac{\sqrt{3}}{2}$
$\cos 30 = \frac{\sqrt{3}}{2}$	$\cos 45 = \frac{1}{\sqrt{2}}$	$\cos 60 = \frac{1}{2}$
$\tan 30 = \frac{1}{\sqrt{3}}$	$\tan 45 = 1$	$\tan 60 = \sqrt{3}$

$$\sin(a \pm b) = \sin a \cos b \pm \cos a \sin b$$
$$\cos(a \pm b) = \cos a \cos b \mp \sin a \sin b$$
$$\tan(a \pm b) = \frac{\tan a \pm \tan b}{1 \mp \tan a \tan b}$$

$$\sin 2\theta = 2 \sin\theta \cos\theta$$
$$\cos 2\theta = \cos^2\theta - \sin^2\theta$$
$$\tan 2\theta = \frac{2\tan\theta}{1 - \tan^2\theta}$$

$$(\cos\theta + i\sin\theta)^n = \cos n\theta + i \sin n\theta$$

de Moivre's theorem

$$e^{(a+bi)x} = e^{ax}(\cos bx + i \sin bx)$$
$$e^{i\theta} = \cos\theta + i\sin\theta$$

Euler's formula

A.2 Some Standard Integrals

$$\int \sin x dx = -\cos x + C$$

$$\int \cos x dx = \sin x + C$$

$$\int \sec^2 x dx = \tan x + C$$

$$\int \csc^2 x dx = -\cot x + C$$

$$\int \sec x \tan x dx = \sec x + C$$

$$\int \csc x \cot x dx = -\csc + C$$

$$\int \tan x dx = \ln(\sec x) + C$$

$$\int \cot x dx = \ln(\sin x) + C$$

$$\int \sec x dx = \ln(\sec x + \tan x) + C$$

$$\int \csc x dx = \ln(\csc x - \cot x) + C$$

$$\int \frac{1}{\sqrt{a^2 - x^2}} dx = \sin^{-1}\frac{x}{a} + C$$

$$\int \frac{-1}{\sqrt{a^2 - x^2}} dx = \cos^{-1}\frac{x}{a} + C$$

$$\int \frac{1}{a^2 + x^2} dx = \frac{1}{a}\tan^{-1}\frac{x}{a} + C$$

$$\int \frac{1}{x\sqrt{x^2 - a^2}} dx = \frac{1}{a}\sec^{-1}\frac{x}{a} + C$$

$$\int \sin^{-1} x dx = x\sin^{-1} x + \sqrt{1 - x^2} + C$$

$$\int u dv = uv - \int v du$$

$$\int \frac{1}{x} dx = \ln x$$

$$\int \ln x dx = x\ln|x| - x + C$$

$$\int \frac{1}{x\ln x} dx = \ln\ln x + C$$

$$\int e^{ax} dx = \frac{1}{a} e^{x} + C$$

$$\int x e^{ax} dx = \frac{1}{a^2}(ax - 1)e^{ax} + C$$

$$\int \sinh x dx = \cosh x + C$$

$$\int \cosh x dx = \sinh x + C$$

$$\int \tanh x dx = \ln\cosh x + C$$

$$\int \mathrm{sech}^2 dx = \tanh x + C$$

$$\int \mathrm{csch} x \coth x dx = -\mathrm{csch} x + C$$

$$\int \mathrm{sech} x \tanh x dx = -\mathrm{sech} x + C$$

$$\int \mathrm{csch}^2 x dx = -\coth x + C$$

$$\int \frac{1}{\sqrt{x^2 + a^2}} dx = \sinh^{-1}\frac{x}{a} + C$$

$$\int \frac{1}{\sqrt{x^2 - a^2}} dx = \cosh^{-1}\frac{x}{a} + C$$

$$\int \frac{1}{a^2 - x^2} dx = \frac{1}{a}\tanh^{-1}\frac{x}{a} + C$$

$$\int \frac{1}{x\sqrt{a^2 - x^2}} dx = -\frac{1}{a}\mathrm{sech}^{-1}\frac{x}{a} + C$$

A.3 Special Functions

The **gamma function** is defined as: $\Gamma(n)=\int_0^{\infty} x^{n-1}e^{-x}dx.$

Some useful results for Γ are:

$$\Gamma(1)=1$$

$$\Gamma(n+1)=n! \quad \text{for } n \text{ is a positive integer}$$

$$\Gamma\left(\frac{1}{2}\right)=\sqrt{\pi}$$

$$\Gamma(n+1)=\sqrt{2\pi n}n^n e^{-n} \quad \text{for large n}$$

$$\Gamma(n)=\frac{\Gamma(n+1)}{n} \quad \text{for } n<0$$

The **beta function** is defined as: $B(m,n)=\int_0^1 x^{m-1}(1-x)^{n-1}dx$

$$=\frac{\Gamma(m)\Gamma(n)}{\Gamma(m+n)}$$

The **error function** is defined as: $erf(x)=\frac{2}{\sqrt{\pi}}\int_0^x e^{u^2}du$

$$=1-\frac{2}{\sqrt{\pi}}\int_x^{\infty} e^{-u^2}du$$

The **exponential integral** is defined as: $Ei(x)=\int_x^{\infty}\frac{e^{-u}}{u}du$

Bessel's differential equation of order n is $x^2y''+xy'+(x^2-n^2)y=0.$
The solution to this equation has the form:

$$y=\sum_{k=0}^{\infty} C_k x^{k+p}$$

One solution is:

$$y=J_n(x)=\sum_{k=0}^{\infty}\frac{(-1)^k}{k!(n+k)!}\left(\frac{x}{2}\right)^{2k+n} \quad \text{where } n \text{ is an integer,}$$

This is a **Bessel function of the first kind** of order n.

Index

Abbe's theory of image formation 235
acceleration 33, 131
alternating series 70
amplitude 240
amplitude function 227
amplitude modulation 171
amplitude spectrum 166, 169, 228
angular momentum quantum number 255
anti-derivative 28, 31
arbitrary constant 28
arccos function 44
arcsine 44
arctan function 44
area 156
area under the curve 30
arithmetic progression 69
arithmetic sequence 69
asymptote 17
augmented matrix 89
auxiliary equation 102, 104, 240
average 75
average velocity 27

bending moments 215
Bessel function of the first kind 276
beta function 276
binomial 73
binomial distribution 80, 81
binomial theorem 21
binomial theorem 38, 73, 80
Bohr 270
Bohr atom 254
Born 239
bound states 251
boundary conditions 95, 247
bra-ket 241, 257

capacitively reactive 199
Cartesian coordinate systems 6
Cayley–Hamilton theorem 91
central limit theorem 83
centre 13
centripetal 131
chain rule 22, 134
change of variable 33, 51
channels 168
characteristic equation 90, 106
characteristic functions 238
Chebyshev's theorem 75
chi-squared 85
circle 11
circulation 148
closed curve 149
coefficient of variation 75
cofactor 88
combination 76
commutator 259
compatibility 213
complementary function 104
completing the square 55
complex conjugate 4, 5
complex eigenvalues 109
complex function 64

complex number 4
complex relative permittivity 198
complex variable 64
compression 211
compressive stress 207
concavity 25
conductivity 198
conservative field 149, 150
constraint factors 144
continuous 18, 20
continuous random variable 78
converge 57
convergent 57
convolution 171, 224
cosine 9
cosine rule 9
cotangent 9
Cramer's rule 89
critical points 24, 26
cubic spline 189
cumulative distribution function 81
cumulative probability distribution function 78
curl 139, 149
cylindrical coordinate system 6

de Broglie 243, 245
de Moivre's 5
decay 40
definite integral 30, 32
deflection curve 217
degeneracy 258
degree 95
degree of confidence 83
degrees of freedom 84, 85
del 138, 258
density function 153
dependent variable 15
derivative 20, 21
derivative of a constant 21
derivative of e^x 39
determinant 88
deviatoric stresses 210
dielectric 197
dielectric constant 197
difference table 185
differential calculus 20
differential equation 27, 95, 192, 200
differential operator 102, 220, 221, 237
differentials 20, 27
differentiation 190
diffraction grating 229, 230
diffraction pattern 235
digital signal processing 171
Dirac 241, 257
direction angles 122
direction cosines 122, 126
directional derivative 136
directrix 12
discontinuity 57
discrete Fourier transform 169
discriminant 16
distance formula 6
distance from a point to a line 128

distance from a point to a plane 126
divergence 138, 158
divergence theorem 158, 162
divergent 57
divided differences method 185
domain 15
dot product 123, 257
double integral 152, 154
double-slit 231

eccentricity 11, 13
eigenfunctions 238, 242, 246, 247
eigenstate 238
eigenvalues 90, 106, 107, 249
eigenvector 90, 106, 197, 257
elastic curve 217
electron 237
electron tunnelling 252
elementary row transformations 89
ellipse 11
end point extrema 24
energy 240, 241, 243
energy density 239
energy operator 258
equation of a plane 125
equilibrium 213
error function 276
Euler method 192
Euler's formula 5, 64, 109, 166, 233
events 77
exact differential 149
exact equation 99, 100
expectation value 241, 248, 259
expectation value of x 244
expected value 79
exponential integral 276
extent 26
extrema 24

factorial 38, 76
failure 77
finite-element analysis 194
first derivative test 24
flux 156, 158
focus 12
formal definition of a limit 18
Fourier integral 167, 182
Fourier series 166, 232
Fourier transform 167, 224, 228, 229, 230, 232, 235
Fraunhofer diffraction pattern 226, 234
frequency resolution 168
function 15, 172
function of a function 22
fundamental theorem of calculus 31

gamma function 276
Gauss' law 157, 158
Gauss' theorem 158
Gaussian distribution 81
general bridge equation 201
general equation of a circle 11
general form of equation for a straight line 10

general solution 95
geometric progression 69
geometric sequence 69
geometric series 70
geometrical optics 226
gradient 10, 136, 138, 150
Green's theorem 160
ground state 270
growth 40

half life 40
half-range Fourier series 170
Hamiltonian 260, 262, 269
Hamiltonian operator 258
harmonic frequencies 178
harmonic oscillator 262
harmonic oscillator potential 267
harmonic series 70
heat conduction equation 180
Heaviside 115
Heisenberg's uncertainty principle 249
helium atom 269
Hermite polynomial 253
Hermitian 258
histogram 78
homogeneous 97, 102
Hooke's law 206, 209
horizontal asymptote 26
hydrogen atom 269
hyperbola 13, 46
hyperbolic functions 46
hyperbolic sector 46

i unit vector 8
imaginary numbers 3
impedance 199
implicit function 22
improper integral 57
indefinite integral 33
independent variable 15
indeterminate form 56
inductively reactive 199
infinite limit 57
infinite sequence 69
infinite square well potential 246, 257, 261
initial conditions 95
instantaneous acceleration 27
instantaneous rate of change 20, 24
instantaneous velocity 27
integral calculus 30
integral of e^x 39
integral transform 113, 220
integral transform operator 220
integrand 30
integrating factor 99, 100
integration 28, 31, 191
integration by parts 52
integrator 221
intensity 239
intensity distribution 230
intercept 10
interpolating polynomial 185
interval of convergence 71, 72

inverse 87
inverse Fourier transform 167, 233
inverse function 15, 44
inverse hyperbolic functions 49
inverse Laplace transform 113
inverse transform 117, 169, 234, 235
inverse trigonometric function 44
irrational numbers 3

k space 228

L'Hôpital's rule 18, 56
Lagrange multiplier 141
Laplace expansion 88
Laplace operator 113, 116, 220
Laplace transform 113, 114, 220, 221
latus rectum 11, 13
laws of logarithms 37
leakage 198
least squares 142
left limit 17
Leibnitz's rule 134
limit 17, 69
line 124
line integral 147
linear differential equation 100, 102
linear equation 10
linear function 15
linear strain 208
linear system 106
linear system of equations 89
logarithm 36
loss factor 198
loss resistor 198
low pass filter 219

Maclaurin series 72, 110
magnetic quantum number 255
magnitude of a complex number 4
major axis 11
mass 153
matter wave 63
maximum 25
Maxwell's equations 163
mean 75, 79, 80
mean stress 210
mean value theorem 30, 31, 135
median 75
microscope 235
mid-ordinate rule 34
midpoint formula 6
midpoint rule 34, 191
minimum 25
minor 88
mode 75
modulated 231
modulation 171
modulus 4
moment of inertia 156, 215
momentum 237, 241, 243

natural logarithm 37
Newton's method 184
Newton-Gregory forward polynomial 185
nonhomogeneous linear differential equation 104
non-periodic 168
non-periodic signal of finite length 168
normal 207
normal distribution 81, 83
normal equations 186
normal stresses 207
normal vector 137
normalisation 244, 248
normalised 241
numerical aperture 235

oblique asymptotes 26
odd 15, 170
odds 77
one-sided limits 17
operator 15
order 95
ordinary differential equations 95
orthogonal 87, 257, 258
overlap integral 256

p series 70
p test 57
parabola 12, 16
parallel circuit 200
parametric equations 124
partial derivatives 62
partial differential equation 95, 174
partial differentiation 133
partial fraction 54, 117
particle in a box 245
particular solution 95
Pascal's triangle 73
permutation 76
perturbation 260
phase 59, 61, 240
phase angle 198
phase shift 229
physical optics 226
Planck's equation 243
plane 125
plane strain 209
plane stress 209
point of inflection 25
points of inflection 24
Poisson distribution 80
Poisson's ratio 209, 211
polynomial 73
polynomial function 15, 23
population mean 83
position vector 121, 130
potential barrier 252
potential energy 237
power 67
power rule 21, 33
power rule for anti-derivatives 28
power rule for functions 22
power series 72, 110, 253
principal planes of stress 212

principal stresses 212
probability 73
probability density 239
probability density function 78, 239
product rule 22
properties of the definite integral 32
pure bending 215, 216, 217
Pythagoras' theorem 7

quadratic equation 16
quadratic formula 16
quadratic polynomial 55
quantum transition 272
quotient rule 22

radian 42
radius of convergence 71
range 15, 75
rank 210
rate of change 40
ratio test 71
rational number 3
real numbers 3
reciprocal hyperbolic functions 46
reciprocal trigonometric functions 43
reciprocal trigonometric ratios cosecant 9
relative permittivity 197
resonance 200
resonant 199
right limit 17
roots 16
Runge–Kutta method 193

s domain 221
sample mean 83
sample standard deviation 83
sample variance 75
sampling frequency 168
Sandwich theorem 18
scalar 138
scalar field 210
scalar potential 149
scalars 7
Schering bridge 203, 204
Schrödinger equation 237, 245, 251, 254, 258
s-domain analysis 222, 223
secant 9, 20
second derivative 23, 25
second derivative test 25
separation constant 238
separation of variables 96, 176
series 70
shear modulus 211
shear strain 206
shear stress 207
SHM 59
simple harmonic motion 59
simple harmonic oscillator 263, 265
simple harmonic oscillator potential 253
Simpson's rule 34, 191
sine rule 9
single slit 230

singular matrix 87, 88
skew lines in space 129
slope 10
spatial filter 235
spatial frequencies 227, 228
spatial frequency 227, 229
spherical coordinate system 6
square well potential 251
standard deviation 75, 79
standard equation of a circle 11
standard equation of a hyperbola 13
standard equation of a parabola 12
standard equation of an ellipse 11
standard error of the mean 83
standard normal distribution 81
standing wave 245, 246, 247, 250
static equilibrium 213
stationary point 24
stationary state 240, 247
step potential 250
stiffness 206
stiffness matrix 194
Stirling's formula 76
strain 206, 208
stress 206
stress deviations 210
stress equilibrium 213
stress tensor 210
substitution 51
success 77
sums and differences : 22
superposition 62, 171, 242, 256
surface integral 154, 155, 156
symmetry 26

tangent 9, 20
tangent plane 137
Taylor series 72, 192
t-domain analysis 222, 223
telescope 235
tensile stress 207
tension 211
tensor field 210
theorems on limits 18
time constant 198
time domain 221
time-dependent equation 238
time-independent equation 238
trace 87
transcendental numbers 3
transfer function 219, 222
transformed 220
transpose 87
transverse wave 60
trapezoidal rule 34, 191
travelling wave 227, 243
trial wave function 266
triaxial stresses 211
trigonometric ratios 9

trigonometric substitution
  53
triple integral 153
tunnelling 250
turning points 24

undetermined coefficients
  104
unit matrix 87
unit vector 121, 130

variance 79, 80
variational method 266
vector 138
vector cross product 127,
  128
vector differential operator
  136, 138
vector equation 8
vector field 121, 210
vector function 121, 147
vector functions 156
vectors 7
velocity 33
vertex 12
vertical asymptote 26
vibration of a stretched
  string 175
volume 152

wave equation 62, 175, 240
wave function 62, 237, 238,
  239, 242, 248
wavelength 60
weighting factor 142
work function 251

Young's modulus 211

zero frequency 228, 229
zero point energy 245